Springer Proceedings in Mathematics & Statistics

Volume 350

This book series features volumes composed of selected contributions from workshops and conferences in all areas of current research in mathematics and statistics, including operation research and optimization. In addition to an overall evaluation of the interest, scientific quality, and timeliness of each proposal at the hands of the publisher, individual contributions are all refereed to the high quality standards of leading journals in the field. Thus, this series provides the research community with well-edited, authoritative reports on developments in the most exciting areas of mathematical and statistical research today.

More information about this series at http://www.springer.com/series/10533

Robert L. Devaney · Kit C. Chan ·
P. B. Vinod Kumar
Editors

Topological Dynamics and Topological Data Analysis

IWCTA 2018, Kochi, India, December 9–11

Springer

Editors
Robert L. Devaney
Department of Mathematics
Boston University
Boston, MA, USA

Kit C. Chan
Department of Mathematics and Statistics
Bowling Green State University
Bowling Green, OH, USA

P. B. Vinod Kumar
Department of Mathematics
Rajagiri School of Engineering
and Technology
Cochin, Kerala, India

ISSN 2194-1009 ISSN 2194-1017 (electronic)
Springer Proceedings in Mathematics & Statistics
ISBN 978-981-16-0173-6 ISBN 978-981-16-0174-3 (eBook)
https://doi.org/10.1007/978-981-16-0174-3

Mathematics Subject Classification: 37Bxx, 37Cxx, 55Mxx, 55Bxx, 55Cxx, 55Dxx, 55Exx, 55Gxx, 55Hxx

This Springer imprint is published by the registered company Springer Nature Singapore Pte Ltd.
The registered company address is: 152 Beach Road, #21-01/04 Gateway East, Singapore 189721, Singapore

Preface

Topological dynamics is an emerging field. Topological data analysis is a young field. Dynamics of data is an interesting zone to stay and watch. Today, the field of topological dynamics and topological data analysis has grown into a respected mathematical discipline with specific concepts and techniques, and with plenty of applications inside and outside mathematics. In December 2018, a workshop and the first international conference on topological dynamics and topological data analysis in India took place at Rajagiri School of Engineering and Technology, Kerala.

In the workshop, from 5th December to 8th December, leading experts from all over the world gave comprehensive survey lectures on the state of the art in their areas. In the coference from 9th December to 11th December, new research results were presented my mathematicians from 14 countries. To name a few—A. N. Sharkovsky, James Yorke, Joseph Auslander, Henk Bruin, Robert Deveney, Saber Elaydi, V. Kannan, G. Rangarajan, Roman Hric, Amit Chattopahyay, Andrei Tetenov, Krzysztof Lesniak, Patrizio Frosini, Dan Burghelea, Dominic Kwietria K, Hisao Kato, Karoly Sumon, Kitchan, Romen Hric, Vijay Natarajan, Anima Nagar, W. J. Charatonik.

This volume contains some invited lectures of the workshop and selected contributions of the conference. Providing readable surveys, it can be used as reference book those who want to start work in the field.

The organizers of the conference would like to thank the management of Rajagiri School of Engineering and Technology, Cochin, Kerala, India, for the inspiration and support provided to conduct the conference.

The organizers acknowledge the financial support given by National Board of Higher Mathematics, India, Dept. of Science and Technology, India, and International Council for Industrial and Applied Mathematics.

Boston, USA — Robert L. Devaney
Bowling Green, USA — Kit C. Chan
Cochin, India — P. B. Vinod Kumar

Contents

An Overview of Unimodal Inverse Limit Spaces

H. Bruin

Abstract An overview of unimodal inverse limit spaces, to support the mini-course "Interval dynamics and Inverse limit spaces", at **IWCTA: International Workshop and Conference on Topology and Applications**, Rajagiri School of Engineering and Technology, Kochi, December 5–8, 2018.

Keywords Inverse limit space · Unimodal map · Tent map · Quadratic map · Embeddings · Endpoints · Folding point · Composant · Ingram conjecture

2000 Mathematics Subject Classification 54H20, 37B45, 37E05

1 Introduction

Unimodal maps are maps of the interval with a single critical point and increasing/decreasing at the left/right of the critical point. The best known examples are quadratic (logistic) maps and tent maps, see Fig. 1.

They are among the simplest maps that, at least for some parameters, are chaotic in every sense that can be given to mathematical chaos. They are not invertible; however, a simple way to make them invertible is by introducing a second coordinate and **thicken** the map:

$$T_a\colon x \mapsto 1 - a|x|, \qquad L_{a,b}\colon (x,y) \mapsto 1 - a|x| + by, x),$$
$$Q_a\colon x \mapsto 1 - ax^2, \qquad H_{a,b}\colon (x,y) \mapsto (1 - ax^2 + by, x).$$

In this way, the tent map becomes a Lozi map and the quadratic map a Hénon map. Figure 2 gives a Lozi attractor (resp. Hénon attractor) obtained as $\cap_{n\geq 0} L^n_{a,b}(U)$ for some well-chosen, forward invariant open disk U. In order to understand the topology of such attractors, unimodal inverse limit spaces (UILs) are a first informative, but

H. Bruin (✉)
Faculty of Mathematics, University of Vienna, Oskar Morgensternplatz 1, 1090 Vienna, Austria
e-mail: henk.bruin@univie.ac.at

R. L. Devaney et al. (eds.), *Topological Dynamics and Topological Data Analysis*, Springer Proceedings in Mathematics & Statistics 350,
https://doi.org/10.1007/978-981-16-0174-3_1

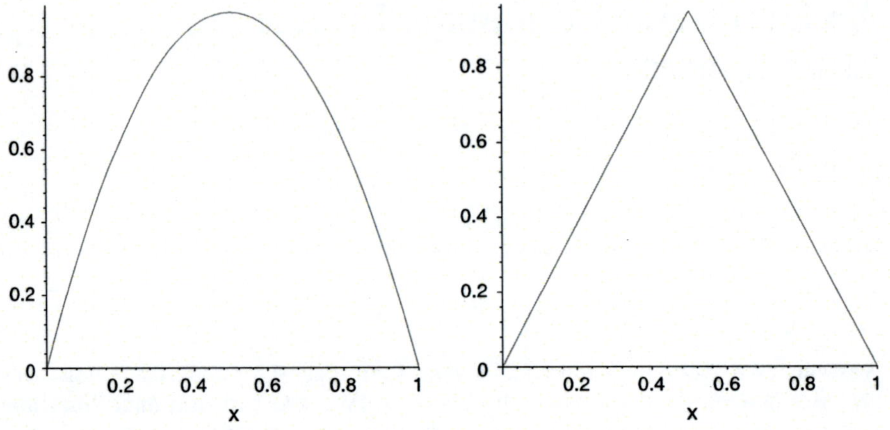

Fig. 1 Unimodal maps: a quadratic map and a tent map

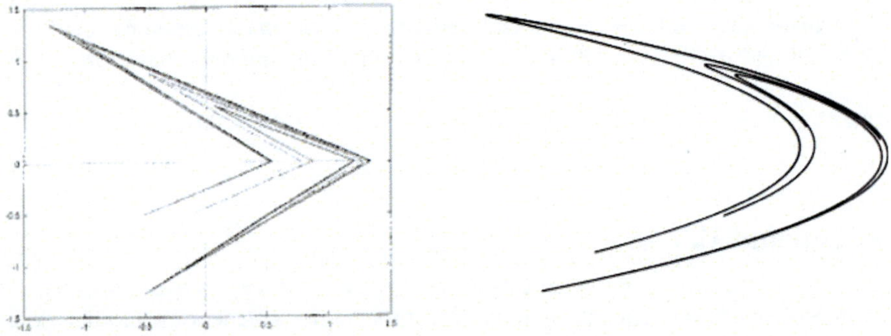

Fig. 2 Lozi and Hénon attractor. The Lozi

certainly not sufficient, step. In fact, all questions asked about UILs in these notes (and more!) can be asked about Lozi attractors and Hénon attractors.

2 Definitions and Notation

Let $\mathbb{N} = \{1, 2, 3, \dots\}$ be the set of natural numbers and $\mathbb{N}_0 = \mathbb{N} \cup \{0\}$. We consider two families of unimodal maps, the family of quadratic maps $Q_a: [0, 1] \to [0, 1]$, with $a \in [2, 4]$, defined as $Q_a(x) = ax(1 - x)$, and the family of tent maps $T_s: [0, 1] \to [0, 1]$ with slope $\pm s$, $s \in [2, 3]$, defined as $T_s(x) = \min\{sx, s(1 - x)\}$. Let f be a map from any of these two families. The **critical** or **turning** point is $c: = 1/2$. Write $c_k: = f^k(c)$. The closed f-invariant interval $[c_2, c_1]$ is called the **core** and denoted as $\varprojlim([c_2, c_1], T)$.

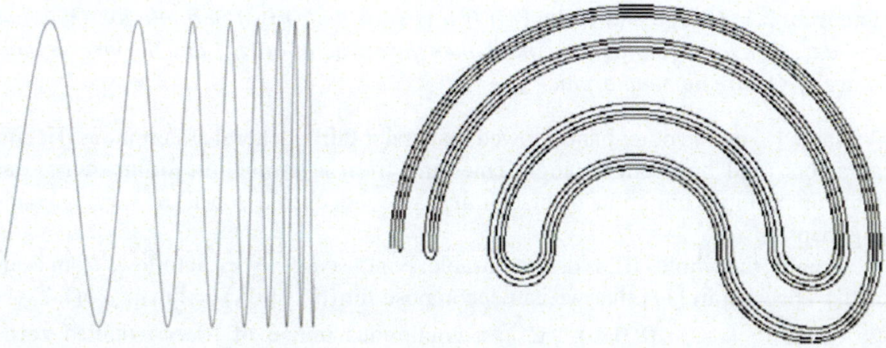

Fig. 3 sin $\frac{1}{x}$ continuum and the Knaster continuum. The sin $\frac{1}{x}$ continuum

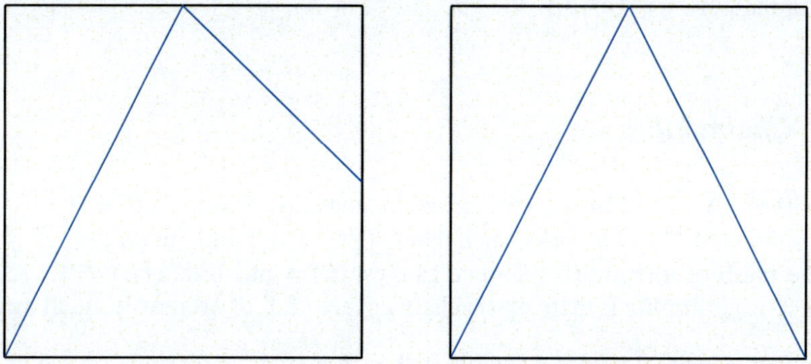

Fig. 4 Maps with sin $\frac{1}{x}$ continuum and the Knaster continuum as inverse limit spaces

The inverse limit space $\varprojlim([0,1], f)$ is the collection of all backward orbits

$$\{x = (\dots, x_{-2}, x_{-1}, x_0) \colon f(x_{-i-1}) = x_{-i} \in [0, c_1] \text{ for all } i \in \mathbb{N}_0\},$$

equipped with metric $d(x, y) = \sum_{i \le 0} 2^i |x_i - y_i|$. The map f is called the **bonding map** of $\varprojlim([0,1], f)$. We define the **induced** or **shift homeomorphism**on $\varprojlim([0,1], f)$ as

$$\sigma(x) := \sigma_f(\dots, x_{-2}, x_{-1}, x_0) = (\dots, x_{-2}, x_{-1}, x_0, f(x_0)).$$

Let $\pi_i \colon \varprojlim([0,1], f) \to [0, c_1]$, $\pi_i(x) = x_{-i}$ be the i-th projection map.

Simple examples of such unimodal inverse limit spaces are the sin $\frac{1}{x}$-continuum and the Knaster continuum (bucket handle) shown in Figs. 3 and 4.

The similarity between a Hénon attractors and the Knaster continuum may suggest that inverse limit spaces are homeomorphic to Hénon attractors in some generality, but in fact, the generality is very limited.

Theorem 2.1 (Barge and Holte [8]) *If a is such that 0 is a periodic for $Q_a(x) = 1 - ax^2$, then for $|b|$ sufficiently small, then the attractor of $H_{a,b}$ and the inverse limit space of Q_a are homeomorphic.*

Barge [6] on the other hand showed that under fairly general assumptions, Hénon attractors (and homoclinic tangle emerging from a homoclinic bifurcations) are homeomorphic to unimodal inverse limit spaces, not even if you allow varying bonding maps.

Usually, the whole UIL is decomposable: For the case $c \le c_1$, it follows from Bennett's Theorem in [11] that we can decompose $\varprojlim([0,1], T_s) = \varprojlim([c_2, c_1], T_s) \cup \mathfrak{C}$, where $\bar{0} := (\dots, 0, 0, 0) \in \mathfrak{C}$ is a continuous image of $[0, \infty)$ (called **zero-composant**) which compactifies on $\varprojlim([c_2, c_1], T_s)$. Inverse limit space of tent map $\varprojlim([c_2, c_1], T_s)$ obtained from the forward invariant interval $[c_2, c_1]$ is called the **core** of the UIL.

2.1 Chainability

Definition 2.2 Let X be a metric space. A **chain in** X is a set $\mathcal{C} = \{\ell_1 \dots, \ell_n\}$ of open subsets of X called **links**, such that $\ell_i \cap \ell_j \neq \emptyset$ if and only if $|i - j| \le 1$.

The **mesh of a chain** $\mathcal{C}$ is defined as $\text{mesh}(\mathcal{C}) = \max\{\text{diam}\, \ell_i : i = 1, \dots, n\}$. A space X is **chainable** if there exists chain covers of X of arbitrarily small mesh.

A corollary of X being chainable is that X contains no **triods** (homeomorphic copies of the letter Y) or circles. All unimodal inverse limit spaces are chainable, and all chainable continua can be **embedded** in the plane, i.e., there is a continuous injection $h: X \to \mathbb{R}^2$ (called **embedding**) such that $h(X)$ and X are homeomorphic. They also possess the **fixed point property**: every continuous map $f: X \to X$ has a fixed point.

Definition 2.3 A point $a \in X \subset \mathbb{R}^2$ is **accessible** if there exists an arc $A = [x, y] \subset \mathbb{R}^2$ such that $a = x$ and $A \cap X = \{a\}$.

Unimodal inverse limit spaces can therefore be embedded in the plane, but in general, there are many (in fact uncountably non-homotopic) ways to do so. There are two **standard ways** that yield an embedding very much like the Lozi attractor (or Hénon attractor) with $b > 0$ (orientation reversing, making the composant $\mathfrak{R}$ of the fixed point $p = (\dots, r, r, r)$ accessible, see [17]) and $b < 0$ (orientation preserving, making the zero-composant accessible, see [16]), respectively.

The result of Anušić et al. gives an idea how much variety there is in embeddings.

Theorem 2.4 (Anušić et al. [2]) *For every point $a \in X$, there exists an embedding of X in the plane such that a is accessible.*

2.2 *Symbolic Dynamics*

We can extend the Milnor–Thurston [26] **kneading theory** to UILs, as done originally in [16]. The symbolic itinerary of the critical value $c_1 \in [0, 1]$ under the action of T is called the **kneading sequence**, and we denote it as $\nu = \nu_1\nu_1\nu_3\dots$, where $\nu_i = 0$ if $c_i < c$ and $\nu_i = 1$ if $c_i > c$. Analogously, to each $x \in \varprojlim([0, 1], T)$, we can assign a symbolic sequence $\overleftarrow{x}.\overrightarrow{x} = \dots s_{-1} \in \{0, \frac{0}{1}, 1\}^{-\mathbb{N}}$ where

$$s_{-i} = \begin{cases} 0 & \pi_i(x) < c, \\ \frac{0}{1} & \pi_i(x) = c, \\ 1 & \pi_i(x) > c, \end{cases} \qquad i \geq 0.$$

Here, $\frac{0}{1}$ means that both 0 and 1 are assigned to x. If c is non-periodic, this can happen only once, i.e., to every point, we assign at most two symbolic itineraries. If c is periodic, say of period n, then we need to make a consistent choice, usually such hat $s_{i+1}\dots s_{i+n}$ contains an even number of 1s.

For a fixed left-infinite sequence $s = \dots s_{-2}s_{-1}s_0 \in \{0, 1\}^{\mathbb{N}_0}$, the subset

$$A(s) := \overline{\{x \in X : \overleftarrow{s} \in \overleftarrow{x}\}}$$

of X is called a **basic arc**. It can be shown that $A(\overleftarrow{x})$ is the maximal closed arc A containing x such that $\pi_0 : A \to I$ is injective. In [17, Lemma 1], it was observed that $A(\overleftarrow{x})$ is indeed an arc (but it can be degenerated, i.e., a single point).

For every basic arc $A(\overleftarrow{x})$, we define

$$N_L(\overleftarrow{x}) := \{n > 1 : s_{-(n-1)}\dots s_{-1} = \nu_1\nu_2\dots\nu_{n-1}, \#_1(\nu_1\dots\nu_{n-1}) \text{ odd}\},$$
$$N_R(\overleftarrow{x}) := \{n \geq 1 : s_{-(n-1)}\dots s_{-1} = \nu_1\nu_2\dots\nu_{n-1}, \#_1(\nu_1\dots\nu_{n-1}) \text{ even}\}.$$

and

$$\tau_L(\overleftarrow{x}) := \sup N_L(\overleftarrow{x}) \quad \text{and} \quad \tau_R(\overleftarrow{x}) := \sup N_R(\overleftarrow{x}).$$

We can construct a model of the inverse limit space $\varprojlim([0, 1], f)$ by gluing basic arcs $A(\overleftarrow{x})$ to $A(\overleftarrow{y})$ at their left (resp. right) end points if and only if $\overleftarrow{x}$ and $\overleftarrow{y}$ agree up to one index, and this index is exactly $\tau_L(\overleftarrow{x}) = \tau_L(\overleftarrow{y})$ (resp. $\tau_R(\overleftarrow{x}) = \tau_R(\overleftarrow{y})$).

3 End points and Folding Points

Definition 3.1 A point x in a chainable continuum is called **end point** if for every two subcontinua $A, B \subset X$, $A \subset B$ or $B \subset A$. We denote the set of end points by $\mathcal{E}$.

As an example, $X = [0, 1]$ has end points 0 and 1 according to this definition. But the triod would have four end points (the branch point too!), which speaks against our intuition. Therefore, we required X to be chainable.

A geometric description of end points (using the notion of **crooked** graphs is due to Barge and Martin [5]. Here, we give a symbolic classification of end points, following [17, Sect. 2].

Lemma 3.2 (Bruin [17], Lemmas 2 and 3) *If* $A(\overleftarrow{x}) \in \{0, 1\}^{\mathbb{N}}$ *is such that* $\tau_L(\overleftarrow{x}), \tau_R(\overleftarrow{x}) < \infty$, *then*

$$\pi_0(A(\overleftarrow{x})) = [T^{\tau_L(\overleftarrow{x})}(c), T^{\tau_R(\overleftarrow{x})}(c)].$$

Without the restriction that $\tau_L(\overleftarrow{x}), \tau_R(\overleftarrow{x}) < \infty$, *we have*

$$\sup \pi_0(A(\overleftarrow{x})) = \inf\{c_n : n \in N_R(\overleftarrow{x})\},$$
$$\inf \pi_0(A(\overleftarrow{x})) = \sup\{c_n : n \in N_L(\overleftarrow{x})\}.$$

This gives the following symbolic characterization of end points.

Proposition 3.3 Bruin [17, Proposition 2] *A point* $x \in X$ *such that* $\pi_i(x) \neq c$ *for every* $i < 0$ *is an end point of* X *if and only if* $\tau_L(\overleftarrow{x}) = \infty$ *and* $\pi_0(x) = \inf \pi_0(A(\overleftarrow{x}))$ *or* $\tau_R(\overleftarrow{x}) = \infty$ *and* $\pi_0(x) = \sup \pi_0(A(\overleftarrow{x}))$.

Definition 3.4 A **folding point** in the core of a unimodal inverse limit is any point that does not have a neighborhood homeomorphic to a Cantor set of open arcs. We denote this set by $\mathcal{F}$.

The omega-limit set of a point is defined as the set of adherence points of its forward orbit:

$$\omega(x) = \{y : \exists\, n_i \to \infty\ T^{n_i}(x) \to y\} = \cap_{j \in \mathbb{N}} \overline{\cup_{i > j}\{T^j(x)\}}.$$

The following characterization of folding points is due to Raines.

Proposition 3.5 (Theorem 2.2 in [28]) *A point* $x \in \varprojlim([c_2, c_1], T)$ *is a folding point if and only if* $\pi_n(x)$ *belongs to* $\omega(c)$ *for every* $n \in \mathbb{N}$.

Theorem 3.6 *The core* $\varprojlim([c_2, c_1], T)$ *contains exactly* N *end points if and only if* c *is periodic of period* N.

The core $\varprojlim([c_2, c_1], T)$ *contains exactly* N *non-end folding points if and only if* c *is* ***pre****periodic of period* N.

Proof This is a special case of the theory developed above (Proposition 3.3) (Fig. 5).

Theorem 3.7 *If* c *is not recurrent, then the core* $\varprojlim([c_2, c_1], T)$ *contains no end points, but folding points do exist.*

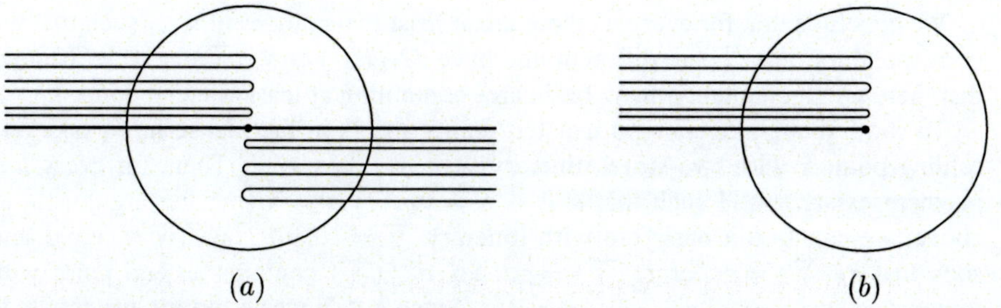

Fig. 5 Neighborhoods of non-end folding point (a) and an end point (b)

Proof Since $\omega(c) \neq \emptyset$, there must be folding points, see Proposition 3.5. But there cannot be any end points, because every backward itinerary $\overleftarrow{x}$ has $N_L(\overleftarrow{x}), N_L(\overleftarrow{x}) < \infty$, so Proposition 3.3 applies.

The following proposition follows implicitly from the proof of Corollary 2 in [17]. It shows that if c is recurrent, then $\#(\mathcal{E} \cap \varprojlim([c_2, c_1], T)) = N \in \mathbb{N}$ if and only if c is N-periodic, and otherwise $\mathcal{E} \cap \varprojlim([c_2, c_1], T)$ is uncountable. We prove it here for completeness.

Proposition 3.8 *If orb(c) is infinite and c is recurrent, then the core inverse limit space X′ has uncountably many end points. Moreover, $\mathcal{E}$ has no isolated points and is dense in $\mathcal{F}$.*

Proof Since c is recurrent, for every $k \in \mathbb{N}$ there exist infinitely many $n \in \mathbb{N}$ such that $\nu_1 \dots \nu_n = \nu_1 \dots \nu_{n-k}\nu_1 \dots \nu_k$.

Take a sequence $(n_j)_{j \in \mathbb{N}}$ such that $\nu_1 \dots \nu_{n_{j+1}} = \nu_1 \dots \nu_{n_{j+1}-n_j}\nu_1 \dots \nu_{n_j}$ for every $j \in \mathbb{N}$. Then, the basic arc given by the itinerary

$$\overleftarrow{x} := \lim_{j \to \infty} \nu_1 \dots \nu_{n_j},$$

is admissible and $\tau_L(\overleftarrow{x}) = \infty$ or $\tau_R(\overleftarrow{x}) = \infty$. Therefore, $A(\overleftarrow{x})$ contains an end point. Note that, since ν is not periodic, $\overleftarrow{x}$ is also not periodic, and thus, $\sigma^k(\overleftarrow{x}) \neq \overleftarrow{x}$ for every $k \in \mathbb{N}$.

To determine the cardinality of end points, we claim that for every fixed $n \in \mathbb{N}$ there are $m_2 > m_1 > n$ such that

$$\nu_1 \dots \nu_{m_2} = \nu_1 \dots \nu_{m_2-n}\nu_1 \dots \nu_n, \quad \nu_1 \dots \nu_{m_1} = \nu_1 \dots \nu_{m_1-n}\nu_1 \dots \nu_n,$$

but $\nu_1 \dots \nu_{m_1}$ is not a suffix of $\nu_1 \dots \nu_{m_2}$. Indeed, if m_2 did not exist, then

$$\overleftarrow{x} = (\nu_1 \dots \nu_{m_1-n})^{-\infty}\nu_1 \dots \nu_n$$

would have a periodic tail. Since c is not periodic, no end point can have such a tail.

We conclude that for every n_j there are at least two choices of n_{j+1} such that the corresponding tails $\overleftarrow{x}$ are different and have $\#N_L(\overleftarrow{x}) \cup N_R(\overleftarrow{x}) = \infty$. It follows that there are uncountably many basic arcs containing at least one end point.

To show that $\mathcal{E}$ contains no isolated points and is in fact dense in $\mathcal{F}$, take any folding point x with **two-sided** itinerary $\dots s_{-2}s_{-1}s_0.s_1s_2\dots$ Then, for every $k \in \mathbb{N}$, there exists $n \in \mathbb{N}$ such that $s_{-k}\dots s_k = \nu_n \dots \nu_{n+2k}$. Using the arguments as above, we can find a basic arc with itinerary $\overleftarrow{y} = \dots \nu_1 \dots \nu_{n-1}\nu_n \dots \nu_{n+2k}$ and such that $\tau_L(\overleftarrow{y}) = \infty$ or $\tau_R(\overleftarrow{y}) = \infty$. So, $\sigma^{-k}(\overleftarrow{y})$ contains an end point with itinerary $\dots \nu_n \dots \nu_{n+k}.\nu_{n+k+1} \dots \nu_{n+2k} \dots$ Since $k \in \mathbb{N}$ was arbitrary, we conclude that there is some (in fact, uncountably many) end points arbitrarily close to x.

The following result about comparing end points with folding points is due to [1]. We first need a definition, going back to Blokh and Lyubich [14]

Definition 3.9 The critical point c is *reluctantly recurrent* if there is $\varepsilon > 0$ and an arbitrary long (but finite!) backward orbit $\bar{y} = (y_{-m}, \dots, y_{-1}, y_0)$ in $\omega(c)$ such that the ε-neighborhood of $y \in I$ has monotone pull-back along $\bar{y}$. Otherwise, c is *persistently recurrent.*

Theorem 3.10 *In an UIL, $\mathcal{F} = \mathcal{E}$ if and only if c is persistently recurrent.*

4 Composants

Definition 4.1 Let X be a continuum and $x \in X$. The **arc-component** $A(x)$ of x is the union of points y such that there is an arc in X connecting x and y. The **composant** $C(x)$ of a point x is the union of all proper subcontinua of X.

For example, if $X = [0, 1]$, then $A(0) = [0, 1]$, but $C(0) = [0, 1)$ (it does not contain 1 because $[0, 1]$ is not a **proper** subcontinuum of X). Also, $A(\frac{1}{2}) = C(\frac{1}{2}) = [0, 1]$ because $[0, 1] = [0, \frac{1}{2}] \cup [\frac{1}{2}, 1]$.

Two arc-components A and $\tilde{A}$ are asymptotic if there are parametrizations

$$\varphi, \tilde{\varphi}\colon \mathbb{R} \to A, \tilde{A} \qquad \text{such that} \qquad \lim_{t\to\infty} d(\varphi(t), \tilde{\varphi}(t)) = 0.$$

The trivial case when $A = \tilde{A}$ is excluded, but A is **self-asymptotic** if there is a parametrization φ such that

$$\lim_{t\to\infty} d(\varphi(t), \tilde{\varphi}(-t)) = 0.$$

Figure 6 gives the UIL of a tent map with $T^3(c) = c$, for which the fixed composant $\mathfrak{R}$ is self-asymptotic. There is a single infinite Wada channel for which the entire shore is equal to $\mathfrak{R}$.

Theorem 4.2 (Barge et al. [9]) *Every UIL with periodic critical point has at least one asymptotic arc-component.*

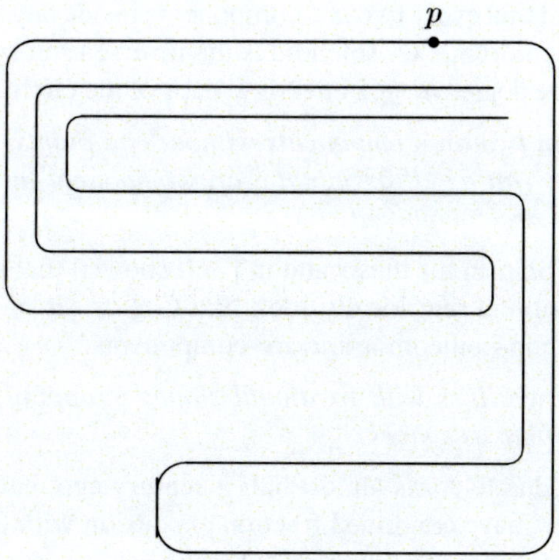

Fig. 6 This representation has a single infinite Wada channel

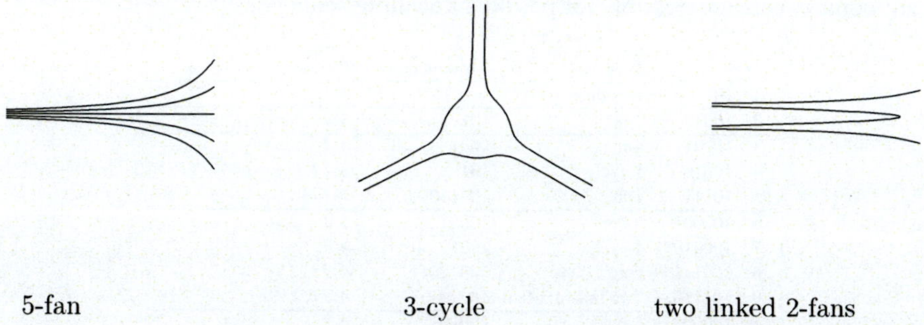

Fig. 7 Configurations of asymptotic arc-components

Proof The proof relies on substitution tilings and the fact that these spaces act as 2-to-1 coverings of inverse limit spaces. In fact, if the period is N, then there are at least $N-1$ and at most $2(N-1)$ "halves" of arc-components asymptotic to some other "halves" of an arc-components.

Conjecture 4.3 *The upper bound is in fact* $2(N-2)$. *Given any two "halves" of arc-components* H *and* H', H *is asymptotic to or coincides with* $\sigma^n(H')$ *for some* $n \in \mathbb{Z}$.

Question 4.4 *If c is non-recurrent, then there are no asymptotic arc-components, see [19], but what is the situation of asymptotic arc-components when c is non-periodic but recurrent?*

In the Knaster continuum, it was shown by Bandt [4] that every two arc-components without not containing the end point are homeomorphic. More generally,

De Man [21] showed that every two arc-components inside any two one-dimensional solenoids are homeomorphic. (A solenoid is the inverse limit space of circles where the bonding maps are degree $n_i \geq 2$ covering maps of the circle as bonding maps f_i.)

Question 4.5 *Given two arc-components without end points, are they homeomorphic? In particular, can a self-asymptotic arc-component be homeomorphic to a non-self-asymptotic arc-component?*

In contrast, Fokkink (in his thesis and in [22]) showed that among all **matchbox manifolds** (i.e., continua that locally look like Cantor set of open arcs) there are uncountably many non-homeomorphic arc-components.

Question 4.6 *Are two lines with irrational slopes wrapping for ever around the torus be homeomorphic as spaces?*

This question is due to Aarts almost half a century ago, but beyond the fact that if the slopes θ and θ' have continued fraction expansion with the same tail then the lines are indeed homeomorphic, nothing is known.

Below, we gave a full list (take from [19]) of what configurations asymptotic arc-components are possible for periodic kneading sequences (Fig. 7)

	ν	Type	Periodic tail(s)	k	Case
1	101	1-cycle	1	2	*I*
2	1001	3-fan	101	3	*I*
3	10001	4-fan	1001	4	*I*
4	10010	3-cycle	101	3	*II*
5	10111	Three 2-fans	101110	3	*III*
6	100001	5-fan	10001	5	*I*
7	100010	4-cycle	1001	4	*II*
8	100111	Four 2-fans	10010011	4	*III*
9	101110	Two linked 3-fans	10, 1	4	*II*, *IV*
10	1000001	6-fan	100001	6	*I*
11	1000010	5-cycle	10001	5	*II*
12	1000111	Five 2-fans	1000100011	5	*III*
13	1000100	Four 2-fans (l.i.p.)	10, 1001	4	*II*
14	1001101	Four 2-fans	10011010	4	*III*
15	1001110	Five 2-fans	10010, 10111	5	*II*
16	1001011	Five 2-fans	1001011011	5	*III*
17	1011010	5-cycle	10111	5	*II*
18	1011111	Five 2-fans	1011111110	5	*III*
19	10000001	7-fan	1000001	7	*I*
20	10000010	6-cycle	100001	6	*II*
21	10000111	Six 2-fans	100001110000	6	*III*
22	10000100	Five 2-fans	10001, 10010	5	*II*
23	10001101	Five 2-fans	1000110100	5	*III*
24	10001110	Six 2-fans	100010, 100111	6	*II*
25	10001011	Six 2-fans	100010110011	6	*III*
26	10011010	Six 2-fans (l.i.p.)	101, 100111	6	*II*
27	10011111	Six 2-fans	100111110110	6	*III*
28	10011100	Five 2-fans	10010, 10111	5	*II*
29	10010101	Five 2-fans	1001010111	5	*III*
30	10010110	Six 2-fans (l.i.p.)	100, 101110	6	*II*
31	10110111	Three 3-cycles	101101110	3	*III*
32	10111110	Two linked 4-fans	101110, 1	5	*II*, *IV*

(l.i.p = linked in pairs.)

5 Ingram Conjecture

In the early 90s, a classification problem that became known as the **Ingram Conjecture** was posed:

> If $1 \le s \le \tilde{s} \le 2$, then the inverse limit spaces $\varprojlim([0,1], T_s)$ and $\varprojlim([0,1], T_{\tilde{s}})$ are not homeomorphic.

In the "Continua with the Houston problem book" in 1995 [24, page 257], Ingram writes

> The [...] question was asked of the author by Stu Baldwin at the [...] summer meeting of the AMS at Orono, Maine, in 1991. ... There is a related question which the author has considered to be of interest for several years. He posed it at a problem session at the 1992 Spring Topology Conference in Charlotte for the special case (that the critical point has period) $n = 5$.

After partial results [7, 12, 18, 25, 29, 31, 33], the Ingram Conjecture was finally answered in affirmative by Barge et al. in [10]. In addition (Bruin & Štimac [32]).

Proposition 5.1 (Rigidity) *If $h : \varprojlim([0,1], T) \to \varprojlim([0,1], T)$ is a homeomorphism, then it is isotopic σ^n for some $n \in \mathbb{Z}$. In fact, if $\omega(c) = [c_1, c_2]$, then $h|_{\varprojlim([c_2,c_1],T)} = \sigma^n$.*

However, the proof presented in [10] crucially depends on using the zero-composant $\mathfrak{C}$, so the core version of the Ingram Conjecture still remains open. For Hénon maps, $\mathfrak{C}$ plays the role of the unstable manifold of the saddle point outside the Hénon attractor; it compactifies on the attractor, but it is somewhat unsatisfactory to have to use this (and the embedding in the plane that it presupposes) for the topological classification. It is not possible to derive the core version directly from the non-core version, because it is impossible to reconstruct $\mathfrak{C}$ from the core. This is for instance illustrated by the work of Minc [27] showing that in general there are many non-equivalent rays compactifying on the Knaster bucket handle.

Question 5.2 *Does the Core Ingram Conjecture hold? And the core rigidity proposition?*

Partial results here are by [25, 31] (because their proofs work without the zero-composant) and [3, 20, 23]. In short, the Core Ingram Conjecture holds if c is (pre)periodic or non-recurrent or is persistently recurrent with so-called "Fibonacci-like" combinatorics, but all other cases remain unproved.

Question 5.3 *Does the Ingram Conjecture hold in the multimodal setting, e.g., for cubic maps?*

References

1. L. Alvin, A. Anušić, H. Bruin, Činč, *Folding points of unimodal inverse limit spaces*. This paper has appeared in nonlinearity **33**, 224–248 (2019)
2. A. Anušić, H. Bruin, Činč, Uncountably many planar embeddings of unimodal inverse limit spaces, Discrete Contin. Dyn. Syst. 37 (2017), 2285–2300
3. A. Anušić, H. Bruin, Činč, The core Ingram conjecture for non-recurrent critical points. Nonlinear. **33**, 224–248 (2020)
4. C. Bandt, Composants of the horseshoe. Fund. Math. **144**, 231–241 (1994). And Erratum to the paper: "Composants of the horseshoe". Fund. Math. **146**, 313 (1995)
5. M. Barge, J. Martin, *Endpoints of inverse limit spaces and dynamics*, Continua (Cincinnati, OH, 1994), vol. 170. Lecture Notes in Pure and Applied Mathematics, pp. 165–182. Dekker, New York, 1995
6. M. Barge, Homoclinic intersections and indecomposability, Proc. Amer. Math. Soc. 101 (1987), 541–544
7. M. Barge, B. Diamond, Inverse limit spaces of infinitely renormalizable maps, Topology Appl. 83 (1998), 103–108
8. M. Barge, S. Holte, Nearly one-dimensional Hénon attractors and inverse limits, Nonlinearity 8 (1995), 29–42
9. M. Barge, B. Diamond, C. Holton, Asymptotic orbits of primitive substitutions, Theoret. Comput. Sci. 301 (2003), 439–450
10. M. Barge, H. Bruin, S. Štimac, The Ingram Conjecture, Geom. Topol. 16 (2012), 2481–2516
11. R. Bennett, *On Inverse Limit Sequences*, Master's Thesis, University of Tennessee, 1962
12. L. Block, S. Jakimovik, J. Keesling, L. Kailhofer, On the classification of inverse limits of tent maps, Fund. Math. 187 (2005), no. 2, 171–192
13. L. Block, J. Keesling, B. E. Raines, S. Štimac, Homeomorphisms of unimodal inverse limit spaces with a non-recurrent critical point, Topology Appl. 156 (2009), 2417–2425
14. A. Blokh, L. Lyubich, Measurable dynamics of S-unimodal maps of the interval, Ann. Sci. Ec. Norm. Sup. 24 (1991), 545–573
15. K. Brucks, H. Bruin, Subcontinua of inverse limit spaces of unimodal maps, Fund. Math. 160 (1999) 219–246
16. K. Brucks, B. Diamond, A symbolic representation of inverse limit spaces for a class of unimodal maps, Continuum Theory and Dynamical Systems. Lecture Notes in Pure Appl. Math. **149**, 207–226 (1995)
17. H. Bruin, Planar embeddings of inverse limit spaces of unimodal maps, Topology Appl. 96 (1999), 191–208
18. H. Bruin, Inverse limit spaces of post-critically finite tent maps, Fund. Math. 165 (2000) 125–138
19. H. Bruin, Asymptotic arc-components of unimodal inverse limit spaces, Topology Appl. 152 (2005) 182–200
20. H. Bruin, S. Štimac, Fibonacci-like unimodal inverse limit spaces and the core Ingram conjecture, Topol. Methods Nonlinear Anal. 47 (2016), 147–185
21. R. de Man, On composants of solenoids, Fund. Math. 147 (1995), 181–188
22. R. Fokkink, There are uncountably many homeomorphism types of orbits in flows, Fund. Math. 136 (1990), 147–156
23. C. Good, B. Raines, Continuum many tent maps inverse limits with homeomorphic postcritical ω-limit sets, Fund. Math. 191 (2006), 1–21
24. W.T. Ingram, *Inverse limits on [0, 1] using tent maps and certain other piecewise linear bounding maps, Continua with the Houston problem book (Cincinnati, OH, 1994)*, ed. by H. Cook et al. Lecture Notes in Pure and Appl. Math.170, Dekker (1995), pp. 253–258
25. L. Kailhofer, A classification of inverse limit spaces of tent maps with periodic critical points, Fund. Math. 177 (2003), 95–120

26. J. Milnor, W. Thurston, On iterated maps of the interval: I, II, Preprint 1977, Published in Lecture Notes in Mathematics, vol. 1342. Springer, Berlin New York, 1988, pp. 465–563 (1977)
27. P. Minc, 2_0^{α} ways of approaching a continuum with $[1, \infty)$. Topol. Appl. **202**, 4–54 (2016)
28. B. Raines, Inhomogeneities in non-hyperbolic one-dimensional invariant sets, Fund. Math. 182 (2004), 241–268
29. B. Raines, S. Štimac, A classification of inverse limit spaces of tent maps with non-recurrent critical point. Algebraic and Geometric Topology **9**, 1049–1088 (2009)
30. S. Štimac, Structure of inverse limit spaces of tent maps with finite critical orbit, Fund. Math. 191 (2006), 125–150
31. S. Štimac, A classification of inverse limit spaces of tent maps with finite critical orbit. Topology Appl. **154**, 2265–2281 (2007)
32. H. Bruin, S. Štimac, On isotopy and unimodal inverse limit spaces. Dis. Continu. Dyn. Systs. - Series A **32**, 1245–1253 (2012)
33. R. Swanson, H. Volkmer, Invariants of weak equivalence in primitive matrices. Ergodic Theory Dynam. Systems **20**, 611–626 (2000)

Dimension Theory of Some Non-Markovian Repellers Part I: A Gentle Introduction

Balázs Bárány, Michał Rams, and Károly Simon

Abstract Michael Barnsley introduced a family of fractals sets which are repellers of piecewise affine systems. The study of these fractals was motivated by certain problems that arose in fractal image compression, but the results we obtained can be applied for the computation of the Hausdorff dimension of the graph of some functions, like generalized Takagi functions and fractal interpolation functions. In this paper, we introduce this class of fractals and present the tools in the one-dimensional dynamics and nonconformal fractal theory that are needed to investigate them. This is the first part in a series of two papers. In the continuation, there will be more proofs and we apply the tools introduced here to study some fractal function graphs.

Keywords Self-affine measures · Self-affine sets · Hausdorff dimension.

2010 Mathematics Subject Classification Primary 28A80 · Secondary 28A78

The research of Bárány and Simon was partially supported by the grant OTKA K123782. Bárány acknowledges support also from NKFI PD123970, the János Bolyai Research Scholarship of the Hungarian Academy of Sciences and the New National Excellence Program of the Ministry of Human Capacities ÚNKP-18-4-BME-385. Michał Rams was supported by National Science Centre grant 2014/13/B/ST1/01033 (Poland). This work was partially supported by the grant 346300 for IMPAN from the Simons Foundation and the matching 2015–2019 Polish MNiSW fund.

B. Bárány · K. Simon (✉)
Department of Stochastics, MTA-BME Stochastics Research Group, Budapest University of Technology and Economics, P.O. Box 91, 1521 Budapest, Hungary
e-mail: simonk@math.bme.hu

M. Rams
Institute of Mathematics, Polish Academy of Sciences, ul. Śniadeckich 8, 00-656 Warszawa, Poland
e-mail: rams@impan.pl

R. L. Devaney et al. (eds.), *Topological Dynamics and Topological Data Analysis*, Springer Proceedings in Mathematics & Statistics 350,
https://doi.org/10.1007/978-981-16-0174-3_2

1 Introduction

This is a paper in the intersection of fractal geometry and dynamical systems. Dynamical systems provide us with beautiful and interesting examples of sets, fractal geometry gives us the language to describe them, and both theories give us tools. Tools to understand the geometric properties of those sets, tools to understand the dynamical properties, and most interesting of all—the relations between the two.

This is a paper about tools. Yeah, sure, we will prove some theorem eventually (in the second part of this paper)—but it is just a pretext. Our real goal is to describe the process of understanding the geometric behaviour of a dynamical system, starting from understanding the simplest possible models (conformal uniformly hyperbolic iterated function systems with separation properties) and then throwing out the training wheels, until we get to piecewise affine maps with quite general symbolic description (not necessarily subshifts of finite type).

And, most of all, this is a survey. While the simple models are in the books (the classical positions by Falconer [7] and by Mattila [17]), the modern theory of affine iterated function systems is not in books yet, and neither is Hofbauer's theory. We aren't going to be able to describe all the details, for sure, but we try to at least provide the main ideas and most useful formulas, and also the literature for further reading.

Fine, let us present the hero of our story.

2 Barnsley's Skew Product Maps

In order to define a piecewise affine and piecewise expanding skew product map F on the plane which sends the vertical stripe $D := [0, 1] \times \mathbb{R}$ into itself, first we partition the unit interval $[0, 1] = \bigsqcup_{i=1}^{m} I_i$.
Then we define $F: D \to D$ by

$$F(x, y) := F_i(x, y) \text{ if } (x, y) \in D_i := I_i \times \mathbb{R}, \tag{1}$$

where for all $i = 1, \ldots, m$

$$F_i(x, y) := (f_i(x), g_i(x, y)), \quad \text{for } (x, y) \in D_i \tag{2}$$

and $f_i: I_i \to J_i \subset [0, 1]$ (see Fig. 1) and $g_i: D_i \to \mathbb{R}$ and for $|\lambda_i|, |\gamma_i| > 1$ let

$$f_i(x) := \gamma_i x + v_i, \quad g_i(x, y) = a_i x + \lambda_i y + t_i. \tag{3}$$

Throughout this note we always assume:

Principal assumption The map $f: [0, 1] \to [0, 1]$

$$f(x) := f_i(x), \text{ if } x \in I_i \quad \text{is transitive,} \tag{4}$$

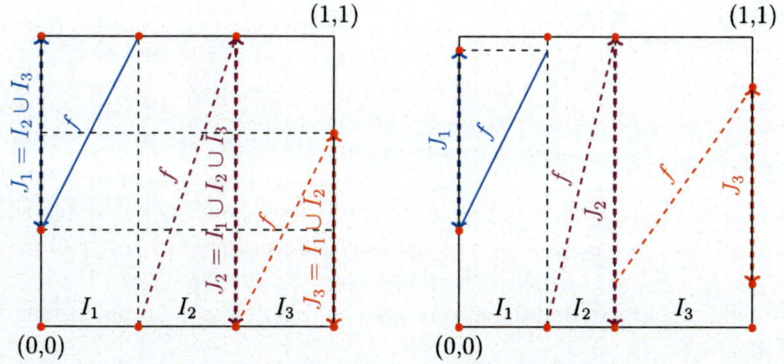

Fig. 1 f is Markov on the left hand-side and non-Markov on the righ-hand side

that is f has an orbit which is dense in [0, 1]. We call the repeller of $F: D \to D$ (which is the graph of a function) Barnsley repeller and we denote it by Λ. We call F Barnsley's skew product map. Let $\mathfrak{S} = \bigcup_{i=1}^{M} \partial I_i$ the singularity set and let $\mathfrak{S}_\infty = \bigcup_{n=0}^{\infty} f^{-n}(\mathfrak{S})$. It was pointed out by Barnsley that Λ is the graph of a function $G: [0, 1] \setminus \mathfrak{S}_\infty: \to \mathbb{R}$ which is defined by

$$G(x) = z, \text{ where } \{F^n(x, z)\}_{n=1}^{\infty} \text{ is bounded.} \tag{5}$$

3 The Hausdorff and Box Dimensions

For a $d \geq 1$ let $A \subset \mathbb{R}^d$ be a set of zero Lebesgue measure and let ν be a measure which is singular with respect to the Lebesgue measure $\mathcal{L}_d$. Then the size of A and ν can be expressed by their fractal dimensions.

3.1 Fractal Dimensions of Sets

The most common fractal dimensions are the Hausdorff and the box dimensions:

Definition 3.1 *(Hausdorff dimension)* Let $A \subset \mathbb{R}^d$ then

$$\dim_{\mathrm{H}} A := \inf \left\{ \alpha : \forall \varepsilon > 0, \exists \{U_i\}_{i=1}^{\infty}, \text{ such that } A \subset \bigcup_{i=1}^{\infty} U_i, \ \sum_{i=1}^{\infty} |U_i|^{\alpha} < \varepsilon \right\}, \tag{6}$$

where $|U_i|$ is the diameter of U.

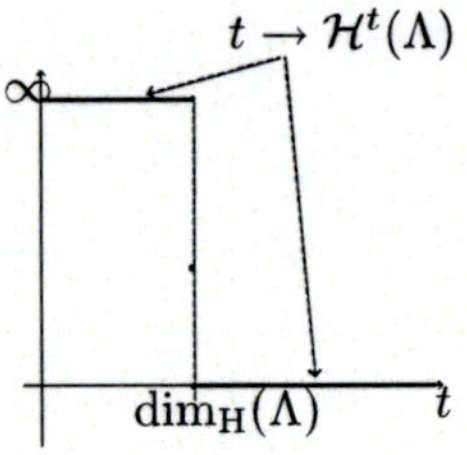

Fig. 2 The definition of the Hausdorff dimension

Equivalently in a more traditional way, we can first define the t-dimensional Hausdorff measure

$$\mathcal{H}^t(A) = \sup_{\delta \to 0} \inf \left\{ \boxed{\sum_{i=1}^{\infty} |E_i|^t} : \Lambda \subset \bigcup_{i=1}^{\infty} E_i,\ |E_i| < \delta \right\}, \tag{7}$$

then we write see (Fig. 2)

$$\dim_{\mathrm{H}} A := \inf \left\{ t : \mathcal{H}^t(A) = 0 \right\} = \sup \left\{ t : \mathcal{H}^t(A) = \infty \right\}. \tag{8}$$

Another very popular notion of fractal dimension is the box dimension:

Definition 3.2 $\dim_{\mathrm{B}} A$

Let $E \subset \mathbb{R}^d$, $E \neq \emptyset$, bounded. $N_\delta(E)$ be the smallest number of sets of diameter δ which can cover E. Then the lower and upper box dimensions of E:

$$\underline{\dim}_{\mathrm{B}}(E) := \liminf_{r \to 0} \frac{\log N_\delta(E)}{-\log \delta}, \tag{9}$$

$$\overline{\dim}_{\mathrm{B}}(E) := \limsup_{r \to 0} \frac{\log N_\delta(E)}{-\log \delta}. \tag{10}$$

If the limit exists then we call it the box dimension of E and we denote it by $\dim_{\mathrm{B}}(E)$.

3.2 *Hausdorff Dimension of Measures*

The Hausdorff dimension of a measure μ is the best lower bound on the Hausdorff dimension of a sets having large μ measures. Depending on what "large" means we define

Definition 3.3 Let μ be a Borel measure on $\mathbb{R}^d$ such that $0 < \mu(\mathbb{R}^d) < \infty$.

(a) Lower Hausdorff dimension of μ is: $\dim_*(\mu) := \inf\{\dim_{\mathrm{H}} A : \mu(A) > 0\}$,
(b) Upper Hausdorff dimension of μ: $\dim^*(\mu) := \inf\{\dim_{\mathrm{H}} A : \mu(A^c) = 0\}$.
(c) The lower and the upper local dimension of the measure μ are:

$$\underline{\dim}(\mu, x) := \liminf_{r\to 0} \frac{\log \mu(B(x,r))}{\log r} \tag{11}$$

and

$$\overline{\dim}(\mu, x) := \limsup_{r\to 0} \frac{\log \mu(B(x,r))}{\log r} \tag{12}$$

We say that the measure μ is exact dimensional if for μ-almost all x $\lim\limits_{r\downarrow 0} \frac{\log \mu(B(x,r))}{\log r}$ exists and equals to a constant.

Lemma 3.4 *Let μ be a measure like in Definition 3.3. Then*

$$\dim_* \mu = \operatorname{essinf}_{x\sim\mu} \underline{\dim}(\mu, x), \quad \dim^* \mu = \operatorname{esssup}_{x\sim\mu} \underline{\dim}(\mu, x) \tag{13}$$

4 Self-similar Sets

From now on we work on $\mathbb{R}^d$. Let $m \geq 2$ and $O_1, \dots, O_m \in O(d)$ orthogonal matrices and $r_1, \dots, r_m \in (0, 1)$ and $t_1, \dots, t_m \in \mathbb{R}^d$. Then

$$\mathcal{S} := \{S_i(x) = r_i \cdot O_i x + t_i\}_{i=1}^m \tag{14}$$

is called a self-similar Iterated Function System on $\mathbb{R}^d$.

Let $B := B(x, R)$ be a closed ball, where R is large. Then

$$\forall i = 1, \dots, m: \quad S_i(B) \subset B. \tag{15}$$

Hence the the following is a nested sequence of compact sets:

$$\left\{ \bigcup_{i_1 \dots i_n} S_{i_1 \dots i_n} B \right\}_{n=1}^{\infty},$$

where we use throughout the paper the notation: $S_{i_1 \dots i_n} := S_{i_1} \circ \cdots \circ S_{i_n}$. The attractor of our IFS $\mathcal{S}$ is

$$\Lambda := \bigcap_{n=1}^{\infty} \bigcup_{i_1 \dots i_n} S_{i_1 \dots i_n} B, \tag{16}$$

which is independent of B as long as B satisfies (15).

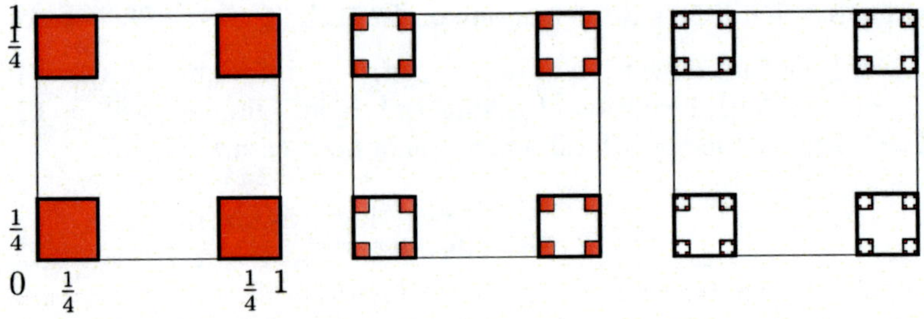

Fig. 3 The four-corner cantor set $C\left(\frac{1}{4}\right)$

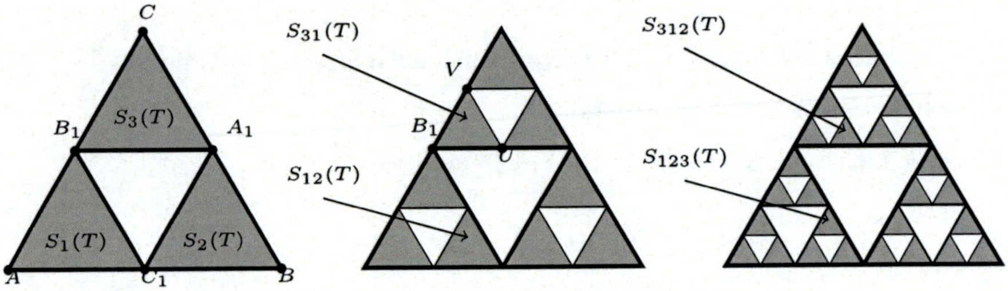

Fig. 4 The Sierpiński gasket: $S_{312}(x) := S_3 \circ S_1 \circ S_2(x) = S_3(S_1(S_2(x)))$

Example 4.1 (Four Corner Set). Figure 3 shows the first three iterations of a famous self-similar set, called the Four Corner Cantor set. Here $B = [0, 1]^2$ and

$$S_i(x, y) = \frac{1}{4}(x, y) + \mathbf{t}_i, \text{ for } \mathbf{t}_1 = (0, 0),\ \mathbf{t}_2 = \left(\frac{3}{4}, 0\right), \mathbf{t}_3 = \left(\frac{3}{4}, \frac{3}{4}\right), \mathbf{t}_3 = \left(0, \frac{3}{4}\right).$$

In the general case, we code the points of the attractor by the elements of the symbolic space:

$$\Sigma := \{1, \dots, m\}^{\mathbb{N}}. \tag{17}$$

The natural projection is $\Pi \colon \Sigma \to \Lambda$:

$$\Pi(\mathbf{i}) := \lim_{n\to\infty} S_{i_1\dots i_n}(0). \tag{18}$$

On Figs. 4 and 5 we indicate how this coding works.

S_i are translations of the appropriate homothety-transformatons of the form:

$$S_i(x) = \frac{1}{2}x + t_i.$$

The sets $\{S_i(T)\}_{i=1}^3$ in the previous examples ar the first cylinders, the sets $\left\{S_{i,j}(T)\right\}_{i,j=1}^3$ are the second cylinders an so on.

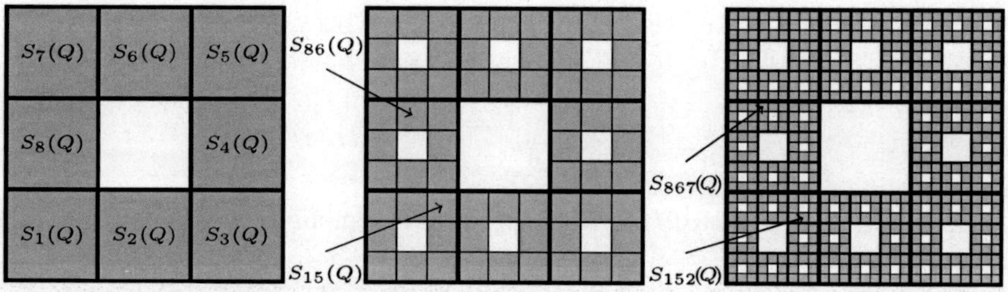

Fig. 5 The third approximation of the Sierpiński carpet

In both of the previous examples, the cylinders were not disjoint, but their interiors were disjoint. This results that the cylinders are well-separated.

Definition 4.2 *(SSP, OSC, SOSC)*. Here, we define three important separation conditions. These will be used in much more general setup then the self-similar IFS.

(a) If $S_i(\Lambda) \cap S_j(\Lambda) = \emptyset$ for all $i \neq j$ the we say that the Strong Separation Property (SSP) holds. (Like in the case of the Four Corner Cantor set.)
(b) If there exists a bounded open set V such that
    (1) $S_i(V) \subset V$ for all $i = 1, \dots, m$
    (2) $S_i(V) \cap S_j(V) = \emptyset$ for all $i \neq j$ then we say that the Open Set Condition (OSC) holds like in the case of the Sierpiński gasket and Sierpiński carpet. Here V is the interior of the right triangle and the unit square respectively.
    (c) If the OSC holds with an open set V satisfying $V \cap \Lambda \neq \emptyset$, where Λ is the attractor, then we say that the Strong Open Set Condition (SOSC) holds.

The OSC and SOSC are equivalent for self-similar (and also for self-conformal) IFS.

Now, we present a heuristic argument in order to guess the Hausdorff dimension of the attractor Λ in the case when the cylinders are disjoint (that is when SSP holds):

We will use the following fact: it is immediate from the definition that for any $r > 0$ we have:

$$\mathcal{H}^s(r \cdot A) := r^s \cdot \mathcal{H}^s(A). \tag{19}$$

Since this is only a heuristic argument we assume that for the appropriate s, (that is for the s satisfying $s = \dim_{\mathrm{H}} \Lambda$) the s-dimensional Hausdorff measure of the attractor Λ has positive and finite. Then,

$$\begin{aligned}\mathcal{H}^s(\Lambda) &= \sum_{i=1}^{m} \mathcal{H}^s(S_i \Lambda) \\ &= \sum_{i=1}^{m} r_i^s \mathcal{H}^s(\Lambda).\end{aligned}$$

By the assumption above, we can divide by $\mathcal{H}^s(\Lambda)$. This yields that:

$$\sum_{i=1}^{m} r_i^s = 1. \tag{20}$$

Even if $\mathcal{S}$ does not satisfy any of the previous assumptions we can define s as the solution of (20).

Definition 4.3 Let $\mathcal{S}$ be a self-similar IFS of the form (14). The similarity dimension $\dim_S(\Lambda) := s$ where s is the unique solution of (20). That is $\sum_{i=1}^{m} r_i^s = 1$. Sometimes, we also say that s is the similarity dimension of the attractor.

Clearly,

$$\dim_H(\Lambda) \leq \dim_S(\Lambda). \tag{21}$$

However "=" does not always hold:

Let $\Lambda_{1/3}$ be the attractor the $\mathcal{S}^{1/3}$ from (24):

$$\mathcal{S}^{1/3} = \mathcal{S} := \left\{ \frac{1}{3}x, \frac{1}{3}x + 1, \frac{1}{3}x + 3 \right\}.$$

Then

$$\dim_B(\Lambda_{1/3}) < 0.9 < 1 = \dim_S(\Lambda_{1/3}). \tag{22}$$

This is so because in this case

$$S_0^{1/3} \circ S_3^{1/3} \equiv S_1^{1/3} \circ S_0^{1/3}$$

so there is an exact overlap.

Theorem 4.4 (Hutchinson's-Moran Theorem [18] and [13]) *Let $\mathcal{S} := \{S_1, \ldots, S_m\}$ be a self-similar IFS on $\mathbb{R}^d$ with contraction ratios $r_1, \ldots, r_m$ and similarity dimension s. We assume that the OSC (Open Set Condition) holds.*
then

(a) $\dim_H \Lambda = s$, *even we have*
(b) $0 < \mathcal{H}^s(\Lambda) < \infty$,
(c) $\mathcal{H}^s\left(S_i(\Lambda) \cap S_j(\Lambda)\right) = 0$ *for all* $i \neq j$.

Theorem 4.5 (Falconer) *The Hausdorff- and box-dimensions are the same for any self-similar set.*

The following problem is one of the most interesting open problems in Fractal Geometry:

Conjecture 4.6 *(Complete Overlap Conjecture)* Let s be the similarity dimension and let Λ be the attractor of a self-similar IFS $\mathcal{S} = \{S_i\}_{i=1}^m$ on $\mathbb{R}$. Then

$$\dim_{\mathrm{H}}(\Lambda) < \min\{d, s\} \Longleftrightarrow \exists \mathbf{i}, \mathbf{j} \in \Sigma^*, \mathbf{i} \neq \mathbf{j} \text{ s.t. } S_{\mathbf{i}} \equiv S_{\mathbf{j}}. \tag{23}$$

In $\mathbb{R}^2$ the conjecture does not hold. The following example was introduced by Keane et al. [15] and played a very important role in the study of self-similar fractals with overlapping construction.

Example 4.7 For every $\lambda \in (\frac{1}{4}, \frac{2}{5})$ consider the following self-similar set:

$$\widetilde{\Lambda}_\lambda := \left\{ \sum_{i=0}^{\infty} a_i \lambda^i : a_i \in \{0, 1, 3\} \right\}.$$

Then $\widetilde{\Lambda}_\lambda$ is the attractor of the one-parameter (λ) family IFS:

$$\mathcal{S}^\lambda := \left\{ S_i^\lambda(x) := \lambda \cdot x + i \right\}_{i=0,1,3} \tag{24}$$

To normalize it, we write $\Lambda_\lambda := \frac{1-\lambda}{3} \cdot \widetilde{\Lambda}_\lambda$. It was proved by Solomyak [21] that for Lebesgue almost all $\lambda > \frac{1}{3}$ (that is when the similarity dimension is greater than one) we have

$$\dim_{\mathrm{H}} \Lambda_\lambda = 1. \tag{25}$$

Fix a λ slightly greater than $1/3$ for which (25) holds and consider the product set $C_\lambda := \Lambda_\lambda \times [0, 1]$ (see Fig. 6). Then for $\lambda \in \left(\frac{1}{3}, \frac{1}{\sqrt{6}}\right)$ we have

$$\dim_{\mathrm{H}} C_\lambda = 1 + \frac{\log 2}{-\log \lambda} < \min\left\{2, \frac{\log 6}{-\log \lambda}\right\} = \min\{2, \dim_{\mathrm{Sim}}(\mathcal{S})\}.$$

Since there are uncountably many λ like this, and complete overlap can happen only for countably many λ, we get that dimension drop occur in higher dimension not only when we have complete overlaps.

4.1 Self-similar Measures

Analogously to the self-similar sets, we can define the self-similar measures:

Definition 4.8 Given an $m \geq 2$, $\mathcal{S} = \{S_1, \ldots, S_m\}$ self-similar IFS on $\mathbb{R}^d$ with contraction ratios: $r_1, \ldots, r_m$ and we are given a probability vector $\mathbf{p} = (p_1, \ldots, p_m)$. Now we define the self-similar measure $\nu = \nu_{\mathcal{S},\mathbf{p}}$ which corresponds to $\mathcal{S}$ and $\mathbf{p}$:

$$\nu_{\mathcal{S},\mathbf{p}} := \Pi_* \left(\mathbf{p}^{\mathbb{N}}\right) := \mu \circ \Pi^{-1}. \tag{26}$$

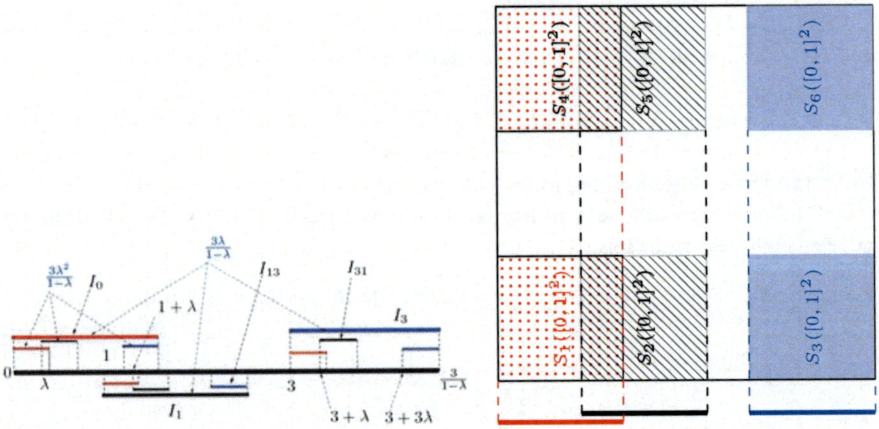

Fig. 6 $\widetilde{\Lambda}_\lambda$ and $C_\lambda := \Lambda_\lambda \times [0, 1]$

Then $\nu_{\mathcal{S},\mathbf{p}}$ is the unique probability Borel measure satisfying

$$\nu_{\mathcal{S},\mathbf{p}}(H) = \sum_{k=1}^{m} p_i \cdot \nu_{\mathcal{S},\mathbf{p}}\left(S_i^{-1}(H)\right), \tag{27}$$

for every Borel set H.

Let $\nu := \nu_{\mathcal{S},\mathbf{p}}$ be the invariant measure for the self-similar IFS on $\mathbb{R}^d$:

$$\mathcal{S} := \{S_i(x) = r_i \cdot O_i x + t_i\}_{i=1}^{m}. \tag{28}$$

Below we give a heuristic argument to show that if the OSC holds then the Hausdorff dimension of ν is equal to the similarity dimension of ν, which is defined by:

$$\dim_{\mathrm{Sim}} \nu := \frac{\sum_{i=1}^{m} p_i \log p_i}{\sum_{i=1}^{m} p_i \log r_i} = \frac{\text{entropy}}{\text{Lyapunov exponent}}. \tag{29}$$

Lemma 4.9 *$\mathcal{S}$ and* $\mathbf{p}$ *as above and we assume that the OSC holds. Then*

$$\dim_{\mathrm{H}} \nu = \dim_{\mathrm{Sim}} \nu. \tag{30}$$

Proof (Heuristic Proof) Let I be a large interval such that $S_i(I) \subset I$ for all $i = 1, \ldots, m$ and we write $I_{i_1 \ldots i_n} := S_{i_1 \ldots i_n} I$ for the level n cylinder intervals. It follows from Birkhoff's Ergodic Theorem that in this case the limit in (11) and (12) exist. That is, Lemma 3.4 indicates that for a ν-typical $x = \Pi(\mathbf{i})$, $\mathbf{i} \in \Sigma$:

$$\dim_{\mathrm{H}} \nu = \lim_{n\to\infty} \frac{\log \nu(I_{i_1\dots i_n})}{\log |I_{i_1\dots i_n}|} \overset{\text{def}}{=} \lim_{n\to\infty} \frac{\log p_{i_1\dots i_n}}{\log r_{i_1\dots i_n}}$$

$$= \frac{\lim\limits_{n\to\infty} \frac{1}{n} \sum\limits_{k=1}^{n} \log p_{i_k}}{\lim\limits_{n\to\infty} \frac{1}{n} \sum\limits_{k=1}^{n} \log r_{i_k}} \overset{\text{LLN}}{=} \frac{\sum\limits_{i=1}^{m} p_i \log p_i}{\sum\limits_{i=1}^{m} p_i \log r_i} = \dim_{\mathrm{Sim}} \nu,$$

where LLN means Law of Large Numbers. Here, we used the notations: $p_{i_1\dots i_n} := p_{i_1} \cdots p_{i_n}$ and $r_{i_1\dots i_n} := r_{i_1} \cdots r_{i_n}$

4.1.1 Hochman Theorem

Let $\mathcal{S} = \{S_i\}_{i=1}^m$ be a self-similar IFS on $\mathbb{R}$ with contraction ratios $\{r_i\}_{i=1}^m$. Let $\Delta_n(\mathcal{S})$ be the smallest distance between the left end points of two level n cylinders having the same length. More formally, $\Delta_n(\mathcal{S})$ is the minimum of $\Delta(\boldsymbol{\omega}, \boldsymbol{\tau})$ for distinct $\boldsymbol{\omega}, \boldsymbol{\tau} \in \Sigma_n$, where

$$\Delta(\boldsymbol{\omega}, \boldsymbol{\tau}) = \begin{cases} \infty & S'_{\boldsymbol{\omega}}(0) \neq S'_{\boldsymbol{\tau}}(0) \\ |S_{\boldsymbol{\omega}}(0) - S_{\boldsymbol{\tau}}(0)| & S'_{\boldsymbol{\omega}}(0) = S'_{\boldsymbol{\tau}}(0). \end{cases}$$

Condition 4.10 *(HESC)* We say that the self-similar IFS $\mathcal{S}$ satisfies Hochman's exponential separation condition (HESC) if there exists an $\varepsilon > 0$ and an $n_k \uparrow \infty$ such that

$$\Delta_{n_k} > \varepsilon^{n_k}. \tag{31}$$

Hochman proved the following very important assertion in [9, Theorem 1.1].

Theorem 4.11 (Hochman) *Assume that $\mathcal{S} = \{S_i\}_{i=1}^m$ is a self-similar IFS on $\mathbb{R}$ which satisfies Hochman's exponential separation condition. Let $\mathbf{p} = (p_1, \dots, p_N)$ be an arbitrary probability vector. Then*

$$\dim_H \left(\nu_{\mathcal{S},\mathbf{p}}\right) = \min \{1, \dim_{\mathrm{Sim}} \nu\}, \tag{32}$$

Remark 4.12 (Relation to the Compete Overlaps Conjecture) Although Hochman's Theorem does not solve the Compete Overlaps Conjecture (Conjecture 4.6) but it makes a very significant progress towards it.

- Exact overlap means that $\Delta_n = 0$ for some n.
- If the OSC holds then $\Delta_n \to 0$ exactly exponentially fast.
- $\Delta_n \to 0$ at least exponentially fast always holds. Namely: $\#\{|\mathbf{i}| = n\} = m^n$. On the other hand: $\#\{r_{\mathbf{i}} \colon |\mathbf{i}| = n\}$ is polynomially large ($r_{\mathbf{i}}$ was the contraction ration of $S_{\mathbf{i}}$). So, there exist distinct $\mathbf{i}, \mathbf{j}$ of length n with $r_{\mathbf{i}} = r_{\mathbf{j}}$ and with with exponentially small $\left|S_{\mathbf{i}}(0) - S_{\mathbf{j}}(0)\right|$.

- However, in case of a dimension drop, that is, if we can find a probability vector $\mathbf{p}$ such that $\dim_{\mathrm{H}} \nu_{\mathcal{S},\mathbf{p}} < \min\{1, \dim_{\mathrm{S}} \nu\}$ then $\Delta_n \to 0$ super exponentially fast. That is

$$\lim_{n\to\infty} -\frac{1}{n} \log \Delta_n = \infty.$$

The following theorem shows that Hochman's theorem solves the complete overlap conjecture in some cases:

Theorem 4.13 (Hochman) *For an self-similar IFS on the line with algebraic parameters we have either exact overlaps, or no dimension drop:* $\dim_{\mathrm{H}} \Lambda = \min\{1, \dim_{\mathrm{S}} \Lambda\}$.

5 Dimension of the Self-conformal Sets and Measures When OSC Holds

We can extend a large part of the dimension theory of self-similar sets to the so called self-conformal ones by using the notion of the topological pressure.

Definition 5.1 *(Conformal IFS on the line)* Let $\eta > 0$ and $m > 1$. We are given $f_1, \dots, f_m \colon [0, 1] \to [0, 1]$ satisfying the following conditions:

(a) $f_i \in C^{1+\eta}[0, 1]$ for all $i = 1, \dots, m$,
(b) $\exists\, 0 < c_1, c_2 < 1$ such that $c_1 < |f_i'(x)| < c_2$ holds for all $i = 1, \dots, m$ and all $x \in [0, 1]$.

Then we say that

$$\mathcal{F} := \{f_1, \dots, f_m\} \tag{33}$$

is a self-conformal IFS. We can define the attractor, the symbolic space and the natural projection analogously as we did in (16), (17) and (18), respectively.

A very important property of the self-conformal IFS the following:

Theorem 5.2 (Bounded Distortion Property) *Let $\mathcal{F}$ be as in Definition 5.1. Then there exist $0 < c_3 < c_4$ such that for all n and for all $(i_1, \dots, i_n) \in (1, \dots, m)^n$ and for all $x, y \in [0, 1]$ we have*

$$c_3 < \frac{f'_{i_1,\dots,i_n}(x)}{f'_{i_1,\dots,i_n}(y)} < c_4, \tag{34}$$

The proof is available in [19]. Our aim is to calculate the Hausdorff dimension of the attractor.

5.1 Hausdorff Dimension of Self-conformal Sets When OSC Is Assumed

Theorem 5.3 *Let $\mathcal{F}$ be a conformal IFS on $\mathbb{R}$ as in definition 5.1 and we assume that the OSC holds. Let s_0 be the root of the pressure formula that is we assume that (82) holds. Then*

$$\dim_{\rm H} \Lambda = s_0. \tag{35}$$

Proof First, we prove that $\dim_{\rm H} \Lambda \leq s_0$. This is so, since the system of level n cylinder intervals $\mathcal{I}_n := \left\{ f_{i_1 \dots i_n}([0,1]) \right\}_{(i_1 \dots i_n) \in (1,\dots,m)^n}$ gives a cover of as small diameter as we want if n is large enough. Moreover, by Lagrange Theorem for suitable $x_\omega \in [0,1]$

$$\sum_{I \in \mathcal{I}_n} |I|^{s_0} = \sum_{|\omega|=n} |f'_\omega(x_\omega)|^{s_0} \leq \frac{1}{c_1 c_3} \sum_{|\omega|=n} \mu(\omega) = \frac{1}{c_1 c_3}.$$

That is $\mathcal{H}^{s_0}(\Lambda) < \infty$ consequently $\dim_{\rm H} \Lambda \leq s_0$.

Now we prove that $\dim_{\rm H} \Lambda \geq s_0$. Let μ be the Gibbs measure for the potential ϕ_{s_0} [defined in (78)]. Fix an arbitrary $\mathbf{i} \in \Sigma$. Then, putting together (77), (82) and (83) we obtain the following limit exists

$$\lim_{n \to \infty} \frac{\log \Pi_* \mu(I_{i_1 \dots i_n})}{\log |I_{i_1 \dots i_n}|} \equiv s_0.$$

That is the local dimension of the measure $\Pi_* \mu$ is equal to s_0 at all points of the attractor Λ. Hence $\dim_{\rm H} \Pi_* \mu = s_0$. This implies that $\dim_{\rm H} \Lambda \geq s_0$.

We say that the measure μ in the previous proof is the natural measure for the IFS $\mathcal{F}$.

5.2 Hausdorff Dimension of an Invariant Measure and Lyapunov Exponents

Now, we present the Lyapunov exponents for the classes of maps that occur in this paper.

Ergodic measures for a piecewise monotone map on the interval. Let η be an ergodic measure for a $T\colon [0,1] \to [0,1]$ piecewise monotonic map. Then, the Lyapunov exponent $\chi(\eta) = \int \log |T'| d\eta$. It follows from Hoffbauer and Raith [11, Theorem 1] that

$$\dim_{\rm H} \eta = \frac{h(\mu)}{\chi(\eta)} \quad \text{if } \chi(\eta) > 0. \tag{36}$$

6 The Hausdorff Dimension of Self-affine Sets

Definition 6.1 *(Self-affine IFS and self-affine measures)* We say that

$$\mathcal{F} := \{f_1(x) = A_1x + t_1, \ldots, f_m(x) = A_mx + t_m\} \tag{37}$$

is a self-affine IFS on $\mathbb{R}^d$ for a $d \geq 2$ if $A_1, \ldots, A_m$ are contractive non-singular $d \times d$ matrices and $t_1, \ldots, t_m \in \mathbb{R}^d$. The natural projection Π from the symbolic $\Sigma := \{1, \ldots, m\}^{\mathbb{N}}$ space to the attractor Λ [which is defined as in (16)] is defined as in the self-similar case: $\Pi(\mathbf{i}) := \lim_{n\to\infty} f_{i_1} \circ \cdots \circ f_{i_n}(0)$. The attractors of self-affine IFS are called self-affine sets. The computation of the dimension of the self-affine sets is much more difficult. Namely, in the self-similar case if the cylinders are well-separated that is OSC holds (see Definition 4.2) then

(a) The Hausdorff dimension of the attractor is equal to the similarity dimension s, which can be calculated merely from the contraction ratios (20), regardless the translations, as long as the cylinders remain well-separated.
(b) The appropriate dimensional Hausdorff measure of the attractor is positive and finite.
(c) The Hausdorff and the box dimensions of self-similar sets are the same.

In the self-affine case, we will define the affinity dimension which replaces the similarity dimension. However, not any of the assertions (a)–(c) hold for all self-affine sets with disjoint cylinders.

Example 6.2 On the left-hand side Fig. 7 we see three copies of the unit square. Focus on the one which is on the left-hand side. It contains six shaded rectangles of size $\frac{1}{3} \times \frac{1}{5}$. Denote their left bottom corners by $t_1, \ldots, t_6$ in any particular order. Then, we define the IFS

$$\mathcal{F}^l := \left\{ f_i(x) = \begin{pmatrix} \frac{1}{3} & 0 \\ 0 & \frac{1}{5} \end{pmatrix} \cdot x + t_i \right\}_{i=1}^{6}.$$

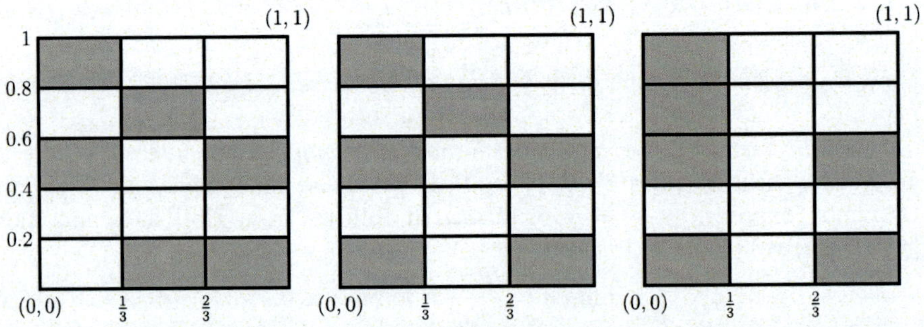

Fig. 7 Left: $\dim_H \Lambda^l = \dim_B \Lambda^l = \dim_{Aff} \Lambda$ middle: $\dim_H \Lambda^m < \dim_B \Lambda^m = \dim_{Aff} \Lambda^m$ right: $\dim_H \Lambda^r < \dim_B \Lambda^r < \dim_{Aff} \Lambda^r$

Let Λ^l be the attractor of $\mathcal{F}^l$. Clearly, the first cylinders of $\mathcal{F}^l$ are the shaded rectangles on Figure. We say that $\mathcal{F}^l$ and Λ^l are generated by the left hand-side of the Fig. 7. We define $\mathcal{F}^m$, Λ^m and $\mathcal{F}^r$, Λ^r, respectively, generated by the rectangles in the middle and right-hand side unit squares on Fig. 7. These self affine sets belongs to the family of Bedford-McMullen carpets (see [7] for more details). The linear parts are the same in each of the three systems they differ only in the translation vectors. However, $\dim_{\mathrm{H}} \Lambda^l = \dim_{\mathrm{B}} \Lambda^l = \dim_{\mathrm{Aff}} \Lambda^l$, $\dim_{\mathrm{H}} \Lambda^m < \dim_{\mathrm{B}} \Lambda^m = \dim_{\mathrm{Aff}} \Lambda^m$ and $\dim_{\mathrm{H}} \Lambda^r < \dim_{\mathrm{B}} \Lambda^r < \dim_{\mathrm{Aff}} \Lambda^r$, where the affinity dimension $\dim_{\mathrm{Aff}}$ plays the same rolle here as the similarity dimension in the case of self-similar sets and it will be defined in Sect. 6.1.

Moreover, if d^l, d^m and d^r are the Hausdorff dimension of Λ^l, Λ^m and Λ^r respectively, then

$$0 < \mathcal{H}^{d^l}(\Lambda^l) < \infty, \quad \mathcal{H}^{d^m}(\Lambda^m) = \mathcal{H}^{d^r}(\Lambda^r) = \infty.$$

For simplicity, here we explain everything on the plane but the definitions and discussions in $\mathbb{R}^d$ for $d \geq 3$ are similar (see e.g. [7, Sect. 9.4] for the introduction in higher dimension).

We can define the self-affine measures exactly as we defined self-similar measures in Sect. 4.1. That is for a probability vector $\mathbf{p} = (p_1, \ldots, p_m)$ the self-affine measure corresponding to $\mathcal{F}$ and $\mathbf{p}$ is

$$\nu = \nu_{\mathcal{F},\mathbf{p}} := \Pi_*(\mathbf{p}^{\mathbb{N}}). \tag{38}$$

6.1 Singular Value Function, Affinity Dimension, Falconer's Theorem

Most of the basic concepts of this field were introduced by Falconer [8]. The *singular value function* $\phi^s(A)$ of a matrix A is defined by

$$\phi^s(A) = \begin{cases} \alpha_{\lceil s \rceil}(A)^{s-\lfloor s \rfloor} \prod_{j=1}^{\lfloor s \rfloor} \alpha_j(A) & \text{if } 0 \leq s \leq \operatorname{rank}(A), \\ |\det(A)|^{s/\operatorname{rank}(A)} & \text{if } \operatorname{rank}(A) < s, \end{cases} \tag{39}$$

where $\alpha_i(A)$ denotes the ith singular value of A. On the plane, for a non-singular matrix A this is simply

$$\phi^s(A) := \begin{cases} \alpha_1(A), & \text{if } s \leq 1; \\ \alpha_1(A)\alpha_2^{s-1}(A), & \text{if } 1 \leq s \leq 2; \\ (\alpha_1(A)\alpha_2(A))^{s/2}, & \text{if } s \geq 2. \end{cases} \tag{40}$$

Using the singular value function Falconer [8] defined the affinity dimension $\dim_{\mathrm{Aff}} \Lambda$ as the root of the subadditive pressure formula

$$P_{A_1,\ldots A_m}(\dim_{\mathrm{Aff}} \Lambda) = 0, \tag{41}$$

where the function $s \mapsto P_{A_1,\ldots A_m}(s)$ is defined in the Appendix Example B.3. This is the value of the Hausdorff dimension of Λ in most of the cases.

Theorem 6.3 (Falconer) *Fix the $d \times d$ non-singular matrices $A_1, \ldots, A_m$ in any particular ways satisfying $\max_{1\le i\le m} \|A_i\| < 1/2$. For every $\mathbf{t} = (t_1, \ldots, t_m) \in \mathbb{R}^{md}$ we consider the following self-affine IFS on $\mathbb{R}^d$: $\mathcal{F}^{\mathbf{t}} := \{f_i(x) := A_i x + t_i\}_{i=1}^m$, where the translations $\mathbf{t} = (t_1, \ldots, t_m)$ are considered as parameters. Then,* $\dim_{\mathrm{H}} \Lambda = \dim_{\mathrm{B}} \Lambda = \dim_{\mathrm{Aff}} \Lambda$ *for Lebesgue almost all choices of $(t_1, \ldots, t_m) \in \mathbb{R}^{dm}$.*

7 Ergodic Measures for a Self-affine IFS

Let $\mathcal{F}$ be a self-affine IFS as in Definition 6.1. Then for an arbitrary ergodic measure ν on Σ we have

$$\chi_k(\nu) := \chi_k(\Pi_*\nu) := \lim_{n\to\infty} \frac{1}{n} \log \alpha_k(A_{i_1} \cdots A_{i_n}). \tag{42}$$

where $\alpha_k(B)$ is the k-th singular value of the matrix B.

In high generality, we know only almost all type formulas for the Hausdorff dimension of $\Pi_*\nu$. Namely, we consider the translations $\mathbf{t} = (t_1, \ldots, t_m)$ as parameters (as in Theorem 6.3) in the self affine IFS of the form (37) and we write $\mathcal{F}^{\mathbf{t}}$ instead of $\mathcal{F}$, $\Pi^{\mathbf{t}}$ instead of Π and $\Pi^{\mathbf{t}}_*\nu$ instead of $\Pi_*\nu$. Then [14, Theorem 1.9] gives an analogous assertion to Falconer's theorem (Theorem 6.3) for self-affine measures instead of self-affine sets:

Theorem 7.1 (Jordan Pollicott and Simon) *Let ν be an arbitrary ergodic measure on $\Sigma = \{1, \ldots, m\}^{\mathbb{N}}$. If $\max_{1\le i\le m} \|A_i\| < 1/2$ then for almost all $\mathbf{t}$ (w.r.t. the $m \cdot d$-dimensional Lebesgue measure) we have*

$$\dim_{\mathrm{H}}(\Pi^{\mathbf{t}}_*\nu) = \min\{d, D(\nu)\}, \tag{43}$$

where $D(\nu)$ is the Lyapunov dimension for the ergodic measure ν defined below.

Definition 7.2 Let $\mathcal{F}$ be a self-affine IFS as in Definition 6.1. Then, for an arbitrary ergodic measure ν on Σ

$$D(\nu) := k + \frac{h(\nu) + \chi_1(\nu) + \cdots + \chi_k(\nu)}{-\chi_{k+1}(\nu)}, \tag{44}$$

if $k = k(\nu) = \max\{i : 0 < h(\nu) + \chi_1(\nu) + \cdots + \chi_i(\nu)\} \le d$. On the other hand, if $0 < h(\nu) + \chi_1(\nu) + \cdots + \chi_d(\nu)$ then we define

$$D(\nu) := d \cdot \frac{h(\nu)}{-(\chi_1(\nu) + \cdots + \chi_d(\nu))}. \tag{45}$$

We call $D(\nu)$ the Lyapunov dimension of the measure ν.

Example 7.3 In this paper, we mostly work on the plane ($d = 2$). In this case

$$D(\nu) = \begin{cases} \frac{h(\nu)}{|\chi_1(\nu)|}, & \text{if } h(\nu) \leq |\chi_1(\nu)| ; \\ 1 + \frac{h(\nu) - |\chi_1(\nu)|}{|\chi_2(\nu)|}, & \text{if } |\chi_1(\nu)| \leq h(\nu) \leq |\chi_1(\nu)| + |\chi_2(\nu)|; \\ 2 \cdot \frac{h(\nu)}{|\chi_1(\nu)| + |\chi_2(\nu)}|, & \text{if } |\chi_1(\nu)| + |\chi_2(\nu)| \leq h(\nu). \end{cases} \tag{46}$$

Recently, there have been a number of very significant achievements on this field. Here, we mention only one of them. Bárány, Hocfhman and Rapaport [1, Theorem 1.2] computed the Hausdorff dimension of self-affine measures under some mild conditions. They obtained this by combining the entropy growth theorem by Hochman [9] with the method of Bárány and Käenmäki [2] about the dimension of the projections of self-affine measures, that they got by an application of the Furstenberg measures.

7.1 *Self-affine Measures*

Definition 7.4 Let $\mathcal{F} := \{f_i(x) := A_i x + t_i\}_{i=1}^m$ be a self-affine IFS on $\mathbb{R}^d$ and let $\mathbf{p}$ be a probability vector. Then, the corresponding self-affine measure can be defined exactly as we defined the self-similar measures. That is

$$\nu = \nu_{\mathcal{F},\mathbf{p}} := \Pi_* \left(\mathbf{p}^{\mathbb{N}}\right), \tag{47}$$

In their very recent seminal paper Bárány, Hochman and Rapaport [1, Theorems 1.1 and 1.2] proved the following

Theorem 7.5 (Bárány, Hochman and Rapaport) *Let $\mathcal{F} := \{f_i(x) := A_i x + t_i\}_{i=1}^m$ be a self-affine IFS on $\mathbb{R}^2$ which satisfies both of the following conditions:*

(a) the strong open set condition (see Definition 4.2) and
(b) The normalized linear parts $\left\{A_i/\sqrt{|\det A_i|}\right\}_{i=1}^m$ generate a non-compact and totally irreducible subgroup of $GL_2(\mathbb{R}^d)$ (that is they do not preserve any finite union of non-trivial linear spaces,)

Then for an arbitrary probability vector, $\mathbf{p}$ *we have*

$$\dim_{\mathrm{H}} \nu_{\mathcal{F},\mathbf{p}} = D(\nu_{\mathcal{F},\mathbf{p}}) \text{ and } \dim_{\mathrm{H}} \Lambda = \dim_{\mathrm{B}} \Lambda = \dim_{\mathrm{Aff}} \Lambda, \tag{48}$$

where Λ is the attractor of $\mathcal{F}$ and we remind the reader that the affinity dimension $\dim_{\mathrm{Aff}}$ *was defined in* (41).

This theorem does not cover the case of those self affine IFS for which all of the mappings have lower triangular linear parts. However, the same authors proved in [1, Proposition 6.6]

Theorem 7.6 (Bárány, Hochman and Rapaport) *Let* $\mathcal{F} := \{f_i(x) := A_i x + t_i\}_{i=1}^m$ *be a self-affine IFS on* $\mathbb{R}^2$ *which satisfies both of the following conditions:*

(c) The linear parts of all of the mapping of $\mathcal{F}$ *are lower triangular:* $A_i = \begin{pmatrix} a_i & 0 \\ b_i & c_i \end{pmatrix}$ *for* $i = 1, \dots, m$ *and*

(d) $a_i < c_i$ *for all* $i = 1, \dots, m$.

Then, for an arbitrary probability vector $\mathbf{p}$ *we have*

$$\dim_{\mathrm{H}} \nu_{\mathcal{F},\mathbf{p}} = D(\nu_{\mathcal{F},\mathbf{p}}) \text{ and } \dim_{\mathrm{H}} \Lambda = \dim_{\mathrm{B}} \Lambda = \dim_{\mathrm{Aff}} \Lambda, \tag{49}$$

where Λ *is the attractor of* $\mathcal{F}$.

8 Ergodic Measures for Barnsley's Skew Product Maps

We use the notation of Sect. 2. Let μ be an ergodic measure for the Barnsley's skew product map F, which was defined in Sect. 2. The two Lyapunov exponents $\chi_1(\mu)$ and $\chi_2(\mu)$ of F are

$$\chi_x(\mu) = \int \log \|D_{\mathrm{proj}(\mathbf{x})} f\| \mathrm{d}\mu(\mathbf{x}) = \sum_{i=1}^{m} \mu(I_i \times \mathbb{R}) \log \gamma_i \text{ and}$$

$$\chi_y(\mu) = \int \log \|\partial_2 g(\mathbf{x})\| \mathrm{d}\mu(\mathbf{x}) = \sum_{i=1}^{m} \mu(I_i \times \mathbb{R}) \log \lambda_i,$$

where $\mathrm{proj}(\mathbf{x})$ is the orthogonal projection of an $\mathbf{x} \in D$ to the x-axis and ∂_2 means the derivative with respect to the second coordinate.

Remark 8.1 If $0 < \chi_x(\mu) \le \chi_y(\mu)$ then

$$\dim \mu = \frac{h(\mu)}{\chi_x(\mu)},$$

Namely, the upper bound is trivial, and the lower bound follows from the fact that $\mathrm{proj}_* \mu$ is f-invariant and ergodic and the result of Hofbauer and Raith [11, Theorem 1] [see (36)]. That is why we can restrict ourselves to the case when

$$\chi_1(\mu) := \chi_x(\mu) = \sum_{i=1}^{m} \mu(I_i \times \mathbb{R}) \log \gamma_i > \chi_2(\mu) := \chi_y(\mu) = \sum_{i=1}^{m} \mu(I_i \times \mathbb{R}) \log \lambda_i > 0. \tag{50}$$

In this case, the best guess for the dimension of the μ is the so-called Lyapunov dimension to be defined below.

Definition 8.2 Let $\mu \in \mathcal{E}_F(\Lambda)$ satisfying $\chi_x(\mu) > \chi_y(\mu) > 0$. We define the Lyapunov dimension

$$D(\nu) := \begin{cases} \frac{h(\nu}{\chi_y(\nu)}, & \text{if } h(\nu) \leq \chi_y(\nu)\ ; \\ 1 + \frac{h(\nu)-\chi_y(\nu)}{\chi_x(\nu)}, & \text{if } \chi_y(\nu) \leq h(\nu) \leq \chi_x(\nu) + \chi_y(\nu); \\ 2 \cdot \frac{h(\nu)}{\chi_x(\nu)+\chi_y(\nu)}, & \text{if } \chi_x(\nu) + \chi_y(\nu) \leq h(\nu). \end{cases} \tag{51}$$

9 Hofbauer's Pressure

In the previous sections (and in the appendix), we presented the dimension theory for the self-affine iterated function systems. However, the principal distinction of the Barnsley's maps from the iterated function systems lies in the fact that the symbolic space for the Barnsley's skew product map is not a full shift. In this section, we will present the most general version of thermodynamical formalism theory, developed in a series of papers by Franz Hofbauer with his co-authors. This theory is not completely general, it assumes the system comes form piecewise monotone maps of the interval, but this assumption is satisfied in our situation.

Let us remind the notations. Our base map $f : [0, 1] \to [0, 1]$ is piecewise monotone: we can divide the interval $[0, 1]$ into finitely many closed intervals with disjoint interiors $[0, 1] = \bigcup_1^m I_i$. We denote by $\mathfrak{S}$ the set of endpoints of intervals I_i. We assume that $f|_{I_i^o}$ is continuous and monotone (strictly increasing or strictly decreasing) on I_i^o. We define f_i as the extension of $f|_{I_i^o}$ by continuity to the endpoints of I_i.

In order that the symbolic expansion of the system (to be defined below) is compact, we need to take a formal modification of the maps. We would like to consider f_i as the restriction of f to I_i. Naturally, such a definition can in general lead to the map being doubly defined on some points in $\mathfrak{S}_\infty$, but this set is countable. Formally speaking, if for a point $x \in \mathfrak{S}$ the left and right limits of f disagree then we define $f(x_-) = \lim_{z\nearrow x} f(z)$ and $f(x_+) = \lim_{z\searrow x} f(z)$. We then proceed to inductively double all the preimages of x. For a point $y \in f^{-1}(x)$, $y \notin \mathfrak{S}$ we define: if f is increasing at y then $f(y-) = x_-$ and $f(y_+) = x_+$, otherwise $f(y-) = x_+$ and $f(y_+) = x_-$. And for a point $y \in f^{-1}(x)$, $y \in \mathfrak{S}$: if $\lim_{z\nearrow y} f(z) = x$ and f is increasing in $(y - \varepsilon, y)$ then $f(y_-) = x_-$, if it is decreasing then $f(y_-) = x_+$, if $\lim_{z\searrow y} f(z) = x$ and f is increasing in $(y, y + \varepsilon)$ then $f(y_+) = x_+$, if it is decreasing then $f(y_+) = x_-$. We set the natural topology: at each doubled point x $\lim_{z\nearrow x} z = x_-$, $\lim_{z\searrow x} z = x_+$. We also redefine the partition intervals: if $I_i = [x, y]$ and one or both of the endpoints are doubled then we set $I_i = [x_+, y_-]$.

Observe that the resulting set is not an interval anymore, but a Cantor set - but with a natural projection onto the interval, which is 2-1 on a countable set and 1-1 elsewhere. The well-known special case of this construction: consider the interval $[0, 1]$ with the map $f(x) = 2x(\text{mod}1)$ and divide each dyadic point into two. That is, $1/2 = 0.10000..._2 = 0.01111..._2$, we formally define $(1/2)_- = 0.01111..._2$ and $(1/2)_+ = 0.10000..._2$ – and the same for all the other dyadic points. The result is a full shift on two symbols, which is conjugate (modulo a countable set) to the original map.

Note that for the piecewise monotone map the minimal possible partition is given by the intervals of monotonicity of f, but we can freely subdivide the intervals I_i further, and the resulting maps will also belong to considered class. In particular, we can freely demand that for any given continuous potential $\varphi\colon [0, 1] \to \mathbb{R}$ its variation $\sup \varphi - \inf \varphi$ is arbitrarily small on each I_i.

Let A be a compact, f-invariant, f-transitive set. For the rest of the section, our dynamical system will be the restriction of f to A.

Let $\widetilde{\Sigma} \subset \{1, \ldots, m\}^{\mathbb{N}}$ be the symbolic system of our dynamics, defined as the set of sequences $\omega \in \{1, \ldots, m\}^{\mathbb{N}}$ such that there exists $x \in A$ such that for $n = 0, 1, \ldots$

$$f^n(x) \in I_{\omega_n}.$$

One can check that $\widetilde{\Sigma}$ is a *subshift*, that is a σ-invariant and closed subset of $\{1, \ldots, f\}^{\mathbb{N}}$. The sequence ω will be called *symbolic expansion* of x, x will be called *representation* of ω. We will write $x = \pi(\omega)$. We will assume the partition $\{I_i\}$ is *generating*, that is each $\omega \in \widetilde{\Sigma}$ has unique representation. This always holds if f is expanding.

For any finite word $\tau^n \in \{1, \ldots, m\}^n$ denote by $C[\tau^n]$ the set of points $x \in A$ such that $\pi^{-1}(x)$ begins with τ^n. This set will be called n-th level *cylinder*. The set of n-th level cylinders will be denoted D_n. For $x \in A$, let $C_n(x)$ be the n-th level cylinder containing x. Denote $d_n(x) = \text{diam} C_n(x)$ and $\varphi_n(x) = \sup\{\varphi(y) - \varphi(z); y, z \in C_n(x)\}$. We have

$$\lim_{n\to\infty} d_n(x) = \lim_{n\to\infty} \varphi_n(x) = 0.$$

Definition 9.1 We say that A is *Markov* if there exists such partition $\{I_i\}$ and such n that for every n-th level cylinder $C[\tau^n]$ its image $T(C[\tau^n])$ is a union of n-th level cylinders. Equivalently, A is Markov if for some partition $\{I_i\}$ the subshift $\widetilde{\Sigma}$ is a *subshift of finite type*, that is, a subshift defined as all the infinite words $\omega \in \{1, \ldots, m\}^{\mathbb{N}}$ that do not contain any word from some finite list of finite words.

9.1 Pressure and Markov Sets

Let $\varphi\colon [0, 1] \to \mathbb{R}$ be a piecewise continuous potential, with the set of discontinuities contained in $\mathfrak{S}$. For the Markov systems we can define the pressure in the usual way:

$$P(A, \varphi) = \lim_{n\to\infty} \frac{1}{n} \log \sum_{C[\omega^n]\in D_n} \exp(\sup_{x\in C[\omega^n]} S_n\varphi(x)), \tag{52}$$

compare (76). For the non-Markov systems the right hand side of this equation is still well-defined, but is considered too large for applications in dimension theory. Let us give a short explanation.

In the year 1973 Bowen [3] gave the following definition of topological entropy: given a continuous map $f : X \to X$, where X is any f-invariant set (not necessarily compact), let X_n be the n-th level cylinders, then

$$h_{\mathrm{top}}(f, X) = \inf\{s;\ \inf_{X\subset\bigcup E_i} \sum e^{-sn(E_i)} = 0\},$$

where the sum is taken over covers of X with cylinders and for a cylinder E $n(E)$ denotes its level. Geometrically, the Bowen's definition of topological entropy is similar to the Hausdorff dimension as the usual definition (67) is similar to the box counting dimension—or more precisely, the Bowen's definition is the Hausdorff dimension and (67) is the box counting dimension, both calculated in a special metric (so-called dynamical metric). Still, Bowen proved that for compact X the two definitions are equal, while for noncompact the Bowen's definition gives in general a smaller number. For example, for a countable set X the Bowen's entropy is always 0.

Our set A is compact, so there is no disagreement about what $h_{\mathrm{top}}(f, A)$ is. However, even though the pressure is heuristically a very similar object to the topological entropy (in both cases we are just counting how many trajectories the system has, except in the case of pressure we count the trajectories with some weights, given by the potential), there is no analogue of Bowen's theorem. Thus, we can always define the pressure by formula (52), but it is only an upper bound for the correct formula – which we do not know.

Except for the Markov systems. For a Markov system each n-th level cylinder is *large*, in the sense that there exists $\delta > 0$ such that for every $C[\omega^n] \in D_n$ we have

$$\mathrm{diam} f^n(C[\omega^n]) > \delta.$$

It is not necessarily so for non-Markov systems: some n-th level cylinders might be very tiny (they will be not only n-th level cylinders but also $n+1, \ldots, n+\ell$-th level cylinders, for some possibly large ℓ). As the result, the sum on the right hand side of (52) overstates their importance (counting them as n-th level cylinders, while they would be counted as $n+\ell$-th level cylinders by Bowen). Thus, Franz Hofbauer in [10] gave a better definition of pressure:

$$P(A, \varphi) = \sup_{B\subset A, B \mathrm{Markov}} P(B, \varphi), \tag{53}$$

where $P(B, \varphi)$ is given by (52). For Markov A (53) gives the same value as (52). We note that it is still an open question whether the formula (53) can be strictly smaller than (52) for non-Markov A.

9.2 Conformal Measure and Small Cylinders

We finish the section with two more important results of Franz Hofbauer. The first of them was obtained together with Mariusz Urbański [12]. We will call a probabilistic measure μ defined on A *conformal* for the potential φ if for every n and for every $C[\omega^n] \in D_n$ we have

$$\mu(TC[\omega^n]) = \int_{C[\omega^n]} e^{P(A,\varphi)-\varphi} \mathrm{d}\mu.$$

As the partition is generating, this formula can be iterated:

$$\mu(T^n C[\omega^n]) = \int_{C[\omega^n]} e^{nP(A,\varphi)-S_n\varphi} \mathrm{d}\mu.$$

Theorem 9.2 (Hofbauer, Urbański) *Let A be topologically transitive, compact, T-invariant set of positive entropy. Then, for every piecewise continuous potential φ there exists a nonatomic conformal measure $\mu(A, \varphi)$ with support A.*

The second result of Hofbauer, from [10], provides a way of estimating the set of points $x \in A$ such that for every n the cylinder $C_n(x)$ is not large. Denote

$$N_\rho(A, \mu) = \{x \in A; \limsup_{n\to\infty} \mu(T^n C_n(x)) \leq \rho\}.$$

Denote also by $D(\alpha)$ the set of points $x \in A$ with Lyapunov exponent α. We remind that $\varphi_1(x)$ denotes the variation of potential φ in first level cylinder containing x.

Lemma 9.3 (Hofbauer) *For every $\alpha > \sup_x(\log|F'|)_1(x)$,*

$$\lim_{\rho\to 0} \dim_H(N_\rho \cap D(\alpha)) = 0.$$

We note that $\sup_x(\log|F'|)_1(x)$ can be arbitrarily decreased by considering sub-partitions of $\{I_i\}$.

10 The Dimension of Barnsley's Repellers

First, we recall the basic definitions.

10.1 The Basic Definitions

First, we recall the definition of Barnsley's skew product maps: Given $\{I_i\}_{i=1}^m$ which is a partition of $[0, 1]$. Let $D_i := I_i \times \mathbb{R}$. For $(x, y) \in D_i$ we defined $F_i(x, y) := (f_i(x), g_i(x, y))$, where $f_i : I_i \to J_i \subset [0, 1]$ onto, and

$$f_i(x) := \gamma_i x + v_i, \;\; g_i(x, y) = a_i x + \lambda_i y + t_i, \;\; |\lambda_i|, |\gamma_i| > 1, \quad t_i, v_i \in \mathbb{R}. \tag{54}$$

Also recall that we define $f(x) := f_i(x)$ if $x \in I_i$. The set of admissible words is defined as

$$X := \text{cl}\left\{(i_1, i_2, \dots) \in \Sigma : \exists x \in I \text{ such that } \forall n \geq 0,\ f^n(x) \in I_{i_n}^o\right\}, \tag{55}$$

where $\text{cl}(A)$ is the closure of the set $A \subset \Sigma := \{1, \dots, m\}^{\mathbb{N}}$ in the usual topology on Σ.

Definition 10.1 We say that f is *Markov* if $f(\overline{I_i})$ is equal to a finite union of elements in $\{\overline{I_i}\}_{i=1}^m$ for every $i = 1, \dots, m$.

10.2 Diagonal and Essentially Non-diagonal System

Since, the maps F_i are affine the derivatives DF_i are constant lower triangular matrices

$$DF_i := \begin{pmatrix} \gamma_i & 0 \\ a_i & \lambda_i \end{pmatrix}.$$

However, it is very important if the derivative matrices are diagonal or essentially non diagonal along the dynamics since the proofs that work for the essentially non-diagonal case do not work for the diagonal ones and we need different assumptions in these different cases.

Definition 10.2 We say that

(a) F is diagonal if all the matrices DF_i are diagonal.
(b) F is essentially diagonal if the system of matrices $\{DF_i\}_{i=1}^m$, simultaneously diagonizable. This holds if

$$\frac{\gamma_i - \lambda_i,}{a_i} = \frac{\gamma_j - \lambda_j}{a_j}, \quad \forall i, j \in \{1, \ldots, m\}. \tag{56}$$

(c) F is essentially non-diagonal along the dynamics if there are admissible words $\boldsymbol{\omega}, \boldsymbol{\tau}, \in X$ and another word $\boldsymbol{\eta}$ such that $\boldsymbol{\omega\eta\tau} \in X$ such that

(1) both $f_{\boldsymbol{\omega}}$ and $f_{\boldsymbol{\tau}}$ have fixed points

(2) $\{DF_{\boldsymbol{\omega}}, DF_{\boldsymbol{\tau}}\}$ are not simultaneously diagonizable. That is for

$$DF_{\boldsymbol{\omega}} = \begin{pmatrix} \gamma_{\boldsymbol{\omega}} & 0 \\ a_{\boldsymbol{\omega}} & \lambda_{\boldsymbol{\omega}} \end{pmatrix} \quad \text{and} \quad DF_{\boldsymbol{\tau}} = \begin{pmatrix} \gamma_{\boldsymbol{\tau}} & 0 \\ a_{\boldsymbol{\tau}} & \lambda_{\boldsymbol{\tau}} \end{pmatrix}$$

we have

$$\frac{\gamma_{\boldsymbol{\omega}} - \lambda_{\boldsymbol{\omega}}}{a_{\boldsymbol{\omega}}} \neq \frac{\gamma_{\boldsymbol{\tau}} - \lambda_{\boldsymbol{\tau}}}{a_{\boldsymbol{\tau}}}.$$

The reason for this restrictive definition in (c) is that during the proof we approximate by Markov sub-systems and we need to guarantee that even the approximating Markov sub-system remains essentially non-diagonal.

10.3 Markov Pressure and Hofbauer Pressure

Using the notation of (3), we introduce potential:

$$\varphi^s(x) = \begin{cases} -s \log |\lambda_i| & \text{if } 0 \leq s \leq 1, \\ -(\log |\lambda_i| + (s-1) \log |\gamma_i|) & \text{if } 1 < s \leq 2. \end{cases} \tag{57}$$

Definition 10.3 $[P(s, B)]$ Let $s > 0$ and $B \subset [0, 1]$ be a Markov subset. Recall that in (52) we defined the pressure $P(B, \varphi)$ for Markov subset $B \subset [0, 1]$ and potential φ. Using this definition we can define

$$P(s, B) := P(B, \varphi^s). \tag{58}$$

The following lemma helps to get better understanding:

Lemma 10.4 *Assume that $B \subset [0, 1]$ is Markov of type-1 set. That is for every $i, j \in \{1, \ldots, m\}$ either $I_j \cap B \subset f(I_i \cap B)$ or $(I_j \cap B) \cap f(I_i \cap B) = \emptyset$. Then*

$$A_{i,j}^{(s)} = \begin{cases} (1/\lambda_i) \cdot (1/\gamma_i)^{s-1} & \text{if } I_j \cap B \subseteq f(I_i \cap B) \\ 0 & \text{otherwise.} \end{cases}$$

Then $P(s, B) = \log \rho(A^{(s)})$, where $\rho(A)$ denotes the spectral radius of A.

We remark that every subshifts of type-n can be corresponded to a type-1 subshift by defining a new alphabet, and subdividing the monotonicity intervals into smaller intervals.

Definition 10.5 $[P_{Mar}(s), P_{\mathrm{Hof}}(s)]$ Now we define the functions $s \mapsto P_{\mathrm{Mar}}(s)$ and $s \mapsto P_{\mathrm{Hof}}(s)$ as follows:

(a) If f is Markov then we write $P_{\mathrm{Mar}}(s) := P(s, [0, 1])$
(b) If f is none Markov then we write

$$P_{\mathrm{Hof}}(s) := \sup_{B \subset [0,1],\ B \text{ Markov}} P(s, B). \tag{59}$$

10.4 The Main Results

Theorem 10.6 *Suppose that*

(a) F is essentially diagonal,
(b) $\gamma_i > \lambda_i$ for every $i = 1, \ldots, m$,
(c) The self-similar IFS $\{g_i^{-1}(y) = \frac{y-t_i}{\lambda_i}\}_{i=1}^{M}$ satisfies HESC (see Condition 4.10)

then

$$\dim_H \Lambda = \dim_B \Lambda = \sup_{\mu \in \mathcal{M}_{\mathrm{erg}}(\Lambda)} D(\mu) = s_0,$$

where s_0 is the unique number such that

- $P_{\mathrm{Mar}}(s_0) = 0$ *if f is Markov, otherwise*
- $P_{\mathrm{Hof}}(s_0) = 0$.

Theorem 10.7 *Assume that F is essentially non-diagonal and f is a topologically transitive. If $\gamma_i > \lambda_i$ for every $i = 1, \ldots, m$ then*

$$\dim_H \Lambda = \dim_B \Lambda = \sup_{\mu \in \mathcal{M}_{\mathrm{erg}}(\Lambda)} D(\mu) = s_0,$$

where s_0 is the unique number such that

- $P_{\mathrm{Mar}}(s_0) = 0$ *if f is Markov, otherwise*
- $P_{\mathrm{Hof}}(s_0) = 0$.

Appendix 1. Thermodynamical Formalism

First we introduce the subshift of finite type.

1.1 Subshift of Finite Type

Let $\Sigma = \{1, \ldots, m\}^{\mathbb{N}}$ be endowed with the usual topology, which generated by the distance $\mathrm{dist}(\mathbf{i}, \mathbf{j}) := m^{-|\mathbf{i} \wedge \mathbf{j}|}$, where

$$|\mathbf{i} \wedge \mathbf{j}| = \max\{n: \forall|\ell| \leq n, i_\ell = j_\ell\}.$$

For some $k < r$ we write $[\mathbf{i}]_{k,r} = \{\mathbf{j} \in \Sigma : i_\ell = j_\ell,\ \forall \ell \in \{k, \ldots, r\}\}$ for the (k, r) cylinder sets. If $k = 1$ then we write simply $[\mathbf{i}]_r$. Similarly,

$$[i_1, \ldots, i_n] := \{\mathbf{j} \in \Sigma: i_k = j_k, \forall k = 1, \ldots, n\}.$$

For an $\mathbf{i} \in \Sigma$ we write

$$\mathbf{i}|_n := (i_1, \ldots, i_n) \in (1, \ldots, m)^n =: \Sigma_n. \tag{60}$$

Definition A.1 *(subshift of finite type)* Given an $m \times m$ matrix A of 0's and 1's. Let $\Sigma_A := \left\{\mathbf{i} \in \Sigma: A_{i_k, i_{k+1}} = 1,\ \forall k \in \mathbb{N}\right\}$ and let σ be the left shift on Σ_A. That is $\sigma(i_1, i_2, i_3 \ldots) := (i_2, i_3 \ldots)$ for every $(i_0, i_1, i_2, \ldots) \in \Sigma_A$. Clearly, $\sigma(\Sigma_A) = \Sigma_A$ and $\sigma|_{\Sigma_A}$ is a homeomorphism on Σ_A. Sometimes we call $\sigma|_{\Sigma_A}$ *topological Markov chain.*

We always assume that for every $k \in \{1, \ldots, m\}$ there exist some $\mathbf{i} \in \Sigma_A$ such that $i_0 = k$. From now on we call

- (Σ, σ) a full shift and
- (Σ_A, σ) as subshift of finite type.

Also for the rest of this Section we assume that A is an $m \times m$ primitive matrix.

$$\Sigma_{A,n} := \{\mathbf{i} = (i_1, \ldots, i_n): [i_1, \ldots, i_n] \cap \Sigma_A \neq \emptyset\}.$$

1.2 Ergodic Measures

Given a measurable self-map T of a measurable space $(X, \mathcal{B})$. That is $T: X \to X$ and $T^{-1}B \in \mathcal{B}$ for every $B \in \mathcal{B}$. We write

- $\mathcal{M}(X)$ for the set of Borel probability measures on $(X, \mathcal{B})$,
- $\mathcal{M}_T(X)$ for the set of invariant measures. That is

$$\mathcal{M}_T(X) = \left\{\mu \in \mathcal{M}(X) \colon \mu(A) = \mu(T^{-1}A),\ \forall A \in \mathcal{B}\right\},$$

- $\mathcal{E}_T(X)$ for the ergodic measures. That is

$$\mathcal{E}_T(X) = \left\{\mu \in \mathcal{M}_T(X) \colon A = T^{-1}A \Longrightarrow \text{ either } \mu(A) = 0, \text{ or } \mu(A) = 1\right\}.$$

We frequently use Birkhoff's Ergodic Theorem.

Theorem A.2 [Birkhoff's Ergodic Theorem] *Let $\mu \in \mathcal{E}_T(X)$ and let $f \in L^1(X, \mu)$. Then for μ-almost all $x \in X$ the ergodic averages converge both in L^1 and pointwise:*

$$\lim_{n\to\infty} \frac{1}{n} \sum_{k=0}^{n-1} f(T^k(x)) = \int f(x)\mathrm{d}\mu(x). \tag{61}$$

1.3 Entropy

One of the basic concepts of the thermodynamical formalism is the entropy. There is measure theoretical and topological entropy. Here, we just present the definitions and a basic property. For further reading we recommend [4, 22] and a very detailed introduction is given in [20].

1.3.1 Measure Theoretical Entropy on (Σ_A, σ) for an Ergodic Measure

First, we define the measure theoretical entropy on Σ_A for an ergodic (with respect to the left shift σ) measure. (We always assume that A is a primitive matrix.)

Definition A.3 *[Entropy (measure theoretical)]* Let μ be an ergodic measure on Σ_A. We can define the entropy of μ as

$$h(\mu) := \lim_{n\to\infty} \frac{1}{n} \sum_{\boldsymbol{\omega}\in\Sigma_{A,n}} \mu([\boldsymbol{\omega}]) \log \mu([\boldsymbol{\omega}]). \tag{62}$$

Theorem A.4 [Shannon Breiman McMillian Theorem] *Let $\mu \in \mathcal{E}_\sigma(\Sigma)$. Then for μ-almost all $\mathbf{i} \in \Sigma_A$ we have*

$$\lim_{n\to\infty} \frac{1}{n} \log \mu[\mathbf{i}|_n] = h(\mu). \tag{63}$$

For the proof see [4].

Example A.5 (a) *Bernoulli shift*. Given a probability vector $\mathbf{p} := (p_1, \ldots, p_m)$, where p_i and $\sum_{i=1}^{m} p_i = 1$. Then, we say the $\mu := \mathbf{p}^{\mathbb{N}}$ is the *Bernoulli measure* corresponding to $\mathbf{p}$. It is easy to see that

$$h(\mu) = -\sum_{i=1}^{m} p_i \log p_i. \tag{64}$$

(b) *Markov Shift* Given a stochastic matrix $P = (p_{i,j})_{1 \le i,j \le m}$. That is $\sum_{j=1}^{m} p_{i,j} = 1$, $p_{i,j} \ge 0$. We assume that P is primitive (it was enough to assume less). Then by Perron Frobenius Theorem, there exists a left eigenvector $\mathbf{p} = (p_1, \ldots, p_m)$ which is a probability vector, such that $\mathbf{p}^T \cdot P = \mathbf{p}^T$, ($\mathbf{p}$ is considered as a column vector). We define the *Markov measure* μ on Σ corresponding to $(\mathbf{p}, P)$ by $\mu([\boldsymbol{\omega}]) := p_{\omega_1} \cdot p_{\omega_1,\omega_2} \cdots p_{\omega_{n_1},\omega_{\omega_n}}$, where $\boldsymbol{\omega} \in \Sigma_n$ and $\boldsymbol{\omega} = (\omega_1, \ldots, \omega_n)$. Then,

$$h(\mu) = -\sum_{i,j=1}^{m} p_i p_{i,j} \log p_{i,j} \tag{65}$$

(c) *Parry measure* Let $A = (a_{i,j})_{1 \le i,j \le m}^{m}$ be an primitive matrix (to assume irreducibility was enough again) whose entries belong to $\{0, 1\}$. Then, we define the canonical Markov measure as follows: Let λ be the largest (Perron-Frobenius) eigenvalue. Let $\mathbf{u} := (u_1, \ldots, u_m)$ and $\mathbf{v} := (v_1, \ldots, v_m)$ be the left and right (positive) eigenvectors satisfying $\sum_{i=1}^{m} u_i = 1$ and $\sum_{i=1}^{m} u_i v_i = 1$ (see [22, p. 16]). Then we define

$$p_i := u_i v_i \text{ and } \quad p_{i,j} := \frac{a_{i,j} v_j}{\lambda v_i} \tag{66}$$

Let μ be the Markov measure corresponding to $(\mathbf{p}, P)$. Then, the unique measure on Σ_A with maximal entropy is μ and $h(\mu) = \log \lambda$.

1.3.2 Topological Entropy on Compact Metric Spaces for Continuous Mappings

Now, we give the definition of the topological entropy in a more general setup (see e.g. [5, pp. 165–170]).

Definition A.6 *(*Topological entropy) Given a homeomorphism T of the compact metric space (X, d). For $\varepsilon > 0$ we say the orbits of length n

$$x, T(x), \ldots, T^{n-1}(x) \text{ and } y, T(y), \ldots, T^{n-1}(y)$$

are the same with ε-precision if

$$d(T^i(x), T^i(y)) < \varepsilon, \quad \forall i = 0, \ldots, n-1.$$

Fix an $\varepsilon > 0$ and an $n \in \mathbb{N}$. Let $s_n(x, \varepsilon)$ be the maximal number of n-orbits which are different with ε-precision. Then, we define the topological pressure of T by

$$h_{\text{top}}(T) = \lim_{\varepsilon \to 0} \limsup_{n \to \infty} \frac{1}{n} \log s_n(\varepsilon) \tag{67}$$

We remark that this is not the most common way to define the topological entropy.

Theorem A.7 *Let $T: X \to X$ be a contiuous map of a compact metric space. then $h_{\text{top}}(T) = \sup\{h_T(\mu): \mu$ is an invariant measure for $T\}$.*

We defined the measure theoretical entropy only on subshift of finite type. The definition in the general case is similar see, e.g., [4] and [22]. Before we give some examples we need the following definition that will also be used later.

Definition A.8 Let $T: I \to I$, where $I \subset \mathbb{R}$ is an interval.

- We say that T is a piecewise monotone map if there is a finite partition of I such that on every class of this partition the map T is monotone.
- Let T be a piecewise monotone map. The the lap number $\ell(T)$ is the number of maximal monotonicity intervals of T.

Example A.9 (a) For a subshift of finite type (Σ_A, σ) the topological entropy of σ is $\log \lambda$, where λ is the largest eigenvalue of the primitive 0, 1 matrix A.

(b) Here we use the notation of Example A.9. It follows from a theorem of Misiurewicz and Szlenk that for a piecewise monotone map T, we have

$$h(T) = \lim_{n \to \infty} \frac{1}{n} \log \ell(T^n), \tag{68}$$

where T^n is the n-fold composition of T. In particular, $h(T) \leq \ell(T)$. Moreover, if T is piecewise affine and its the slope of $\pm s$ at every point (except the turning points) then $h(T) = \max\{0, \log s\}$. (see [5] for the proofs.).

1.4 Lyapunov Exponent

To define the Lyapunov exponents, we need Oseledec Theorem. The following version of Oseledec Theorem is from Krengel's book [16, pp. 42–47] where the proof is also presented. Given a finite measure space $(\Omega, \mathcal{A}, \mu)$ and $\tau: \Omega \to \Omega$ measure preserving. Further, M denotes the set of $r \times r$ matrices. Put

$$P_n(A, \omega) := A(\tau^{n-1}\omega) \cdots A(\tau\omega)A(\omega).$$

Theorem A.10 [Oseledec] *Legyen $A: \Omega \to M$ be measurable and we assume that*

$$\log^+ \|A(\cdot)\| \in L_1(\mu). \tag{69}$$

Then, there exists an invariant $\Omega' \subset \Omega$ which has full μ-measure such that

1.

$$\lim_{n\to\infty}\left(P_n^*(A,\omega)\cdot P_n(A,\omega)\right)^{1/2n} =: \Lambda(\omega)$$

exists and Λ *is a symmetric positive semidefinite matrix.*

2. *Let* $\exp(\lambda_1(\omega)) > \cdots > \exp(\lambda_s(\omega))$ *are the different eigenvalues of* Λ *and let* E_ν *be the eigenspace of* Λ *which belongs to* $\exp\lambda_\nu(\omega)$. *Then for*

$$H_\nu(\omega) := E_s(\omega)\bigoplus E_{s-1}(\omega)\bigoplus\cdots\bigoplus E_{s+1-\nu}(\omega)$$

we have

$$\lim_{n\to\infty}\frac{1}{n}\log\|P_n(A,\omega)\mathbf{v}\| = \lambda_{s+1-\nu}(\omega), \qquad \forall\mathbf{v}\in H_\nu(\omega)\setminus H_{\nu-1}(\omega), \tag{70}$$

where $H_0(\omega)\equiv\emptyset$.

3. $\omega\mapsto \dim E_\nu(\omega)$ *and* $\omega\mapsto\lambda_\nu(\omega)$ *are* τ*-invariant maps and we call* $\dim E_\nu(\omega)$ *the multiplicity of* $\lambda_i(\omega)$.

Definition A.11 *(Lyapunov exponenets)* Let μ be an ergodic measure. Then it follows from (3) that for all $i=1,\dots,s$ and for μ-almost all $\omega\in\Omega$, $\lambda_i(\omega)$ and $\dim E_\nu(\omega)$ are constants that we call λ_i and d_i, respectively, for $1,\dots,s$. We partition the index set

$$\{1,\dots,r\} = \bigsqcup_{k=1}^{s}\mathcal{I}_k,\quad \mathcal{I}_k := \{d_1+\cdots+d_{k-1}+1,\cdots,d_1+\cdots+d_{k-1}+d_k\} \tag{71}$$

Then, we define the Lyapunov exponents $\chi_1\geq\chi_2\geq\cdots\geq\chi_r$ as follows:

$$\underbrace{\chi_1=\cdots=\chi_{d_1}}_{:=\lambda_1} > \underbrace{\chi_{d_1+1}=\cdots=\chi_{d_1+d_2}}_{:=\lambda_2} > \underbrace{\chi_{d_1+d_2+1}=\cdots=\chi_{d_1+d_2+d_3}}_{:=\lambda_3} > \cdots$$
$$> \underbrace{\chi_{d_1+\cdots+d_{s-2}+1}=\cdots=\chi_{d_1+\cdots+d_{s-2}+d_{s-1}}}_{:=\lambda_{s-1}} > \underbrace{\chi_{d_1+\cdots+d_{s-1}+1}=\cdots=\chi_{d_1+\cdots+d_{s-1}+d_s}}_{:=\lambda_s}. \tag{72}$$

1.5 Topological Pressure and Gibbs Measure

In this section, we always assume that A is a primitive $m\times m$ matrix and we consider the topological Markov chain (or subshift of finite type) (σ,Σ_A) as defined in Definition A.1

Definition A.12 (Hölder continuity) We say that a function $\phi\colon\Sigma_A\to\mathbb{R}$ is Hölder continuous if there exists $b>0$ and $\alpha\in(0,1)$ such that

$$\mathrm{var}_k\phi := \sup\{|\phi(\mathbf{i}) - \phi(\mathbf{j})| : |\mathbf{i} \wedge \mathbf{j}| \geq k\} \leq b\alpha^k. \tag{73}$$

The set of Hölder continuous functions on Σ_A is denoted by $\mathcal{F}_A$. For a $\phi \in \mathcal{F}_A$ and $\boldsymbol{\omega} = (\omega_1, \dots, \omega_n) \in \{1, \dots, m\}^n$

$$S_n\phi(\boldsymbol{\omega}) := \sup\left\{\sum_{\ell=0}^{n-1} \phi(\sigma^\ell \mathbf{j}) : \mathbf{j} \in [\boldsymbol{\omega}] \cap \Sigma_A\right\}. \tag{74}$$

First observe that for any $\phi \in \mathcal{F}_A$ satisfying (73): and for any $\mathbf{j}, \mathbf{j}' \in [\boldsymbol{\omega}]$, where $\boldsymbol{\omega} = (\omega_1, \dots, \omega_n) \in \Sigma_{A,n}$ we have

$$\left|\sum_{\ell=0}^{n-1} \phi(\sigma^\ell \mathbf{j}) - \sum_{\ell=0}^{n-1} \phi(\sigma^\ell \mathbf{j}')\right| \leq \frac{b}{1-\alpha} \tag{75}$$

holds for all n and $\boldsymbol{\omega} \in \Sigma_{A,n}$. This yields that the topological pressure of the potential $\boldsymbol{\phi}$ for the topological Markov shift (Σ_A, σ) is

$$P(\boldsymbol{\phi}) := \lim_{n\to\infty} \frac{1}{n} \log\left(\sum_{\mathbf{i}\in\Sigma_{A,n}} \mathrm{e}^{S_n\phi(\mathbf{i})}\right) \tag{76}$$

does not depend on which $\mathbf{j} \in [\mathbf{i}]$ is chosen. Let $\mathcal{M}_\sigma(\Sigma_A)$ denote the σ-invariant probability measures on Σ_A. The so-called Gibbs measure together with the topological pressure play central role in dimension theory:

Theorem A.13 [The Existence of Gibbs Measure Theorem] *Suppose that*

- *A is primitive and*
- *$\phi \in \mathcal{F}_A$.*

Then there exists a unique $\mu \in \mathcal{M}_\sigma(\Sigma_A)$ for which $\exists c_1, c_2 > 0$ such that for $\forall \mathbf{i} \in \Sigma_A$ and $\forall \ell$:

$$c_1 \leq \frac{\mu([\mathbf{i}]_\ell)}{\exp(-\ell \cdot P(\phi) + S_\ell\phi(\mathbf{i}))} \leq c_2, \tag{77}$$

where recall that we defined $[\mathbf{i}]_\ell = \{\mathbf{j} \in \Sigma_A : i_k = j_k,\ \forall k \in \{1, \dots, \ell\}\}$. It can be proved that μ is mixing, consequently ergodic.

We say that μ is the Gibbs measure for the potential ϕ. For the proof see [4].

1.6 The Root of the Pressure Formula

Let $\mathcal{F}$ be a conformal IFS on $\mathbb{R}$ as in definition 5.1 and we assume that the SSP holds. That is $f_i([0, 1]) \cap f_j([0, 1]) = \emptyset$ for all $i \neq j$. Let $\phi_s : \Sigma \to \mathbb{R}$ be

$$\phi_s(\mathbf{i}) := \log |f'_{i_1}(\sigma \mathbf{i})|^s. \tag{78}$$

Then for every $\mathbf{i} \in \Sigma$ and n we have

$$\phi_s(\sigma^{n-1}\mathbf{i}) + \cdots + \phi_s(\sigma \mathbf{i}) + \phi_s(\mathbf{i}) = \log |f'_{i_1 \dots i_n}(\Pi(\sigma^n \mathbf{i}))|^s. \tag{79}$$

Using this and the Bounded Distortion Property, we obtain that for every n and for every $\boldsymbol{\omega} \in \Sigma_n := \{1, \dots, m\}^n$

$$s \log c_1 < \left| S_n \phi_s(\boldsymbol{\omega}) - \log |f'_{i_1 \dots i_n}(\Pi(\sigma^n \mathbf{i}))|^s \right| < s \log c_2. \tag{80}$$

Hence we get

$$P(s) := P(\phi_s) = \lim_{n \to \infty} \frac{1}{n} \log \sum_{|\omega|=n} |f'_{i_1 \dots i_n}(0)|^s, \tag{81}$$

It is easy to see that the function $s \mapsto P(\phi_s)$ is positive at zero, negative at 1, continuous and strictly decreasing. So it has a unique zero in (0, 1). Let us denote this unique zero by s_0. That is

$$P(s_0) = 0. \tag{82}$$

This is the reason that we say that s_0 is the root of the pressure formula.

Let μ be the Gibbs measure for the potential ϕ_{s_0}. Then for every n, $\boldsymbol{\omega} \in \Sigma_n$, and $x \in (0, 1)$ we have

$$c_1 c_3 < \frac{\mu([\boldsymbol{\omega}])}{|f'_{\boldsymbol{\omega}}(x)|^{s_0}} < c_2 c_4. \tag{83}$$

Appendix 2. Subadditive Pressure

Falconer introduced subadditive pressure in [8] and in a more explicit form in [6, Sect. 3].

Definition B.1 *(Subadditive pressure)* Assume that $\psi_n \colon \Sigma_A \to \mathbb{R}$, $n = 1, 2, \dots$ satisfy the following three conditions:

(a) $\psi_{n+m}(\mathbf{i}) \le \psi_n(\mathbf{i}) + \psi_m(\sigma^m \mathbf{i})$, $n, m \in \mathbf{N}$
(b) There exists an $a > 0$ such that $\left|\frac{1}{n}\psi_n(\mathbf{i})\right| \le a$, for all $\mathbf{i} \in \Sigma_A$, $n \in \mathbf{N}$
(c) There exists an $a > 0$ such that $|\psi_n(\mathbf{i}) - \psi_n(\mathbf{j})| \le b$ for all $n \in \mathbb{N}$ and $\mathbf{i}, \mathbf{j} \in \Sigma_A$.

Foe every $\boldsymbol{\omega} \in \Sigma_{A,n}$ we fix an arbitrary $\mathbf{i}_{\boldsymbol{\omega}} \in [\boldsymbol{\omega}]$. Then the subadditive pressure associated to $\{\psi_n\}$ is

$$P(\{\psi_n\}) := \lim_{n \to \infty} \frac{1}{n} \log \sum_{\boldsymbol{\omega} \in \Sigma_{A,n}} \exp(\psi_n(\mathbf{i}_{\boldsymbol{\omega}})) = \inf_n \frac{1}{n} \log \sum_{\boldsymbol{\omega} \in \Sigma_{A,n}} \exp(\psi_n(\mathbf{i}_{\boldsymbol{\omega}})). \tag{84}$$

The the second equality is verified in [6, Sect. 3] is a slightly different setup. The connection to the additive pressure is that

$$P(\{\psi_n\}) = \lim_{N\to\infty} \frac{1}{N} P\left(\sigma^N, \psi_N\right) = \inf_N \frac{1}{N} P\left(\sigma^N, \psi_N\right), \tag{85}$$

where $P\left(\sigma^N, \psi_N\right)$ is the additive pressure (defined in (76)) for the potential ψ_N on the topological Markov shift (Σ_A, σ^N).

Most commonly we use this in the following special case:

Example B.2 In the case of the additive pressure $\psi_n(\mathbf{i}) = \sum_{k=0}^{n-1} f(\sigma^n \mathbf{i})$ for a continuous function $f : \Sigma_A \to \mathbb{R}$.

Example B.3 Given contracting non-singular $d \times d$ matrices $A_1, \ldots, A_m$ (the linear part of a self-affine IFS of the form 37). Then for every $s \geq 0$ we define

$$\psi_n^s : \Sigma_A \to \mathbb{R}, \qquad \psi_n^s(\mathbf{i}) := \log \phi^s(A_{i_1} \cdots A_{i_n}) \text{ and } P(s) := P_{A_1 \ldots A_n}(s) := P\left(\left\{\psi_n^s\right\}\right). \tag{86}$$

where ϕ^s is the singular value function defined in (40). It is immediate that the function $s \mapsto P_{A_1 \ldots A_n}(s)$ is strictly decreasing, continuous, positive at zero and negative at any s which is large enough. So, it has a unique zero $s_{A_1 \ldots A_n} > 0$. That is

$$P_{A_1 \ldots A_n}(s_{A_1 \ldots A_n}) = 0. \tag{87}$$

References

1. B. Bárány, M. Hochman, A. Rapaport, Hausdorff dimension of planar self-affine sets and measures. Invent. Math. **216**(3), 601–659 (2019). arXiv:1712.07353
2. B. Bárány, Antti Käenmäki, Ledrappier-young formula and exact dimensionality of self-affine measures. Adv. Math. **318**, 88–129 (2017)
3. R. Bowen, Topological entropy for noncompact sets. Trans. Am. Math. Soc. **184**, 125–136 (1973)
4. R. Bowen, *Equilibrium states and the ergodic theory of Anosov diffeomorphisms*, vol. 470, Lecture Notes in Mathematics (Springer-Verlag, Berlin-New York, 1975)
5. W. De Melo, S. Van Strien, *One-dimensional dynamics*, vol. 25 (Springer Science & Business Media, 2012)
6. K.J. Falconer, Bounded distortion and dimension for nonconformal repellers. Math. Proc. Cambridge Philos. Soc. **115**(2), 315–334 (1994)
7. K. Falconer, *Fractal geometry: mathematical foundations and applications* (John Wiley & Sons, 2004)
8. K.J. Falconer, The Hausdorff dimension of self-affine fractals. Math. Proc. Camb. Phil. Soc. **103**(3), 339–350 (1988)
9. M. Hochman, On self-similar sets with overlaps and inverse theorems for entropy. Ann. Math. **180**(2), 773–822 (2014)
10. F. Hofbauer, Multifractal spectra of Birkhoff averages for a piecewise monotone interval map. Fundamenta Mathematicae **208**, 95–121 (2010)
11. F. Hofbauer, P. Raith, The Hausdorff dimension of an ergodic invariant measure for a piecewise monotonic map of the interval. Canad. Math. Bull. **35**(1), 84–98 (1992)

12. F. Hofbauer, M. Urbański, Fractal properties of invariant subsets for piecewise monotonic maps on the interval. Trans. Am. Math. Soc. **343**(2), 659–673 (1994)
13. J.E. Hutchinson, Fractals and self-similarity. Indiana Univ. Math. J. **30**(5), 713–747 (1981)
14. T. Jordan, M. Pollicott, K. Simon, Hausdorff dimension for randomly perturbed self affine attractors. Comm. Math. Phys. **270**(2), 519–544 (2007)
15. M. Keane, M. Smorodinsky, B. Solomyak, On the morphology of expansions with deleted digits. Trans. Am. Math. Soc. **347**(3), 955–966 (1995)
16. U. Krengel, *Ergodic theorems*, vol. 6 (Walter de Gruyter, 2011)
17. P. Mattila, *Geometry of sets and measures in Euclidean spaces: fractals and rectifiability*, vol. 44 (Cambridge University Press, 1999)
18. P.A.P. Moran, Additive functions of intervals and Hausdorff measure. Proc. Cambridge Philos. Soc. **42**, 15–23 (1946)
19. Y.B. Pesin, *Dimension theory in dynamical systems*. Chicago Lectures in Mathematics (University of Chicago Press, Chicago, IL, 1997). Contemporary views and applications
20. F. Przytycki, M. Urbański, *Conformal fractals: ergodic theory methods*, vol. 371 (Cambridge University Press, 2010)
21. B. Solomyak, Measure and dimension for some fractal families, in textitMathematical Proceedings of the Cambridge Philosophical Society, vol. 124, pp. 531–546 (Citeseer, 1998)
22. P. Walters, *An introduction to ergodic theory*, vol. 79, Graduate Texts in Mathematics (Springer-Verlag, New York-Berlin, 1982)

Dimension Theory of Some Non-Markovian Repellers Part II: Dynamically Defined Function Graphs

Balázs Bárány, Michał Rams, and Károly Simon

Abstract This is the second part in a series of two papers. Here, we give an overview on the dimension theory of some dynamically defined function graphs, like Takagi and Weierstrass function, and we study the dimension of Markovian fractal interpolation functions and generalized Takagi functions generated by non-Markovian dynamics.

Keywords Self-affine measures · Self-affine sets · Hausdorff dimension

2010 Mathematics Subject Classification 28A80 (Primary) · 28A78 (Secondary)

The research of Bárány and Simon was partially supported by the grant OTKA K123782. Bárány acknowledges support also from NKFI PD123970 and the János Bolyai Research Scholarship of the Hungarian Academy of Sciences and the New National Excellence Program of the Ministry of Human Capacities ÚNKP-18-4-BME-385. Michał Rams was supported by National Science Centre grant 2014/13/B/ST1/01033 (Poland). This work was partially supported by the grant 346300 for IMPAN from the Simons Foundation and the matching 2015–2019 Polish MNiSW fund.

B. Bárány (✉) · K. Simon
Department of Stochastics, MTA-BME Stochastics Research Group, Budapest University of Technology and Economics, P.O. Box 91, Budapest 1521, Hungary
e-mail: simonk@math.bme.hu

M. Rams
Polish Academy of Sciences, Institute of Mathematics, ul. Śniadeckich 8, 00-656 Warszawa, Poland
e-mail: rams@impan.pl

R. L. Devaney et al. (eds.), *Topological Dynamics and Topological Data Analysis*, Springer Proceedings in Mathematics & Statistics 350,
https://doi.org/10.1007/978-981-16-0174-3_3

1 The Weierstrass and Takagi Functions

The study of the geometric properties of the graphs of real functions goes back to the nineteenth century. Karl Weierstrass introduced in 1872 a function, which is continuous but nowhere differentiable. That was one of the first examples of such functions and for nowadays, became a famous example:

$$W_{\alpha,b}(x) = \sum_{n=0}^{\infty} \alpha^n \cos(2\pi b^n x), \tag{1}$$

where $b > 1$ and $1/b < \alpha < 1$. In fact, Weierstrass proved the non-differentiability for some values of parameters, and the proof for all parameters was given by Hardy [14] in 1916.

Takagi [25] published his simple example of a continuous but nowhere differentiable function in 1901,

$$T(x) = \sum_{n=0}^{\infty} 2^{-n} \psi(2^n x), \tag{2}$$

where $\psi(x) = \text{dist}(x, \mathbb{Z})$. Unlike for the Weierstrass function, it is easy to show that T has at no point a finite derivative, which proof is due to Billingsley [10]. For further properties and historical background of the functions above, see the survey papers of Allaart and Kawamura [1] and Barański [2].

Later, starting from the work of Besicovitch and Ursell [9], the graphs of $W_{\alpha,b}$ and related functions were studied from a geometric point of view as fractal curves in the plane. In general, let

$$G_{\alpha,b}(x) = \sum_{n=0}^{\infty} \alpha^n \phi(b^n x) \tag{3}$$

for $x \in \mathbb{R}$, where $b \in \mathbb{N}$, $1/b < \alpha < 1$ and $\phi\colon \mathbb{R} \mapsto \mathbb{R}$ is a non-constant $\mathbb{Z}$-periodic Lipschitz continuous piecewise C^1 function. Kaplan et al. [17] proved that a function of the form (3) is either piecewise C^1 smooth or the box dimension of the graph is equal to

$$D = 2 + \frac{\log \alpha}{\log b}. \tag{4}$$

This fact is related to the Hölder continuity of the function $G_{\alpha,b}$. In fact, if the function $g\colon [0, 1] \mapsto \mathbb{R}$ is Hölder continuous with exponent α then

$$\overline{\dim}_B\{(x, g(x)) : x \in [0, 1]\} \leq 2 - \alpha.$$

For instance, the case of smoothness of $G_{\alpha,b}$ happens if $\phi(x) = \alpha h(bx) - h(x)$ for some smooth function h.

The problem of determining the value of the Hausdorff dimension turned out to be much more complicated. Mandelbrot formulated the conjecture in 1977 [20] that the Hausdorff dimension of the graph of $W_{\alpha,b}$ equals to D, but this has been solved only recently.

Ledrappier [19] gave a sufficient condition in order to determine the Hausdorff dimension of the graph. In details, let $\underline{\xi} = \{\xi_i, i = 1, 2, \ldots\}$ be a sequence of independent Bernoulli variables with values $0, \ldots, b-1$ and with probabilities $\mathbb{P}(\xi_j = k) = 1/b$. If the distribution of the random variable

$$Y_x(\underline{\xi}) = \sum_{n=1}^{\infty} (b\alpha)^{-n} \phi' \left(\frac{x}{b^n} + \frac{\xi_1}{b^n} + \frac{\xi_2}{b^{n-1}} + \cdots + \frac{\xi_n}{b} \right) \tag{5}$$

has dimension 1 for Lebesgue almost every x then

$$\dim_H\{(x, G_{\alpha,b}(x)) : x \in [0, 1]\} = D.$$

This condition relies on the so-called Ledrappier-Young formula.

Although for the first sight this condition may seem very restrictive, it turned out that it is widely applicable. In the case of Weierstrass functions (1), Barański et al. [3] showed that for every $b \geq 2$ integers there exists $\alpha_b \in [1/b, 1)$ such that for every $\alpha \in (\alpha_b, 1)$,

$$\dim_H\{(x, W_{\alpha,b}(x)) : x \in [0, 1]\} = D.$$

Recently, Shen [23] proved that $\alpha_b = 1/b$.

In the case of Takagi function, the distribution of the random variable $Y_x(\underline{\xi})$ is independent of x and it is the Bernoulli convolution, related to Erdős' problem [11, 12]. For simplicity denote T_α the function $G_{\alpha,2}$ with $\psi(x) = \mathrm{dist}(x, \mathbb{Z})$. It is easy to see that $Y_x(\underline{\xi}) = \sum_{n=0}^{\infty} (\delta_{\xi_n,0} - \delta_{\xi_n,1})(2\alpha)^{-n}$ in (5), where $\delta_{i,j} = 1$ if $i = j$ and 0 otherwise. Using this phenomena, Solomyak [24] showed that for Lebesgue almost every $\alpha \in (1/2, 1)$,

$$\dim_H\{(x, T_\alpha(x)) : x \in [0, 1]\} = D. \tag{6}$$

Applying the result of Hochman [15], [5, Theorem 4.11], there exists a set $E \subset (1/2, 1)$ such that $\dim_H E = 0$ and for every $\alpha \in (1/2, 1) \setminus E$, (6) holds. Recently, Varjú [26] showed that the distribution of $Y(\underline{\xi})$ has dimension 1 if $(2\alpha)^{-1}$ is a transcendental number (which is transcendental if and only if α is transcendental), and hence (6) holds.

However, it is well known that for Pisot numbers (for instance $(2\alpha)^{-1} = (\sqrt{5} - 1)/2$ the golden ratio) the distribution of $Y(\underline{\xi})$ is singular and has dimension strictly smaller than 1 and thus, Ledrappier's condition (5) cannot be applied. Recently, with different method, Bárány et al. [4] proved that (6) holds for every $\alpha \in (1/2, 1)$.

2 Dynamically Defined Function Graphs

Let $G_{\alpha,b}$ be the function defined in (3) with $b > 1$ integer, $1/b < \alpha < 1$ and $\phi\colon \mathbb{R} \mapsto \mathbb{R}$ is a non-constant 1-periodic Lipschitz continuous piecewise C^1 function. It is easy to see that $G_{\alpha,b}$ satisfies certain self-similarity equation

$$G_{\alpha,b}(x) = \alpha G_{\alpha,b}(bx) + \phi(x). \tag{7}$$

Since ϕ is 1-periodic and thus, $G_{\alpha,b}$ as well, Eq. (7) implies that $\mathrm{graph}(G_{\alpha,b}) = \{(x, G_{\alpha,b}(x)) : x \in [0,1]\}$ is invariant with respect to the dynamics

$$F(x,y) = \left(bx \mod 1, \frac{y - \phi(x)}{\alpha}\right) \text{ for } (x,y) \in [0,1] \times \mathbb{R},$$

and $\{F^n(x,y)\}$ is bounded if and only if $y = G_{\alpha,b}(x)$.

One can define the local inverses of F such that

$$\tilde{F}_i(x,y) = \left(\frac{x+i}{b}, \alpha y + \phi\left(\frac{x+i}{b}\right)\right) \text{ for } i = 0, \ldots, b-1.$$

Hence, $\mathrm{graph}(G_{\alpha,b}) = \bigcup_{i=0}^{b-1} \tilde{F}_i(\mathrm{graph}(G_{\alpha,b}))$. For a visualization of the local inverses in the cases of $W_{1/2,3}$ and $T_{2/3}$, see Fig. 1.

Observe that for the Takagi function T_α, the function ϕ is piecewise linear; moreover, the singularity occurs exactly at $x = 1/2$. Thus, $\mathrm{graph}(T_\alpha)$ is a self-affine set, see [5, Definition 6.1], with IFS

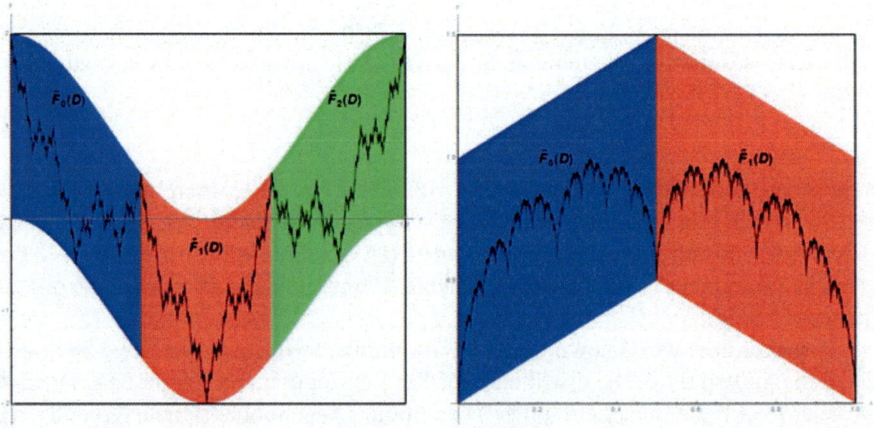

Fig. 1 Graph of $W_{1/2,3}$ and $T_{2/3}$ as repellers

$$\left\{\tilde{F}_0(\underline{x}) = \begin{pmatrix} \frac{1}{2} & 0 \\ 1 & \alpha \end{pmatrix} \underline{x}, \tilde{F}_1(\underline{x}) = \begin{pmatrix} \frac{1}{2} & 0 \\ -1 & \alpha \end{pmatrix} \underline{x} + \begin{pmatrix} \frac{1}{2} \\ 1 \end{pmatrix}\right\}$$

formed by lower triangular matrices.

A wider family of continuous functions, which are attractors of affine IFS, is the fractal interpolation functions, introduced by Barnsley [6]. Let a dataset $\Delta = \{(x_i, y_i) \in [0,1] \times \mathbb{R} : i = 0, 1, \ldots, m\}$ be given so that $0 = x_0 < x_1 < \cdots < x_{m-1} < x_m = 1$. We concern the graphs of continuous functions $G\colon [0,1] \mapsto \mathbb{R}$, which interpolate the data according to $G(x_i) = y_i$ for $i \in \{0, 1, \ldots, m\}$, and graph$(G)$ is the attractor of an IFS, which contains only affine transformations with lower triangular matrices. That is,

$$\left\{\tilde{F}_i\begin{pmatrix} x \\ y \end{pmatrix} = \begin{pmatrix} (x_i - x_{i-1})x + x_{i-1} \\ (y_i - y_{i-1} - \alpha_i(y_m - y_0))x + \alpha_i y + y_{i-1} - \alpha_i y_0 \end{pmatrix}\right\}_{i=1}^{m}$$

where $\alpha_i \in (-1,1) \setminus \{0\}$ are free parameters for $i = 1, \ldots, m$. In other words, the interpolation function G is the repeller of the piecewise linear, expanding map F, where $F(x,y) = F_i(x,y)$ if $x_{i-1} < x < x_i$ and

$$F_i(x,y) = \left(\frac{x - x_{i-1}}{x_i - x_{i-1}}, \frac{y - (y_i - y_{i-1} - \alpha_i(y_m - y_0))\frac{x - x_{i-1}}{x_i - x_{i-1}} - y_{i-1} + \alpha_i y_0}{\alpha_i}\right). \quad (8)$$

For a visualization of a fractal interpolation function, see Fig. 2.

Note that if Δ is collinear then $G_{\underline{\alpha},\Delta}$ is a linear function and thus, its graph has dimension 1. Thus, without loss of generality, the non-collinearity of Δ might be assumed without loss of generality.

Let us introduce the notation $G_{\underline{\alpha},\Delta}$, which denotes the fractal interpolation function for the dataset Δ and free parameters $\underline{\alpha} \in \left((-1,1) \setminus \{0\}\right)^{|\Delta|-1}$.

Barnsley and Harrington [7] calculated the box dimension of graph(G) in a special case. Namely, when $x_i - x_{i-1} = 1/m$ and $\alpha_i = \alpha$ for every $i = 1, \ldots, m$ with $1/m < \alpha$, and the data is not situated on a line. Note that in this case the interpolation function corresponds to $G_{\alpha,m}$ in (3) with

$$\phi(x) = (y_i - y_{i-1} - \alpha(y_m - y_0))\left(mx + \frac{y_{i-1} - \alpha y_0}{y_i - y_{i-1} - \alpha(y_m - y_0)} - (i-1)\right) \quad \text{if } \frac{i-1}{m} \le x < \frac{i}{m}. \quad (9)$$

In this case,

$$\dim_B \operatorname{graph}(G_{\alpha,m}) = 2 + \frac{\log \alpha}{\log m}.$$

This result was later generalized by Bedford [8] for general α_i but with the assumption that $x_i - x_{i-1} = 1/m$ with $\alpha_i > 1/m$ for every $i = 1, \ldots, m$. Ruan et al. [22]

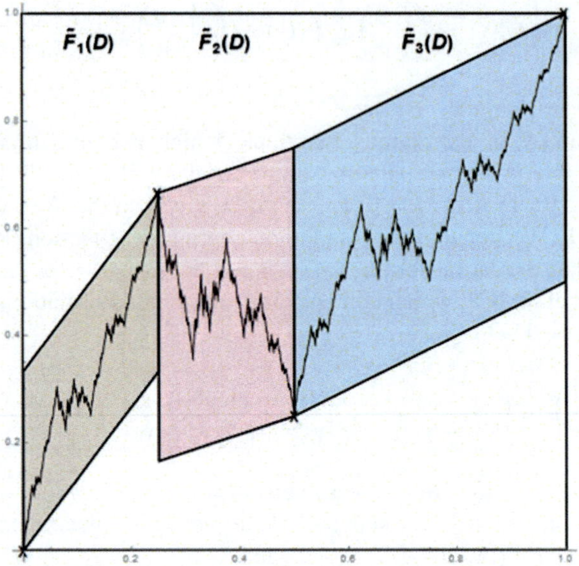

Fig. 2 Fractal interpolation function and its defining dynamics for $\Delta = \{(0,0), (1/4, 2/3), (1/2, 1/4), (1,1)\}$ and $\alpha_1 = 1/3$, $\alpha_2 = -1/2$ and $\alpha_3 = 1/2$

studied the box dimension in further generality. The complete characterization of the box counting dimension follows by Falconer and Miao [13, Corollary 3.1]. Namely, if Δ is not collinear then

$$\dim_B \operatorname{graph}(G_{\underline{\alpha},\Delta}) = \begin{cases} 1 & \text{if } \sum_{i=1}^m |\alpha_i| \leq 1 \text{ and} \\ s & \text{if } \sum_{i=1}^m |\alpha_i| > 1, \end{cases}$$

where $\sum_{i=1}^m |\alpha_i|(x_i - x_{i-1})^{s-1} = 1$.

The following extension for the Hausdorff dimension follows by Bárány, Hochman and Rapaport [4].

Theorem 3.1 *Let the dataset* $\Delta = \{(x_i, y_i) \in [0,1] \times \mathbb{R} : i = 0, 1, \ldots, m\}$ *be given so that* $0 = x_0 < x_1 < \cdots < x_{m-1} < x_m = 1$. *If* $\sum_{i=1}^m |\alpha_i| > 1$ *and there exists* $i \neq j$ *such that*

$$\frac{y_i - y_{i-1} - \alpha_i(y_m - y_0)}{x_i - x_{i-1} - \alpha_i} \neq \frac{y_j - y_{j-1} - \alpha_j(y_m - y_0)}{x_j - x_{j-1} - \alpha_j} \tag{10}$$

then

$$\dim_H \operatorname{graph}(G_{\underline{\alpha},\Delta}) = s, \text{ where } \sum_{i=1}^m |\alpha_i|(x_i - x_{i-1})^{s-1} = 1.$$

The assumption (10) is a little bit stronger than non-collinearity of Δ. That is, if Δ is collinear then (10) does not hold. The condition (10) is equivalent with the condition that the matrices $\{DF_i\}_{i=1}^m$ are not simultaneously diagonalizable.

Note that (10) is a milder condition than Ledrappier's condition (5). For example, suppose that the fractal interpolation function corresponds to a function of the form (3) with a 1-periodic piecewise linear ϕ. That is, the dataset $\Delta = \{(\frac{i}{m}, y_i) : i = 0, \dots, m\}$, $y_0 = y_m = 0$ and $\alpha_1 = \cdots = \alpha_m = \alpha$. Then ϕ is the piecewise linear function, connecting the dataset Δ, i.e.,

$$\phi(x) = (y_i - y_{i-1})(mx - (i-1)) + y_{i-1} \text{ if } \frac{i-1}{m} \le x < \frac{i}{m}$$

for $i = 1, \dots, m$. Then (5) has the form

$$Y(\underline{\xi}) = m \sum_{n=1}^{\infty} (m\alpha)^{-n} (y_{\xi_n} - y_{\xi_n - 1}),$$

where $\{\xi_n\}$ are independent random variables with $\mathbb{P}(\xi_i = k) = 1/m$ for $k = 1, \dots, m$. Ledrappier's condition requires that the distribution of the random variable Y has dimension 1 but the condition (10), i.e., $y_i - y_{i-1} \neq y_j - y_{j-1}$ for some $i \neq j$, is equivalent to that the distribution of the random variable Y has positive dimension.

3 Markovian Fractal Interpolation Functions

Let $\Delta = \{(x_i, y_i) \in [0, 1] \times \mathbb{R} : i = 0, 1, \dots, m\}$ be given so that $0 = x_0 < x_1 < \cdots < x_{m-1} < x_m = 1$, and let $\alpha_i \in (-1, 1) \setminus \{0\}$ for $i = 1, \dots, m$. The expanding dynamics, of which repeller is $\text{graph}(G_{\underline{\alpha}, \Delta})$, has a skew product form. That is, the map $F(x, y)$ has the form

$$F(x, y) = F_i(x, y) = (f_i(x), g_i(x, y)) \text{ for } x \in (x_{i-1}, x_i). \tag{11}$$

Thus, there is a base dynamics $f : [0, 1] \mapsto [0, 1]$, which is a piecewise linear, expanding interval map. In particular, each subinterval (x_{i-1}, x_i) is mapped to the complete interval $(0, 1)$. A natural generalization could be when the base dynamics f is a Markovian expanding map with Markov partition $\{(x_{i-1}, x_i) : i = 1, \dots, m\}$.

That is, for every $i = 1, \dots, m$ let $0 \le \ell(i) < r(i) \le m$ be integers such that $\gamma_i := \frac{x_{r(i)} - x_{\ell(i)}}{x_i - x_{i-1}} > 1$. Then let

$$f(x) = f_i(x) := \frac{x_{r(i)} - x_{\ell(i)}}{x_i - x_{i-1}} (x - x_{i-1}) + x_{\ell(i)} \text{ if } x \in (x_{i-1}, x_i).$$

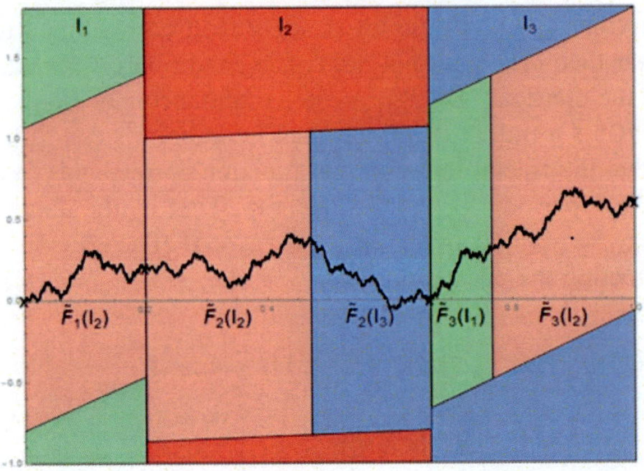

Fig. 3 Markovian fractal interpolation function and its defining dynamics for $\Delta = \{(0,0), (1/5, 1/5), (2/3, 0), (1, 3/5)\}$ and $\alpha_1 = 2/3, \alpha_2 = -2/3$ and $\alpha_3 = 2/3$

By the choice of $\ell(i), r(i)$, the map f is a piecewise linear expanding Markov map, see [5, Definition 10.1].

For each $i = 1, \ldots, m$, let $\alpha_i \in (-1, 1) \setminus \{0\}$ be arbitrary. Then let $g_i(x, y)$ be of the form $g_i(x, y) = \lambda_i y + a_i x + t_i$ such that $\lambda_i = \alpha_i^{-1}$, $g_i(x_{i-1}, y_{i-1}) = y_{\ell(i)}$ and $g_i(x_i, y_i) = y_{r(i)}$. This assumption guarantees that the repeller of F in (11) is a graph of a function G so that $G(x_i) = y_i$ for $i = 0, \ldots, m$. Simple calculations show that

$$a_i = \frac{y_{r(i)} - y_{\ell(i)} - \alpha_i^{-1}(y_i - y_{i-1})}{x_i - x_{i-1}} \text{ and } t_i = y_{\ell(i)} - \alpha_i^{-1} y_{i-1} - a_i x_{i-1}.$$

For a visualization of a Markovian fractal interpolation function, see Fig. 3.

Since the base dynamics is Markov, not all sequences of functions f_i is admissible. We define the following $m \times m$ matrix $A = (A_{i,j})_{i,j=1}^m$ as follows

$$A_{i,j} = \begin{cases} 1 & \text{if } \ell(i) + 1 \leq j \leq r(i), \\ 0 & \text{otherwise.} \end{cases} \tag{12}$$

Hence, an infinite sequence $\mathbf{i} = (i_1, i_2, \ldots)$ is admissible if $A_{i_k, i_{k+1}} = 1$ for every $k = 1, 2, \ldots$. Denote $\Sigma_A \subseteq \{1, \ldots, m\}^{\mathbb{N}}$ the set of all admissible sequences, that is, $\mathbf{i} = (i_1, i_2, \ldots) \in \Sigma_A$ if and only if $A_{i_k, i_{k+1}} = 1$ for every $k \geq 1$. By using the local inverses $\tilde{F}_i$, one can define the natural map from Σ_A to graph(G) as

$$\Pi(\mathbf{i}) = \lim_{n \to \infty} \tilde{F}_{i_1} \circ \cdots \circ \tilde{F}_{i_n}(x_{\ell(i_n)}, y_{\ell(i_n)}). \tag{13}$$

Thus, $\Pi(\mathbf{i})_2 = G(\Pi(\mathbf{i})_1)$, where $\Pi(\mathbf{i})_i$ denotes the ith coordinate of $\Pi(\mathbf{i})$, moreover, $F(\Pi(\mathbf{i})) = \Pi(\sigma\mathbf{i})$, where σ is the left shift on Σ_A (Fig. 4).

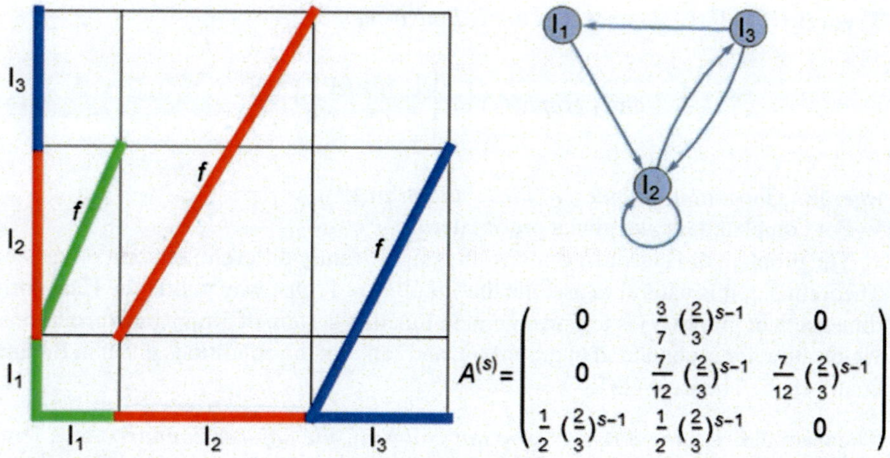

Fig. 4 Base system f, its Markovian structure and the matrix $A^{(s)}$ of the Markovian fractal interpolation function of Fig. 3

Since f is Markov with respect to the intervals $\{[x_{i-1}, x_i]\}_{i=1}^{m}$, one can decompose the intervals into finitely many classes with respect to recurrency. Since the repeller of F restricted to any recurrent class of intervals is graph(G) restricted to the intervals, without loss of generality, we may assume that f is topologically transitive. On the other hand, if the period of f would be $p \geq 2$ then again by decomposing the intervals into finitely many classes, the repeller of F^p restricted to a class is the restriction of graph(G). Thus, without loss of generality, we may assume that f (and the matrix A) is aperiodic, namely there exists a positive $k \geq 1$ such that every element of A^k is positive.

Since the local inverses are strict contractions, there exists an interval $D = [a, b]$ such that $\bigcup_{i=1}^{m} \tilde{F}_i([x_{\ell(i)}, x_{r(i)}] \times D) \subseteq [0, 1] \times D$. In order to determine the box counting dimension of graph(G), it is natural to cover graph(G) with sets of the form $\tilde{F}_{\boldsymbol{\omega}}([x_{\ell(i_{|\omega|})}, x_{r(i_{|\omega|})}] \times D)$. These sets are parallelograms with height parallel to the x-axis $\gamma_{\boldsymbol{\omega}}$ and side length (parallel to the y-axis) $\alpha_{\boldsymbol{\omega}}$.

Let us define the matrix $A^{(s)} = (A_{i,j}^{(s)})_{i,j=1}^{m}$ for $s \in [1, 2]$ as follows

$$A_{i,j}^{(s)} = |\alpha_i|\gamma_i^{-(s-1)} A_{i,j} = \begin{cases} |\alpha_i|\gamma_i^{-(s-1)} & \text{if } \ell(i) + 1 \leq j \leq r(i), \\ 0 & \text{otherwise.} \end{cases} \tag{14}$$

Similarly to Barnsley's fractal interpolation function, we distinguish two cases $\rho(A^{(1)}) \leq 1$ and $\rho(A^{(1)}) > 1$, where $\rho(\cdot)$ denotes the spectral radius. The first case implies that for most of the sets $\tilde{F}_{\boldsymbol{\omega}}([x_{\ell(i_{|\omega|})}, x_{r(i_{|\omega|})}] \times D)$, the component on the x-axis is longer than the component on the y-axis.

Theorem 3.2 *If the dataset* Δ *is not collinear then*

$$\dim_B graph(G) = \begin{cases} 1 & if\ \rho(A^{(1)}) \leq 1, \\ s & if\ \rho(A^{(1)}) > 1, \end{cases} \tag{15}$$

where s *is the unique solution of the equation* $\rho(A^{(s)}) = 1$.

For completeness, we give a proof later.

The problem of Hausdorff dimension is significantly different. In point of view of Theorem 3.2, it is natural to assume that $\rho(A^{(1)}) > 1$. One way to find the Hausdorff dimension of graph(G) is to find a iterated function system of affine transformations, which attractor is contained in graph(G), and satisfies the conditions given in Bárány et al. [4], [5, Theorem 6.3].

Theorem 3.3 *Let the dataset* Δ *be not collinear, the adjacency matrix* A *be irreducible and aperiodic, and* $(\alpha_1, \ldots, \alpha_m) \in ((-1, 1) \setminus \{0\})^m$ *be such that* $\rho(A^{(1)}) > 1$. *Moreover, let us assume that there exist* $\ell \geq 1$, $\boldsymbol{\omega}, \boldsymbol{\tau} \in \Sigma_{A,\ell}$ *such that*

$$\alpha_{\boldsymbol{\omega}} = \alpha_{\boldsymbol{\tau}}, \gamma_{\boldsymbol{\tau}} = \gamma_{\boldsymbol{\omega}}, \omega_1 = \tau_1, \omega_\ell = \tau_\ell \text{ and } D\tilde{F}_{\boldsymbol{\omega}} \neq D\tilde{F}_{\boldsymbol{\tau}}. \tag{16}$$

Then

$$\dim_H \text{graph}(G) = s, \text{ where } s \text{ is the unique solution of } \rho(A^{(s)}) = 1.$$

We remind that $\Sigma_n = \{1, \ldots, m\}^n$ is the collection of words of length n. For $n \in \mathbb{N}$, let $(p_1, \ldots, p_{|\Sigma_n|})$ be a probability vector and let ν be the corresponding Bernoulli measure, living on $(\Sigma_n^{\mathbb{N}}, \sigma_{\Sigma_n})$, where σ_{Σ_n} is the usual left shift but acting on $\Sigma_n^{\mathbb{N}}$. We have a natural isometry between $(\Sigma_n^{\mathbb{N}}, \sigma_{\Sigma_n})$ and (Σ, σ^n), let $\tilde{\nu}$ be the image of ν under this isometry. Finally, let

$$\hat{\nu} = \frac{1}{n} \sum_{i=0}^{n-1} \tilde{\nu} \circ \sigma^{-i}.$$

The measures $\hat{\nu}$ that can be obtained by this construction will be called *n-Bernoulli* measures. Note that the n-Bernoulli measures are ergodic and σ invariant measures on Σ.

Proposition 3.4 *Let* A *be an irreducible and aperiodic adjacency and let* (Σ_A, σ) *be a subshift of finite type and let* μ *be a* σ*-invariant measure supported on* Σ_A. *Then there exists a sequence of n-Bernoulli measures* $\hat{\nu}_n, n \to \infty$ *supported on* Σ_A *and converging to* μ *both in weak-* topology and in entropy.*

Proof Fix k such that all elements of A^k are positive. We choose a pair $(i, j) \in \{1, \ldots, m\}^2$ such that $A_{ij} = 1$. For every $\ell \in \{1, \ldots, m\}$ we can choose a word $\boldsymbol{p}(\ell) \in \Sigma_{A,k}$ such that $p_1 = j$ and $\boldsymbol{p}(\ell)\ell \in \Sigma_{A,k+1}$ and a word $\boldsymbol{s}(\ell) \in \Sigma_{A,k}$ such that $s_k = i$ and $\ell\boldsymbol{s}(\ell) \in \Sigma_{A,k+1}$. For any $n \geq 2k + 1$ and for any word $\boldsymbol{\omega} \in \Sigma_{A,n-2k}$ let

$\hat{\boldsymbol{\omega}} = \boldsymbol{p}(\omega_1)\boldsymbol{\omega}\boldsymbol{s}(\omega_{n-2k})$, denote the set of such words by $\hat{\Sigma}_{A,n}$. Note that $\hat{\Sigma}_{A,n} \subset \Sigma_{A,n}$, moreover each word $\hat{\boldsymbol{\omega}}$ begins with j and ends with i, hence any concatenation of those words is also admissible.

Let us show this construction on the example in Fig. 3. In this case

$$A = \begin{pmatrix} 0 & 1 & 0 \\ 0 & 1 & 1 \\ 1 & 1 & 0 \end{pmatrix}.$$

Choose $(i, j) = (2, 2)$. The matrix A^3 has strictly positive elements, and it is easy to check that choices $\boldsymbol{p}(1) = (2, 3), \boldsymbol{p}(2) = (2, 2), \boldsymbol{p}(3) = (2, 2)$ and $\boldsymbol{s}(1) = (2, 2), \boldsymbol{s}(2) = (2, 2), \boldsymbol{s}(3) = (2, 2)$ are admissible and appropriate.

Let ν_n be the the Bernoulli measure on $(\Sigma_{A,n}^{\mathbb{N}}, \sigma_{\Sigma_n})$ obtained by the probability vector $(p_{\boldsymbol{\tau}})_{\boldsymbol{\tau} \in \Sigma_{A,n}}$, where

$$p_{\boldsymbol{\tau}} = \begin{cases} \mu([\boldsymbol{\omega}]) & \text{if there exists } \boldsymbol{\omega} \in \Sigma_{A,n-2k} \text{such that } \boldsymbol{\tau} = \hat{\boldsymbol{\omega}}, \\ 0 & \text{otherwise.} \end{cases}$$

Let $\tilde{\nu}_n$ be the measure on (Σ_A, σ^n) and let $\hat{\nu}_n$ be the n-Bernoulli measure on (Σ_A, σ) as introduced previously. We need to prove two claims.

Claim 3.5 $h(\hat{\nu}_n) \to h(\mu)$ as $n \to \infty$.

Proof We have

$$h(\mu) = \lim_{n\to\infty} -\frac{1}{n-2k} \sum_{\Sigma_{A,n-2k}} \mu([\boldsymbol{\tau}]) \log \mu([\boldsymbol{\tau}]).$$

At the same time,

$$h(\tilde{\nu}_n, \sigma^n) = -\sum p_{\boldsymbol{\omega}} \log p_{\boldsymbol{\omega}} = -\sum_{\Sigma_{A,n-2k}} \mu([\boldsymbol{\tau}]) \log \mu([\boldsymbol{\tau}]),$$

hence

$$h(\hat{\nu}_n) = -\frac{1}{n} \sum_{\Sigma_{A,n-2k}} \mu([\boldsymbol{\tau}]) \log \mu([\boldsymbol{\tau}]).$$

Claim 3.6 $\hat{\nu}_n \to \mu$ in weak-* topology.

Proof Let $w \colon \Sigma \to \mathbb{R}$ be a continuous function and denote by $\mathrm{var}_\ell(w)$ the supremum of differences $w(x) - w(y)$ over x, y belonging to the same ℓ-th level cylinder. We have

$$\left| \int w d\mu - \int w d\hat{\nu}_n \right| \le \frac{1}{n} \sum_{i=0}^{n-1} \left| \int w d(\mu \circ \sigma^{-i}) - \int w d(\tilde{\nu}_n \circ \sigma^{-i}) \right|$$

(we remind that μ is σ-invariant, hence $\mu = \mu \circ \sigma^{-i}$ for any $i \geq 1$). For any $n - 2k$-th level cylinder set $[\boldsymbol{\omega}]$, $\tilde{\nu}_n(\sigma^{-k}[\boldsymbol{\omega}]) = \tilde{\nu}_n([\boldsymbol{p}(\omega_1)\boldsymbol{\omega}]) = \tilde{\nu}_n([\boldsymbol{p}(\omega_1)\boldsymbol{\omega}\boldsymbol{s}(\omega_{n-k})]) = \mu([\boldsymbol{\omega}])$, hence for $i = k, \ldots, n-k+1$ we have

$$\left|\int wd(\mu \circ \sigma^{-i}) - \int wd(\tilde{\nu}_n \circ \sigma^{-i})\right| \leq \mathrm{var}_{i-k}(w).$$

The other summands can be estimated from above by $\mathrm{var}_0(w)$. Summarizing,

$$\left|\int wd\mu - \int wd\hat{\nu}_n\right| \leq \frac{2k}{n}\mathrm{var}_0(w) + \frac{n-2k}{n}\frac{1}{n-2k}\sum_{i=1}^{n-2k}\mathrm{var}_i(w) \to 0.$$

The combination of Claims 3.5 and 3.6 proves the proposition.

Proof *(Proof of Theorem 3.3)* The strategy of the proof is the following:

(1) Find a σ-invariant ergodic probability measure μ on Σ_A which natural projection is a candidate for achieving the Hausdorff dimension;
(2) find a approximating sequence of n-step Bernoulli measures $\hat{\nu}_n$ such that $\hat{\nu}_n \to \mu$ in weak-* and entropy topology;
(3) show that $\dim_H \Pi_*\hat{\nu}_n \to s$ as $n \to \infty$.

First, we find the measure μ. Let s be such that $\rho(A^{(s)}) = 1$. Since there exists a $k \geq 1$ such that $(A^{(s)})^k$ has strictly positive elements. Then by Perron-Frobenius theorem, there exists a vector $p = (p_1, \ldots, p_m)^T$ with strictly positive elements such that $A^{(s)}p = p$. Let $P_{i,j} = A^{(s)}_{i,j}\frac{p_j}{p_i}$. Then the matrix $P = (P_{i,j})_{i,j=1}^m$ is a probability matrix, which is aperiodic and recurrent. Thus, there exists a unique probability vector $q = (q_1, \ldots, q_m)$ with positive elements such that $qP = q$. Then for a cylinder set $[i_1, \ldots, i_n]$ let

$$\mu([i_1, \ldots, i_n]) = q_{i_1}P_{i_1,i_2}\cdots P_{i_{n-1},i_n}. \tag{17}$$

It is easy to see by the definition of Lyapunov exponents in formula [5, (8.1)] that

$$h(\mu) = -\sum_{i,j} q_i P_{i,j}\log P_{i,j} = -\sum_{i=1}^m q_i \log|\alpha_i|\gamma_i^{-(s-1)} = \chi_2(\mu) + (s-1)\chi_1(\mu).$$

Moreover, since $\frac{h(\mu)}{\chi_1(\mu)} \leq 1 < s$ we have $\chi_2(\mu) < \chi_1(\mu)$, and thus, $D(\mu) = s$ by [5, Definition 8.2].

By Proposition 3.4, for every $\varepsilon > 0$ there exists a sequence of n-step Bernoulli measures $\hat{\nu}_n$ and a $N \geq 1$ such that for every $n \geq N$

$$|h(\mu) - h(\hat{\nu}_n)|, |\chi_2(\mu) - \chi_2(\hat{\nu}_n)|, |\chi_1(\mu) - \chi_1(\hat{\nu}_n)| < \varepsilon.$$

One can choose $\varepsilon < (\chi_1(\mu) - \chi_2(\mu))/100$, so $\chi_2(\hat{\nu}_n) < \chi_1(\hat{\nu}_n)$. Now, we approximate $\hat{\nu}_n$ with a nm-step Bernoulli measure $\overline{\nu}_{n,m}$, which is supported on words $\boldsymbol{\omega} \in (\Sigma_{A,n})^m$ for which $\gamma_{\boldsymbol{\omega}}^{-1} < \alpha_{\boldsymbol{\omega}}$. More precisely, let

$$Y_{m,n} = \{\boldsymbol{\omega} \in \Sigma_{A,nm} : \hat{\nu}_n(C[\omega]) > 0 \text{ and } \gamma_{\boldsymbol{\omega}}^{-1} < \alpha_{\boldsymbol{\omega}}\},$$

and let $\hat{\nu}_{n,m}$ be the Bernoulli measure on $(Y_{m,n})^{\mathbb{N}}$ defined with the probabilities $\left(\hat{\nu}_n(C[\boldsymbol{\omega}])/\hat{\nu}_n(Y_{m,n})\right)_{\boldsymbol{\omega} \in Y_{m,n}}$, and let $\overline{\nu}_{n,m}$ be the corresponding nm-step Bernoulli measure.

By the strong law of large numbers and Egorov's theorem, for every $\varepsilon > 0$ there exists $M = M(n) > 0$ such that for every $m \geq M$

$$|h(\overline{\nu}_{n,m}) - h(\hat{\nu}_n)|, |\chi_2(\overline{\nu}_{n,m}) - \chi_2(\hat{\nu}_n)|, |\chi_1(\overline{\nu}_{n,m}) - \chi_1(\hat{\nu}_n)| < \varepsilon.$$

Thus, $|s - D(\overline{\nu}_{n,m})| < C\varepsilon$ with some constant $C > 0$ independent of n, m.

By definition, $\operatorname{supp}(\Pi_* \overline{\nu}_{n,m}) \subseteq \operatorname{graph}(G)$. Thus, in order to apply [5, Theorem 6.3], it is enough to show that there exists $\boldsymbol{\omega} \neq \boldsymbol{\tau} \in Y_{m,n}$ such that $D\tilde{F}_{\boldsymbol{\omega}}$ and $D\tilde{F}_{\boldsymbol{\tau}}$ are not simultaneously diagonalizable. Let $\ell \geq 1$ and $\boldsymbol{\omega}_1, \boldsymbol{\tau}_1 \in \Sigma_{A,\ell}$ as in (16). Without loss of generality, we may assume that $n - 2k \gg \ell$. Since the first and last symbols of $\boldsymbol{\omega}_1, \boldsymbol{\tau}_1$ are the same, one can choose $\boldsymbol{\upsilon}_1, \boldsymbol{\upsilon}_2$ such that $\hat{\nu}_n(C[\boldsymbol{\upsilon}_1\boldsymbol{\omega}_1\boldsymbol{\upsilon}_2]), \hat{\nu}_n(C[\boldsymbol{\upsilon}_1\boldsymbol{\tau}_1\boldsymbol{\upsilon}_2]) > 0$. By the strong law of large numbers, for every sufficiently large $m \geq 1$ one can find $\boldsymbol{\kappa} \in \Sigma_{A,n(m-1)}$ such that $\boldsymbol{\upsilon}_1\boldsymbol{\omega}_1\boldsymbol{\upsilon}_2\boldsymbol{\kappa}, \boldsymbol{\upsilon}_1\boldsymbol{\tau}_1\boldsymbol{\upsilon}_2\boldsymbol{\kappa} \in Y_{m,n}$. By definition, $\alpha_{\boldsymbol{\upsilon}_1\boldsymbol{\tau}_1\boldsymbol{\upsilon}_2\boldsymbol{\kappa}} = \alpha_{\boldsymbol{\upsilon}_1\boldsymbol{\omega}_1\boldsymbol{\upsilon}_2\boldsymbol{\kappa}}$ and $\gamma_{\boldsymbol{\upsilon}_1\boldsymbol{\tau}_1\boldsymbol{\upsilon}_2\boldsymbol{\kappa}} = \gamma_{\boldsymbol{\upsilon}_1\boldsymbol{\omega}_1\boldsymbol{\upsilon}_2\boldsymbol{\kappa}}$. Thus, $D\tilde{F}_{\boldsymbol{\upsilon}_1\boldsymbol{\tau}_1\boldsymbol{\upsilon}_2\boldsymbol{\kappa}}$ and $D\tilde{F}_{\boldsymbol{\upsilon}_1\boldsymbol{\omega}_1\boldsymbol{\upsilon}_2\boldsymbol{\kappa}}$ are not simultaneously diagonalizable if and only if $D\tilde{F}_{\boldsymbol{\upsilon}_1\boldsymbol{\tau}_1\boldsymbol{\upsilon}_2\boldsymbol{\kappa}} \neq D\tilde{F}_{\boldsymbol{\upsilon}_1\boldsymbol{\omega}_1\boldsymbol{\upsilon}_2\boldsymbol{\kappa}}$. But this is true since $D\tilde{F}_{\boldsymbol{\omega}_1} \neq D\tilde{F}_{\boldsymbol{\tau}_1}$. Hence, by [5, Theorem 6.3]

$$\dim_H \operatorname{graph}(G) \geq \dim_H \Pi_* \overline{\nu}_{n,m} = D(\overline{\nu}_{n,m}) \geq s - C\varepsilon.$$

The statement follows by taking $\varepsilon \to 0$.

Proof *(Proof of Theorem 3.2)* Since the lower box-counting dimension is always an upper bound for the Hausdorff dimension and the upper box-counting dimension is always at most s, in point of view of Theorem 3.3, it is enough to show for diagonal systems. That is, by applying an affine transformation on the dataset Δ, we may assume that $a_i = 0$ for every $i = 1, \ldots, m$. Since Δ is not collinear, $G([0, 1])$ is an interval D with $|D| > 0$. Let $\Sigma_A^{(r)} = \left\{\boldsymbol{\omega} \in \bigcup_{\ell=1}^{\infty} \Sigma_{A,\ell} : \gamma_{\boldsymbol{\omega}}^{-1} \leq r < \gamma_{\boldsymbol{\omega}|_{|\boldsymbol{\omega}|-1}}^{-1}\right\}$. There needed at least $\sum_{\boldsymbol{\omega} \in \Sigma_A^{(r)}} \left\lceil \frac{|D| \cdot \alpha_{\boldsymbol{\omega}}}{\gamma_{\boldsymbol{\omega}}^{-1}} \right\rceil$-many squares of side length r to cover $\operatorname{graph}(G)$. By using the measure μ defined in (17),

$$\sum_{\boldsymbol{\omega} \in \Sigma_A^{(r)}} \left\lceil \frac{|D| \cdot \alpha_{\boldsymbol{\omega}}}{\gamma_{\boldsymbol{\omega}}^{-1}} \right\rceil \geq r^{-s} \sum_{\boldsymbol{\omega} \in \Sigma_A^{(r)}} \frac{|D| \cdot \alpha_{\boldsymbol{\omega}}}{\gamma_{\boldsymbol{\omega}}^{-1}} \gamma_{\boldsymbol{\omega}}^{-s} \geq r^{-s} C \sum_{\boldsymbol{\omega} \in \Sigma_A^{(r)}} \mu([\boldsymbol{\omega}]) = r^{-s} C,$$

where $C = |D| \min_{i,j} p_i/p_j$.

4 Continuous Generalized Takagi Functions

In the previous examples, the base dynamics $f\colon [0,1] \mapsto [0,1]$ was a Markovian expanding, piecewise linear map with Markov partition formed by intervals. For general systems of the form (11), the base dynamics is not Markovian. However, it is hard to get a graph of a continuous function as a repeller of such systems. Finally, we present here a special case, for which the repeller is a continuous function graph but the base dynamics is non-Markovian. This example can be considered as generalized Takagi functions.

Let us recall that the α-Takagi function T_α was defined as $T_\alpha(x) := \sum_{n=1}^{\infty} \alpha^n \psi(2^n \cdot x)$, where we defined $\psi(z) = \mathrm{dist}(z, \mathbb{Z})$.

To define a continuous generalization of this family first we fix the two parameters $\alpha \in (0,1)$ and $\beta \in (1,2)$ such that $\alpha \cdot \beta > 1$. Then we introduce (see Fig. 5) the function $B_\beta : [0,1] \to [0,1]$

$$B_\beta(x) := \begin{cases} \beta x, & \text{if } x \in \left[0, \frac{1}{2}\right]; \\ 1 - \beta(1-x), & \text{if } x \in \left(\frac{1}{2}, 1\right]. \end{cases} \tag{18}$$

This map will be our base dynamics.

Now we define the continuous generalized (α, β)-Takagi function $T_{\alpha,\beta}\colon [0,1] \to \mathbb{R}^+$ as

$$T_{\alpha,\beta}(x) := \sum_{k=0}^{\infty} \alpha^k \cdot \psi\left(B_\beta^n(x)\right). \tag{19}$$

The fact that the function $T_{\alpha,\beta}(x)$ is continuous follows from the fact that for all n the function $x \mapsto \psi\left(B_\beta^n(x)\right)$ is continuous (see the right-hand side of Fig. 5). Indeed, it is easy to see by the symmetry $B_\beta(x) = 1 - B_\beta(1-x)$ that for a continuous function $g\colon [0,1] \mapsto \mathbb{R}$, which is symmetric to the line $x = 1/2$, the map $g \circ B_\beta$ is continuous and symmetric to $x = 1/2$.

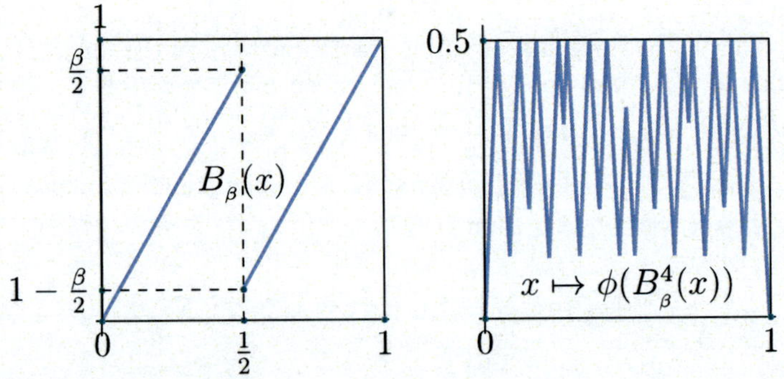

Fig. 5 Functions $B_\beta(x)$ and $\psi\left(B_\beta^4(x)\right)$

The graphs of the functions $T_{\alpha,\beta}(x)$ are not self-affine but the graphs of these functions have a less restrictive weakened form of self-affinity. Namely, we write

$$I_1 := \left[0, \frac{1}{2}\right],\ I_2 := \left[\frac{1}{2}, 1\right] \text{ and } J_1 := \left[0, \frac{\beta}{2}\right],\ J_2 := \left[1 - \frac{\beta}{2}.1\right] \tag{20}$$

and

$$\widetilde{I}_\ell := I_\ell \times [0, M_{\alpha,\beta}] \text{ and } \widetilde{J}_\ell := J_\ell \times [0, M_{\alpha,\beta}], \quad \ell = 1, 2, \tag{21}$$

where $M_{\alpha,\beta} := \max_{x\in[0,1]} T_{\alpha,\beta}(x)$. Then

$$\text{Graph}(T_{\alpha,\beta}) = \widetilde{F}_1\left(\text{Graph}(T_{\alpha,\beta}) \cap \widetilde{J}_1\right) \bigcup \widetilde{F}_2\left(\text{Graph}(T_{\alpha,\beta}) \cap \widetilde{J}_2\right), \tag{22}$$

where

$$\begin{aligned} \widetilde{F}_0(x, y) &:= \left(\frac{1}{\beta} \cdot x, \frac{1}{\beta} \cdot x + \alpha \cdot y\right) \text{ and} \\ \widetilde{F}_1(x, y) &:= \left(1 - \frac{1}{\beta} \cdot (1 - x), \frac{1}{\beta} \cdot (1 - x) + \alpha \cdot y\right). \end{aligned} \tag{23}$$

The union in (22) is almost disjoint, the intersection is the only point of graph$(T_{\alpha,\beta})$ which lies on the vertical line $x = \frac{1}{2}$. This follows from the fact that

$$\text{graph}(T_{\alpha,\beta}) \cap \widetilde{I}_\ell = \widetilde{F}_\ell\left(\text{Graph}(T_{\alpha,\beta}) \cap \widetilde{J}_\ell\right), \quad \ell = 1, 2. \tag{24}$$

See Fig. 6. If we compare this function graph with the graph of the self-affine Takagi map $T_{3/2}$ (see on the right-hand side of Fig. 1) then we can see the difference. Namely,

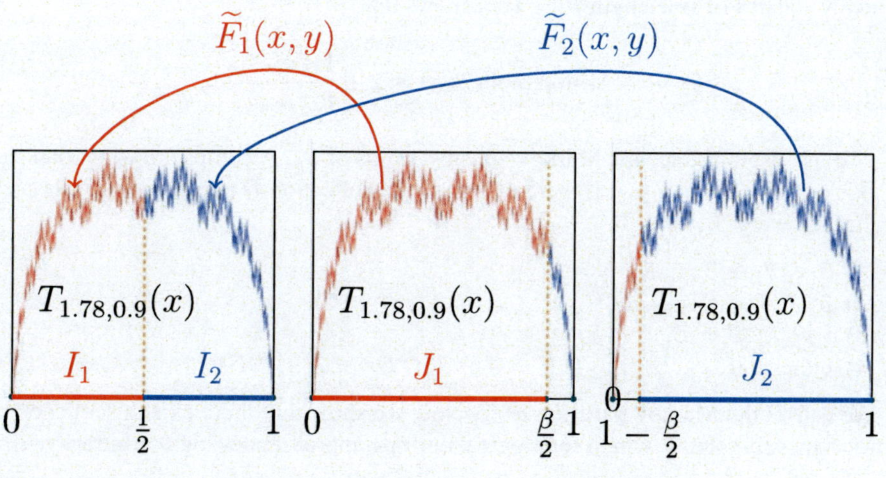

Fig. 6 graph$(T_{1.78,0.9})$ is the union of affine images of *parts* of graph$(T_{1.78,0.9})$

in the case of $T_{3/2}$, both the left- and right-hand sides of graph(T_α) are affine images of the whole graph graph(T_α). As opposed to that in the case of $T_{\alpha,\beta}$ the left- and the right-hand sides: graph$(T_{\alpha,\beta}) \cap \widetilde{I}_1$ and graph$(T_{\alpha,\beta}) \cap \widetilde{I}_2$ are affine images of certain *parts* of graph$(T_{\alpha,\beta})$ and not the whole one. That is why the family of $T_{\alpha,\beta}$ is much more general.

Theorem 3.7 *For every value of* α *and* $1 < \beta \leq 2$ *such that* $\alpha \cdot \beta > 1$

$$\dim_H \text{graph}(T_{\alpha,\beta}) = \dim_B \text{graph}(T_{\alpha,\beta}) = 2 + \frac{\log \alpha}{\log \beta}.$$

In order to calculate $\dim_H$ graph$(T_{\alpha,\beta})$, we give the upper bound by using natural covers and for the lower bound we find "large enough" Markovian subsystems of B_β. The set of admissible sequences is

$$\Sigma_\beta = \{(i_1, i_2, \ldots) : \exists x \in [0, 1] \text{ such that } B_\beta^n(x) \in I_{i_n} \text{ for every } n \geq 1\}.$$

Since the base system B_β is not Markovian for a general value of β, the set of admissible sequences cannot be generated by an adjacency matrix. By Rokhlin's formula, see [18, 21], $\lim_{n\to\infty} \frac{1}{n} \log \sharp\Sigma_\beta^{(n)} = \log \beta$, where $\Sigma_\beta^{(n)} = \{(i_1, \ldots, i_n) : \exists \mathbf{j} \in \Sigma_\beta \text{ such that } j_k = i_k \text{ for } k \geq 1\}$.

For each $\boldsymbol{\omega} \in \Sigma_\beta^{(n)}$, let us define the cylinder sets by induction. Namely, for $n = 1$ let $\mathcal{C}_\omega = \tilde{F}_\omega(\widetilde{J}_\omega)$ the cylinder set corresponding to $\omega \in \Sigma_\beta^{(1)}$. For $n > 1$ and $\boldsymbol{\omega} \in \Sigma_\beta^{(n)}$, let $\mathcal{C}_{\boldsymbol{\omega}} = \tilde{F}_{\omega_1}(\mathcal{C}_{\sigma\boldsymbol{\omega}} \cap \widetilde{J}_{\omega_1})$, where $\sigma\boldsymbol{\omega}$ is the word of length $n-1$ by deleting the first symbol of $\boldsymbol{\omega}$. For each $\boldsymbol{\omega} \in \Sigma_\beta^{(n)}$, the set $\mathcal{C}_\omega$ is a parallelogram with height parallel to the x-axis is at most β^{-n} and side length parallel to the y-axis is $\alpha^n M_{\alpha,\beta}$. Since $\alpha\beta > 1$ we get that the tangent of the angle between the sides is uniformly bounded, denote the bound by C. Thus, graph$(T_{\alpha,\beta})$ can be covered by at most $\sharp\Sigma_\beta^{(n)} \cdot (M_{\alpha,\beta}(\alpha\beta)^n + C)$-many squares of sidelength β^{-n}. This shows that

$$\overline{\dim}_B\text{graph}(T_{\alpha,\beta}) \leq 2 + \frac{\log \alpha}{\log \beta}.$$

Now, we introduce the Markovian subsystems of B_β. A compact B_β-invariant set $\mathcal{B}$ is called *Markov subset* if there exists a finite collection $\mathcal{D}$ of closed intervals such that for every $\mathfrak{I}_1, \mathfrak{I}_2 \in \mathcal{D}$.

(1) $\mathfrak{I}_1 \subseteq I_1$ or $\mathfrak{I}_1 \subseteq I_2$,
(2) $\mathfrak{I}_1^o \cap \mathfrak{I}_2^o = \emptyset$ if $\mathfrak{I}_1 \neq \mathfrak{I}_2$,
(3) $\bigcup_{\mathfrak{I}\in\mathcal{D}} \cap \mathcal{B} = \mathcal{B}$,
(4) either $B_\beta(\mathfrak{I}_1 \cap \mathcal{B}) \cap \mathfrak{I}_2 \cap \mathcal{B} = \emptyset$ or $\mathfrak{I}_2 \cap \mathcal{B} \subseteq B_\beta(\mathfrak{I}_1 \cap \mathcal{B})$.

We call $\mathcal{D}$ the Markov partition of $\mathcal{B}$. Now we show that there exist a sequence of Markov subsystems, which topological entropy approximates $\log \beta$ arbitrarily.

Lemma 3.8 *For every* $\varepsilon > 0$ *there exists* $m \geq 1$, *a Markov subset* $\mathcal{B}_m \subset [0, 1]$ *and* $\mathcal{D}_m$ *Markov partition such that*

$$h_{top}(B_\beta|_{\mathcal{B}_m}) > h_{top}(B_\beta) - \varepsilon.$$

Moreover, we can assume that there exist intervals in $\mathcal{D}_m$ *which contain* 0 *and* 1.

The claim follows from Hofbauer et al. [16, Proposition 1(a, b, c) and Lemma 2].

Similarly to (14), we define a matrix $A^{(s),m}$ for every $m \geq 1$, which gives the dimension of graph$(T_{\alpha,\beta}|_{\mathcal{B}_m})$. Namely, let $A^{(s),m}$ be a $\#\mathcal{D}_m \times \#\mathcal{D}_m$ matrix such that

$$A^{(s),m}_{\mathfrak{I},\mathfrak{J}} = \begin{cases} \alpha\beta^{-(s-1)} & \text{if } \mathfrak{J} \cap \mathcal{B}_m \subseteq B_\beta(\mathfrak{I} \cap \mathcal{B}_m) \text{ for } \mathfrak{I}, \mathfrak{J} \in \mathcal{D}_m \\ 0 & \text{otherwise.} \end{cases}$$

Let s_m be such that $\rho(A^{(s_m),m}) = 1$. For, $\mathfrak{I}, \mathfrak{J} \in \mathcal{D}_m$, let

$$\mathfrak{I} \xrightarrow{n} \mathfrak{J} = \{(\mathfrak{I}_1, \ldots, \mathfrak{I}_n) : \\ \mathfrak{I}_j \in \mathcal{D}_m,\ \mathfrak{I}_1 = \mathfrak{I}, \mathfrak{I}_n = \mathfrak{J},\ B_\beta(\mathfrak{I}_j \cap \mathcal{B}_m) \supseteq \mathfrak{I}_{j+1} \cap \mathcal{B}_m \text{ for } 1 \leq j \leq n-1\big\}.$$

By definition,

$$h_{top}(B_\beta|_{\mathcal{B}_m}) = \lim_{n\to\infty} \frac{\log \# \bigcup_{\mathfrak{I},\mathfrak{J}\in\mathcal{D}_m} \mathfrak{I} \xrightarrow{n} \mathfrak{J}}{n}.$$

But for every $k \geq 1$, and $\mathfrak{I}, \mathfrak{J} \in \mathcal{D}_m$,

$$\left(\left(A^{(s_m)}\right)^k\right)_{\mathfrak{I},\mathfrak{J}} = \left(\alpha\beta^{-(s_m-1)}\right)^k \cdot \#(\mathfrak{I} \xrightarrow{n} \mathfrak{J}).$$

Hence,

$$\log\beta - \varepsilon < h_{top}(B_\beta|_{\mathcal{B}_m}) = \lim_{k\to\infty} \frac{\log \frac{\|(A^{(s_m)})^k\|_1}{(\alpha\beta^{-(s_m-1)})^k}}{k} = -\log\left(\alpha\beta^{-(s_m-1)}\right),$$

which implies that $s_m > 2 + \frac{\log\alpha}{\log\beta} - \varepsilon$. One can decompose $\mathcal{D}_m$ into recurrent and transient classes. It is easy to see that there exists a recurrent class R such that restricting $A^{(s_m),m}$ for R, $\rho(A^{(s_m),m}|_R) = 1$. Denote the Markov subset of $\mathcal{B}_m$ restricted to R by $\mathcal{R}_m$. Similarly to (17), there exists a Markov measure μ_m such that $D(\mu_m) = s_m$. By Proposition 3.4, for every $\varepsilon > 0$ there exists an n-step Bernoulli measure $\nu_{n,m}$ such that $D(\nu_{n,m}) > s_m - \varepsilon$. By [5, Theorem 7.6] , $\dim_H \Pi_*\nu_{n,m} = D(\nu_{n,m})$, which gives the lower bound.

References

1. P.C. Allaart, K. Kawamura, The Takagi function: a survey. Real Anal. Exch. **37**(1), 1–54 (2011/12)
2. K. Barański, Dimension of the graphs of the Weierstrass-type functions, in *Fractal Geometry and Stochastics V*, volume 70 of *Progr. Probab.* (Birkhäuser/Springer, Cham, 2015), pp. 77–91
3. K. Barański, B. Bárány, J. Romanowska, On the dimension of the graph of the classical Weierstrass function. Adv. Math. 265, 32–59 (2014)
4. B. Bárány, M. Hochman, A. Rapaport, Hausdorff dimension of planar self-affine sets and measures. Invent. Math. **216**(3), 601–659 (2019)
5. B. Bárány, M. Rams, K. Simon, Dimension theory of some non-Markovian repellers Part I: a gentle introduction
6. M.F. Barnsley, Fractal functions and interpolation. Constr. Approx. 2(4), 303–329 (1986)
7. M.F. Barnsley, A.N. Harrington, The calculus of fractal interpolation functions. J. Approx. Theor. 57(1), 14–34 (1989)
8. T. Bedford, Hölder exponents and box dimension for self-affine fractal functions. Constr. Approx. **5**(1), 33–48 (1989). Fractal approximation
9. A.S. Besicovitch, H.D. Ursell, Sets of fractional dimensions (v): on dimensional numbers of some continuous curves. J. Lond. Math. Soc. **s1-12**(1), 18–25
10. P. Billingsley, Notes: Van Der Waerden's continuous nowhere differentiable function. Am. Math. Monthly 89(9), 691 (1982)
11. P. Erdös, On a family of symmetric Bernoulli convolutions. Am. J. Math. 61, 974–976 (1939)
12. P. Erdös, On the smoothness properties of a family of Bernoulli convolutions. Am. J. Math. 62, 180–186 (1940)
13. K. Falconer, J. Miao, Dimensions of self-affine fractals and multifractals generated by upper-triangular matrices. Fractals 15(3), 289–299 (2007)
14. G.H. Hardy, Weierstrass's non-differentiable function. Trans. Am. Math. Soc. 17(3), 301–325 (1916)
15. M. Hochman. On self-similar sets with overlaps and inverse theorems for entropy. Ann. Math. 180(2), 773–822 (2014)
16. F. Hofbauer, P. Raith, K. Simon, Hausdorff dimension for some hyperbolic attractors with overlaps and without finite Markov partition. Ergodic Theor. Dyn. Syst. 27(4), 1143–1165 (2007)
17. J.L. Kaplan, J. Mallet-Paret, J.A. Yorke, The Lyapunov dimension of a nowhere differentiable attracting torus. Ergodic Theory Dyn. Syst. 4(2), 261–281 (1984)
18. F. Ledrappier, Some properties of absolutely continuous invariant measures on an interval. Ergodic Theory Dyn. Syst. 1(1), 77–93 (1981)
19. F. Ledrappier, On the dimension of some graphs, in *Symbolic dynamics and its applications (New Haven, CT, 1991)*, volume 135 of *Contemp. Math.* (American Mathematical Society, Providence, RI, 1992), pp. 285–293
20. B.B. Mandelbrot. *Fractals: Form, Chance, and Dimension*, revised edn. (W. H. Freeman and Co., San Francisco, Calif., 1977). Translated from the French
21. V.A. Rohlin, Exact endomorphisms of a Lebesgue space. Izv. Akad. Nauk SSSR Ser. Mat. 25, 499–530 (1961)
22. H.-J. Ruan, S. Wei-Yi, K. Yao, Box dimension and fractional integral of linear fractal interpolation functions. J. Approx. Theory 161(1), 187–197 (2009)
23. W. Shen, Hausdorff dimension of the graphs of the classical Weierstrass functions. Math. Z. 289(1–2), 223–266 (2018)
24. B. Solomyak, Measure and dimension for some fractal families. Math. Proc. Cambridge Philos. Soc. 124(3), 531–546 (1998)
25. T. Takagi, A simple example of the continuous function without derivative. Tokyo Sugaku-Butsurigakkwai Hokoku 1, F176–F177 (1901)
26. P.P. Varjú, On the dimension of bernoulli convolutions for all transcendental parameters. Ann. Math. **189**(3), 1001–1011 (2019)

Iterated Function Systems—A Topological Approach. Attractors

Krzysztof Leśniak

Abstract Various types of basins, attractors and their fiberings are defined and shortly discussed in the realm of iterated function systems on normal topological spaces.

Keywords Iterated function system · Strict attractor · Pointwise basin · Fast basin

2010 Mathematics Subject Classification Primary: 54H20 · Secondary: 47H09

1 Introduction

The aim of this article is to present some topological basics on attractors of IFSs in view of recent advances in the fractal geometry. It is based on the series of articles: [2–6, 8]. We introduce the concepts of basin, pointwise basin, fast basin, strict attractor, pointwise strict attractor, point-fibred attractor, strongly fibred attractor and homoclinic attractor. Relation of these concepts with the chaos game algorithm and fractal manifolds is mentioned in passing. For a thorough discussion of the existence of attractors, invariant sets and measures in contractive and non-contractive IFSs we refer to survey [9].

K. Leśniak (✉)
Faculty of Mathematics and Computer Science, Nicolaus Copernicus University in Toruń, Chopina 12/18, 87-100 Toruń, Poland
e-mail: much@mat.umk.pl

R. L. Devaney et al. (eds.), *Topological Dynamics and Topological Data Analysis*, Springer Proceedings in Mathematics & Statistics 350,
https://doi.org/10.1007/978-981-16-0174-3_4

2 IFS

Throughout the paper, X will be a normal topological space. As usual, $\overline{S}$ stands for the closure and $\mathrm{Int}(S)$ for the interior of $S \subseteq X$.

We distinguish the following collections of sets:

- 2^X, all subsets of X;
- $\mathcal{C}(X)$, nonempty closed sets;
- $\mathcal{CB}(X)$, nonempty bounded closed sets (provided X is a metric space);
- $\mathcal{K}(X)$, nonempty compact sets.

The *Vietoris topology* in $\mathcal{C}(X)$ is generated by subbasic sets of two forms

$$V^+ = \{C \in \mathcal{C}(X) : C \subseteq V\},$$
$$V^- = \{C \in \mathcal{C}(X) : C \cap V \neq \emptyset\},$$

where V runs through all open subsets of X. If X is a metric space, then the Vietoris topology and the Hausdorff metric topology agree on $\mathcal{K}(X)$. If a sequence of closed sets $S_n \subseteq X$ converges to $S \subseteq X$ with respect to the Vietoris topology, then we write $S_n \to S$.

An *iterated function system* $\mathcal{F} = \{w_i : i \in I\}$, *IFS* for short, is a finite collection of maps $w_i : X \to X$. Note that we do not assume continuity of w_i.

The *Hutchinson operator* $\mathcal{F}: 2^X \to 2^X$ induced by the IFS $\mathcal{F}$ is defined as follows

$$\mathcal{F}(S) := \bigcup_{i \in I} \overline{w_i(S)} \quad \text{for} \quad S \subseteq X.$$

Note that, without ambiguity, we denote the IFS and the associated Hutchinson operator by the same symbol $\mathcal{F}$. Symbol $\mathcal{F}^n$ will stand for the n-fold composition of $\mathcal{F}$. (Conveniently $\mathcal{F}^0 = \mathrm{id}$.)

Under additional conditions, we can restrict $\mathcal{F}$ to smaller collections of sets. We shall tacitly assume the following condition

$$w_i(K) \in \mathcal{K}(X) \quad \text{for all} \quad K \in \mathcal{K}(X), i \in I,$$

whenever we write $\mathcal{F}: \mathcal{K}(X) \to \mathcal{K}(X)$. This condition is satisfied when all maps w_i are continuous.

If $\mathcal{F}$ comprises continuous maps, then the Hutchinson operator $\mathcal{F}: \mathcal{C}(X) \to \mathcal{C}(X)$ is continuous with respect to the Vietoris topology. If X is a metric space, then $\mathcal{F}: \mathcal{K}(X) \to \mathcal{K}(X)$ is continuous in both, the Vietoris topology and the Hausdorff metric topology, while $\mathcal{F}: \mathcal{CB}(X) \to \mathcal{CB}(X)$ may fail to be continuous with respect to the Hausdorff metric. See [2] for more information about the continuity of $\mathcal{F}$.

3 Basins and Pointwise Basins

Definition 3.1 (Barnsley et al. [3, 4]) Let $A \in \mathcal{K}(X)$ and $\mathcal{F}$ be an IFS on X. We define the *pointwise basin* of A to be the set

$$\mathcal{B}_1(A) = \{x \in X : \mathcal{F}^n(\{x\}) \to A\},$$

and the *basin* of A to be the set

$$\mathcal{B}(A) = \bigcup \mathcal{U}(A),$$
$$\mathcal{U}(A) = \{U \subseteq X : A \subseteq U - \text{open}, \mathcal{F}^n(S) \to A \text{ for all } S \in \mathcal{K}(U)\}.$$

A nonempty compact set A is

(i) a *pointwise strict attractor* of $\mathcal{F}$, when $\text{Int}(\mathcal{B}_1(A)) \supseteq A$;
(ii) a *strict attractor* of $\mathcal{F}$, when $\mathcal{B}(A) \neq \emptyset$.

Proposition 3.2 (Barnsley et al. [3] Propositions 8 and 11) *(i) If A is a pointwise strict attractor of $\mathcal{F}$, then* $\text{Int}(\mathcal{B}_1(A)) = \mathcal{B}_1(A)$ *and* $\mathcal{F}(\mathcal{B}_1(A)) \subseteq \mathcal{B}_1(A)$.

(ii) If A is a strict attractor, then A is a pointwise strict attractor, and $\mathcal{B}(A) = \mathcal{B}_1(A)$.

The following criterion explains that pointwise strict attractors which are not strict attractors can exist only in highly non-contractive IFSs.

Proposition 3.3 (Barnsley et al. [3] Lemma 10) *Let $\mathcal{F} = \{w_i : i \in I\}$ be an IFS consisting of nonexpansive maps $w_i : X \to X$ acting on a metric space (X, d). If A is a pointwise strict attractor of $\mathcal{F}$, then A is a strict attractor of $\mathcal{F}$.*

We list now a couple of characteristic examples.

Example 3.4 (Strict attractor is a local concept) Let $w : X \to X$ be a continuous map with two attractive fixed points $x_1, x_2 \in X$, i.e. there exist open neighbourhoods $U_l \ni x_l$, $l = 1, 2$, such that $w^n(x) \to x_l$ for $x \in U_l$. Then, $A_l = \{x_l\}$, $l = 1, 2$, are two pointwise strict attractors of the same $\mathcal{F} = \{w\}$. (If w is locally contractive around x_1, x_2 in a complete metric space X, then we get strict attractors.)

In view of the above example and the example below, let us note that a strict attractor A of the IFS comprising global contractions is global in the sense that $\mathcal{B}(A) = X$.

Example 3.5 (Strict attractor is a topological concept) Let $\mathbb{C}$ be the complex plane. We endow $\mathbb{C}$ with two equivalent metrics: $d(z_1, z_2) = |z_1 - z_2|$ for $z_1, z_2 \in \mathbb{C}$ and $d_1 = \frac{d}{1+d}$. Fix three distinct points $a_1, a_2, a_3 \in \mathbb{C}$. Define $w_i(z) = \frac{1}{2} \cdot (z + a_i)$ for $z \in \mathbb{C}, i = 1, 2, 3$ and consider $\mathcal{F} = \{w_i : i \in \{1, 2, 3\}\}$. It is known that the Sierpiński triangle A with vertices a_1, a_2, a_3 is the *Hutchinson attractor* of $\mathcal{F}$ in $(\mathbb{C}, d)$, i.e. for all nonempty closed and bounded subsets S of $(\mathbb{C}, d)$, the set convergence $\mathcal{F}^n(S) \to$

A takes place with respect to the Hausdorff metric d_H in $\mathcal{CB}(\mathbb{C})$ induced by d. The Hausdorff metric induced by d_1 is not equivalent to d_H, because d_1 and d are not uniformly equivalent. Moreover, $\mathcal{F}^n(\mathbb{C}) = \mathbb{C} \neq A$, and the set $\mathbb{C}$ is closed and bounded in $(\mathbb{C}, d_1)$. Therefore, A is not the Hutchinson attractor of $\mathcal{F}$ in $(\mathbb{C}, d_1)$. On the other hand, A is a strict attractor of $\mathcal{F}$ regardless of the choice of equivalent metric in $\mathbb{C}$.

Example 3.6 (Strict attractor in a discontinuous IFS) Let $\mathcal{F} = \{w_i : i \in I\}$ be an IFS comprising continuous maps $w_i : X \to X$. We assume that $\mathcal{F}$ admits a strict attractor, denoted A. Further, assume that A has two disjoint dense subsets $E_m \subseteq A, m = 1, 2$, i.e. $\overline{E_m} = A$, $E_1 \cap E_2 = \emptyset$. Let also $e_m \in E_m$ be two distinguished points. Define for $i \in I, m = 1, 2$

$$\widetilde{w_{i,m}}(x) = \begin{cases} w_i(x), & x \in E_m \cup (X \setminus A), \\ e_m, & x \in A \setminus E_m. \end{cases}$$

Then, the IFS $\widetilde{\mathcal{F}} = \{\widetilde{w_{i,m}} : (i, m) \in I \times \{1, 2\}\}$ is an IFS of discontinuous maps, and A is a strict attractor of $\widetilde{\mathcal{F}}$. (Indeed, the Hutchinson operators associated with $\mathcal{F}$ and $\widetilde{\mathcal{F}}$ coincide.)

Some other notable examples of strict attractors include:

- the Alexandrov double arrow space—a nonmetrizable compact separable space ([3] Example 6);
- the Warsaw sine curve—a non-locally connected continuum ([4] Example 2).

Pointwise strict attractors, despite their generality, offer sufficiently reach theory to be worth of consideration for IFSs. For instance, the probabilistic chaos game algorithm is valid for them, cf. [3].

If A is a strict attractor of the IFS $\mathcal{F}$ comprising continuous maps, then A is an invariant set, i.e. $\mathcal{F}(A) = A$. (Indeed, $\mathcal{F}^{n+1}(A) = \mathcal{F}(\mathcal{F}^n(A)) \to \mathcal{F}(A) = A$ thanks to continuity of $\mathcal{F}$.) We will see later that attractors which are not invariant can exist in discontinuous IFSs and their existence leads to interesting questions.

4 Point-Fibred and Strongly Fibred Attractors

Let I be a finite set (with a discrete topology). The Tikhonov product I^∞ of countably many copies of I is called the *code space*. It is a Cantor space, i.e. a homeomorph of the Cantor ternary set.

Definition 4.1 (Kieninger [7] chap. 4) Let $\mathcal{F} = \{w_i : i \in I\}$ be an IFS comprising continuous maps. Let A be a strict attractor of $\mathcal{F}$. We define the *coding multifunction* $\pi : I^\infty \to \mathcal{K}(A)$ by the following formula

$$\pi(\iota) = \bigcap_{n=1}^{\infty} w_{i_1} \circ \ldots \circ w_{i_n}(A) \quad \text{for } \iota = (i_n)_{n=1}^{\infty} \in I^\infty.$$

The strict attractor A is said to be

- *point-fibred* if π is single-valued, i.e. $\pi(\iota)$ is a singleton for each $\iota \in I^\infty$;
- *strongly fibred* if for every open $V \subseteq X$ with $V \cap A \neq \emptyset$ there exists $\iota \in I^\infty$ such that $\pi(\iota) \subseteq V$.

Note that the coding map π provides a fibering of the attractor A into a nondisjoint union: $A = \bigcup_{\iota \in I^\infty} \pi(\iota)$.

Proposition 4.2 (Barnsley and Leśniak [1] Proposition 1) *The coding multifunction π of a strict attractor A of an IFS $\mathcal{F}$ comprising continuous maps w_i does not depend on the choice of a forward invariant compact cap $C \supseteq A$, $\mathcal{F}(C) \subseteq C$, that is for every forward invariant compact cap $C \subseteq \mathcal{B}(A)$ and every $\iota = (i_n)_{n=1}^\infty$ we have*

$$\pi(\iota) = \bigcap_{n=1}^{\infty} w_{i_1} \circ \ldots \circ w_{i_n}(C).$$

An attractor of an IFS comprising weak contractions is point-fibred. Interestingly, we can construct strongly fibred attractors from point-fibred ones.

Example 4.3 (Strongly fibred attractor which is not point-fibred; [1] Example 2.1, [7] Example 4.3.19) Let $\mathcal{F} = \{w_i : i \in I\}$ be an IFS of at least two continuous maps $w_i : X \to X$ on a compact space X which contains at least two points. Assume that the images of these maps tessalate X: $\bigcup_{i \in I} w_i(X) = X$. (We do not demand Int $(w_i(X))$ to be disjoint.) Define an IFS on $X \times X$:

$$\mathcal{F}_\square = \mathrm{id} \times w_i,\, w_i \times \mathrm{id} : i \in I.$$

If X is a point-fibred strict attractor of $\mathcal{F}$, then $X \times X$ is a strongly fibred strict attractor of $F_\square$, but it is not point-fibred.

Example 4.4 (Non-strongly fibred attractor) Let $w : X \to X$ be a minimal map on a compact metric space X (i.e. $\overline{\{w^n(x) : n \geq 0\}} = X$ for each $x \in X$). Then, X is a strict attractor of $\mathcal{F} = \{\mathrm{id}, w\}$, and X is not strongly fibred.

The interesting fact about strongly fibred strict attractors, aside their mosaic inner structure (e.g. [5]), is that we can derandomize the chaos game algorithm for such attractors, cf. [1].

5 Fast Basins

So far we have considered the basin $\mathcal{B}(A)$ and the pointwise basin $\mathcal{B}_1(A)$ of a set A. These domains have the property that the iterations of the IFS $\mathcal{F} = \{w_i : i \in I\}$ starting there, as well as orbits $x_n = w_{i_n} \circ \ldots \circ w_{i_1}(x_0)$, $i_n \in I$, $n \geq 1$, $x_0 \in \mathcal{B}_1(A)$, are attracted by A. We are going to consider the fast basin $\widehat{\mathcal{B}}(A)$ of A, the domain with the property that all iterations (of orbits) fall into A after finite number of steps.

Definition 5.1 (Barnsley et al. [4, 6]) Let A be a strict attractor of an IFS $\mathcal{F}$. The *fast basin* of A is defined by

$$\widehat{\mathcal{B}}(A) = \{x \in X: \mathcal{F}^n(\{x\}) \cap A \neq \emptyset \text{ for some } n \geq 0\}.$$

We describe below the fast basin of the Sierpiński triangle.

Example 5.2 (Sierpiński wallpaper) Let A be the Sierpiński triangle in the complex plane with vertices $a_1 = 0, a_2 = 1, a_3 = \imath \in \mathbb{C}$, generated by the IFS from Example 3.5. Then, $\widehat{\mathcal{B}}(A) = \bigcup_{k,m\in\mathbb{Z}}(A + k \cdot 1 + m \cdot \imath)$.

It should be noted that in general neither $\widehat{\mathcal{B}}(A) \subseteq \mathcal{B}(A)$ nor $\mathcal{B}(A) \subseteq \widehat{\mathcal{B}}(A)$.

Example 5.3 (Fast basin reaching outside basin; [4] Example 5) Let $X = \mathbb{R} \cup \{\infty\}$. Define $w_1(x) = \frac{x}{2}$ for $x \neq \infty$, $w_1(\infty) = \infty$, $w_2(x) = \frac{x+3}{-2x+6}$ for $x \notin \{3, \infty\}$, $w_2(3) = \infty$, $w_2(\infty) = \frac{-1}{2}$. Then $A = [0, 1]$ is a strict attractor with basin $\mathcal{B}(A) = \left(-\infty, \frac{3}{2}\right)$. It turns out that

$$\{3 \cdot 2^k: k \geq 1\} \subseteq \widehat{\mathcal{B}}(A) \setminus \mathcal{B}(A).$$

Denote

- $\mathcal{F}^{-1}(S) = \bigcup_{i\in I} w_i^{-1}(S)$, the large counter-image of $S \subset X$;
- $\widehat{B}(\vartheta) = \bigcup_{k=0}^{\infty} w_{\theta_k}^{-1}(\ldots w_{\theta_1}^{-1}(A)\ldots)$, the *fractal continuation* of A along $\vartheta = (\theta_1, \theta_2, \ldots) \in I^\infty$.

Proposition 5.4 *(Alternative descriptions of the fast basin; Barnsley et al. [4] Propositions 2 and 3) If A is a strict attractor of $\mathcal{F}$ and $\widehat{B}(A)$ is the fast basin of A, then*

(i) $S = \widehat{B}(A)$ is the smallest (with respect to $\subseteq$) solution of the equation

$$\mathcal{F}^{-1}(S) \cup A = S;$$

(ii) $\widehat{B}(A) = \bigcup_{k=0}^{\infty}(\mathcal{F}^k)^{-1}(A) = \bigcup_{\vartheta\in I^\infty} \widehat{B}(\vartheta)$.

The IFS is said to be *invertible* if it consists of homeomorphisms. The characterization of the fast basin given in Proposition 5.4 is the key to the following theorem.

Theorem 5.5 *Let A be a strict attractor of the invertible IFS $\mathcal{F}$ acting on a normal space X. Let $\widehat{\mathcal{B}}(A)$ be the fast basin of A. Let (P) be any of the following properties of a set:*

(i) the Lebesgue topological dimension of the set equals $\delta \in \{0, 1, 2, \ldots\}$;
(ii) the Hausdorff fractal dimension of the set equals $\delta \in [0, \infty)$;
(iii) the set is connected;
(iv) the set is pathwise connected;
(v) the set is boundary (i.e. it has empty topological interior);

(vi) *the set is σ-porous;*
(vii) *the set is hereditarily disconnected (in particular, it has a tree-like structure and admits ultrametrization).*

If A has property (P), then $\widehat{\mathcal{B}}(A)$ has property (P) too. In (ii) and (vi), we need to assume that X is a metric space and the maps constituting $\mathcal{F}$ are b-Lipschitz. In (v), we need to assume that X is a Baire topological space. For (vii), we assume that X is a locally compact metric space.

The work [4] contains a gallery of fast basis. To unveil a true nature of the fast basin $\widehat{\mathcal{B}}(A)$, one has to introduce inductive topology in a flag of successive enlargements $w_{\theta_k}^{-1}(\ldots w_{\theta_1}^{-1}(A)\ldots)$ (or blow-ups) of A. These blow-ups fill up the fractal continuation $\widehat{B}(\vartheta)$. Properly glued continuations constitute branches (or leaves) of the resulting object called a *fractal manifold*. We refer to [6] for technical details of this construction. A simplistic visualization of this construction in the case of the Sierpiński wallpaper has been offered in [10].

6 Homoclinic Attractors Versus Fast Basins

We are going to address an intricate connection of the existence of non-invariant strict attractors, called *homoclinic attractors*, with the notion of fast basin.

Let $\mathcal{F} = \{w_i : i \in I\}$ be an IFS of continuous maps $w_i : X \to X$. Let A be a strict attractor with a nontrivial basin $\mathcal{B}(A) \neq A$. Fix $b \in \mathcal{B}(A) \setminus A$. Define $\widetilde{w}_i|A \equiv b$, $\widetilde{w}_i = w_i$ outside A, and

$$\widetilde{\mathcal{F}} = \{\widetilde{w}_i : i \in I\}.$$

Then, $\widetilde{\mathcal{F}}$ is a discontinuous modification of $\mathcal{F}$.

The following question arises: Whether/when A persists a strict attractor after the modification of $\mathcal{F}$? We would have then an attractor of $\widetilde{\mathcal{F}}$ which undergoes an expulsion of its content, i.e. $\widetilde{\mathcal{F}}(A) \nsubseteq A$. The answer is that it depends upon the fast basin $\widehat{\mathcal{B}}(A)$ of the original system $\mathcal{F}$.

Proposition 6.1 (Necessary condition for a homoclinic attractor; [8] Proposition 2) *If A is a strict attractor of $\widetilde{\mathcal{F}}$, then $b \notin \widehat{\mathcal{B}}(A)$.*

Theorem 6.2 [Sufficient condition for a homoclinic attractor; [8] Theorem 3] *If $b \notin \widehat{\mathcal{B}}(A)$ and the following nonresonance condition holds: there exists an open neighbourhood $A \subseteq U(A) \subseteq \mathcal{B}(A)$ such that*

$$\kappa(S) := \sup\{k \geq 0 : \mathcal{F}^k(S) \cap (\widehat{\mathcal{B}}(A) \setminus A) \neq \emptyset\} < \infty$$

for all nonempty compact $S \subseteq U(A)$, then A is a strict attractor of $\widetilde{\mathcal{F}}$.

What about more general modifications $\widetilde{\mathcal{F}}$ of $\mathcal{F}$? Say, $\mathcal{F}$ admits a strict attractor A with basin $\mathcal{B}(A)$ and fast basin $\widehat{\mathcal{B}}(A)$, further $\widetilde{\mathcal{F}}$ is such a modification of $\mathcal{F}$ that

$\widetilde{\mathcal{F}}(A) \subseteq \mathcal{B}(A)$ and $\widetilde{\mathcal{F}}(A) \nsubseteq A$. On this level of generality, Proposition 6.1 would sound like: if A is a strict attractor of $\widetilde{\mathcal{F}}$, then $\widetilde{\mathcal{F}}(A) \cap (\widehat{\mathcal{B}}(A) \setminus A) = \emptyset$. We have the following counterexample for such speculations.

Example 6.3 (Leśniak [8] Example 6) Let us consider the IFS $\mathcal{F} = \{w_i : i \in \{1, 2, 3\}\}$ on $\mathbb{C}$ from Example 5.2. Let $\widetilde{w_i} = w_i$ for $i = 2, 3$, and $\widetilde{w_1}(z) = w_1(z)$ for $z \neq 0$, $\widetilde{w_1}(0) = 2$. The Sierpiński triangle A is a strict attractor of $\mathcal{F}$. It turns out that A is a strict attractor of $\widetilde{\mathcal{F}}$, $\widetilde{\mathcal{F}}(A) \nsubseteq A$, and $2 \in (\widehat{\mathcal{B}}(A) \setminus A) \cap \widetilde{\mathcal{F}}(A)$.

We do not know any good criteria for the existence of homoclinic attractors.

References

1. M.F. Barnsley, K. Leśniak, The chaos game on a general iterated function system from a topological point of view. Int. J. Bifurcation Chaos Appl. Sci. Eng. 24(1), 1450139/1–10 (2014)
2. M.F. Barnsley, K. Leśniak, On the continuity of the Hutchinson operator, Symmetry (Basel) 7 (2015), no. 4, 1831–1840
3. M.F. Barnsley, K. Leśniak, M. Rypka, Chaos game for IFSs on topological spaces, J. Math. Anal. Appl. 435 (2016), no. 2, 1458–1466
4. M.F. Barnsley, K. Leśniak, M. Rypka, Basic topological structure of fast basins. Fractals **26**(01), 1850011/1–11 (2018)
5. M.F. Barnsley, A. Vince, Developments in fractal geometry, Bull. Math. Sci. 3 (2013), no. 2, 299–348
6. M.F. Barnsley, A. Vince, Fast basins and branched fractal manifolds of attractors of iterated function systems. SIGMA Symmetry Integrability Geom. Methods Appl. **11**, 084 (2015)
7. B. Kieninger, *Iterated Function Systems on Compact Hausdorff Spaces* (Shaker-Verlag, Aachen, 2002)
8. K. Leśniak, Homoclinic attractors in discontinuo[u]s iterated function systems, Chaos Solitons Fract. 81 (2015), 146–149
9. K. Leśniak, N. Snigireva, F. Strobin, Weakly contractive iterated function systems and beyond: a manualJ. Difference Equ. Appl. **26**(8), 1114–1173 (2020)
10. *Sierpinski Fractal Manifold.* http://superfractals.com/wpfiles/sierpinski-fractal-manifold/ (alternative source https://www.geogebra.org/m/u43mWStD), 2016 last accessed 16 Nov 2018

Zero-Dimensional Covers of Dynamical Systems

Hisao Kato

Abstract In this article, we study the dynamical properties of two-sided zero-dimensional maps. In particular, we show that if $f: X \to X$ is a two-sided zero-dimensional map on an n-dimensional compactum X with zero-dimensional set $P(f)$ of periodic points, then the map f can be covered by a map on a zero- dimensional compactum via an at most 2^n-to-one map.

Keywords Dynamical systems · Covers (extensions) of dynamical systems · Periodic point · Dimension · Cantor sets · General position

1 Introduction

A pair (X, f) is called a *dynamical system* if X is a compact metric space (= compactum) and $f: X \to X$ is a map on X. A dynamical system $(Z, \tilde{f})$ *covers* (X, f) *via* a map $p: Z \to X$ provided that p is an onto map and the following diagram is commutative, i.e., $p\tilde{f} = fp$.

$$\begin{array}{ccc} Z & \xrightarrow{\tilde{f}} & Z \\ \downarrow p & & \downarrow p \\ X & \xrightarrow{f} & X \end{array}$$

Note that (X, f) is also called a *factor* of $(Z, \tilde{f})$ and conversely $(Z, \tilde{f})$ is called a *cover* (or an *extension*) of (X, f). We call the map $p: Z \to X$ a *factor mapping*. If Z is zero-dimensional, then we say that the dynamical system $(Z, \tilde{f})$ is a *zero-dimensional cover* of (X, f). Moreover, if the factor mapping is a finite-to-one map,

H. Kato (✉)
Institute of Mathematics, University of Tsukuba, Tsukuba, Japan
e-mail: hkato@math.tsukuba.ac.jp

R. L. Devaney et al. (eds.), *Topological Dynamics and Topological Data Analysis*, Springer Proceedings in Mathematics & Statistics 350,
https://doi.org/10.1007/978-981-16-0174-3_5

then we say that the dynamical system $(Z, \tilde{f})$ is a *finite-to-one zero-dimensional cover* of (X, f).

The (symbolic) dynamical systems on Cantor sets have been studied by many mathematicians, and also, the strong relations between Markov partitions and symbolic dynamics have been studied (e.g., see [1, 3–5, 11, 17, 19], Proposition 3.19). In [1], R. D. Anderson proved that for any dynamical system (X, f), there exists a zero-dimensional cover $(Z, \tilde{f})$ of (X, f), and moreover in Boyle et al. [4, Theorem A.1] proved that any dynamical system (X, f) has a zero-dimensional cover $(Z, \tilde{f})$ such that the topological entropy $h(f)$ of f is equal to $h(\tilde{f})$, where the factor mappings are not necessarily finite-to-one. In topology, there is a classical theorem by Hurewicz [8] that any compactum X is at most n-dimensional if and only if there is a zero-dimensional compactum Z with an onto map $p: Z \to X$ whose fibers have cardinality at most $n + 1$. In the theory of dynamical systems, we have the related general problem (e.g., see [3, 4, 10, 16]):

Problem 1.1 *What kinds of dynamical systems can be covered by zero-dimensional dynamical systems via finite-to-one maps?*

The motivation for this problem comes from (symbolic) dynamics on Cantor sets. To study dynamical properties of the original dynamics (X, f), the finiteness of the fibers of the factor mapping may be very important, and so, in this article, we focus on the finiteness of fibers of factor mappings. Related to Problem 1.1, first Kulesza [16] proved the following significant theorem:

Theorem 1.2 (Kulesza [16]) *For each homeomorphism f on an n-dimensional compactum X with zero-dimensional set $P(f)$ of periodic points, there is a zero-dimensional cover $(Z, \tilde{f})$ of (X, f) via an at most $(n+1)^n$-to-one map such that $\tilde{f}: Z \to Z$ is a homeomorphism.*

He also showed that Problem 1.1 needs the assumption $\dim P(f) \leq 0$. In fact, for the disk $X = [0, 1]^2$ or some one-dimensional continuum X, there is a dynamical system (X, f) such that $f: X \to X$ is a homeomorphism on X with $\dim P(f) = 1$ and (X, f) has no zero-dimensional cover via a finite-to-one map (see the proof of Example 2.2 and Remark 2.3 of [16]). In [10] Ikegami, Kato and Ueda improved the theorem of Kulesza as follows: The condition of at most $(n+1)^n$-to-one map can be strengthened to the condition of at most 2^n-to-one map.

The aim of this article is to give a partial answer to Problem 1.1. In fact, we show that the above theorem is also true for a class of maps containing two-sided zero-dimensional maps. For the special case that (X, f) is a positively expansive dynamical system with $\dim X = n$, (X, f) can be covered by a subshift (Σ, σ) of the shift map $\sigma: \{1, 2, ..., k\}^\infty \to \{1, 2, ..., k\}^\infty$ via an at most 2^n-to-one map. Also, we study some dynamical zero-dimensional decomposition theorems of spaces related to such maps (see [14]).

2 Preliminaries

In this article, all spaces are separable metric spaces, and maps are continuous functions. Let $\mathbb{N}$ be the set of all natural numbers, i.e., $\mathbb{N} = \{1, 2, 3, \dots\}$, $\mathbb{Z}$ the set of all integers and $\mathbb{Z}_+$ the set of all nonnegative integers, i.e., $\mathbb{Z}_+ = \{0\} \cup \mathbb{N}$ (= $\{0, 1, 2, \dots\}$). Also, let $\mathbb{R}$ be the real line. If K is a subset of a space X, then $\mathrm{cl}(K)$, $\mathrm{bd}(K)$ and $\mathrm{int}(K)$ denote the closure, the boundary and the interior of K in X, respectively. A subset A of a space X is an F_σ-set of X if A is a countable union of closed subsets of X. Also, a subset B of X is a G_δ-set of X if B is an intersection of countable open subsets of X. For a space X, $\dim X$ means the topological (covering) dimension of X (e.g., see [6]). For a collection $\mathcal{C}$ of subsets of X, we put

$$\mathrm{ord}(\mathcal{C}) = \sup\{\mathrm{ord}_x \mathcal{C} \mid x \in X\},$$

where $\mathrm{ord}_x \mathcal{C}$ is the number of members of $\mathcal{C}$ which contains x. A closed set K in X is *regular closed* in X if $\mathrm{cl}(\mathrm{int}(K)) = K$. A collection $\mathcal{C}$ of regular closed sets in X is called a *regular closed partition* of X provided that

$$\bigcup \mathcal{C} \;(= \bigcup \{C \mid C \in \mathcal{C}\}) = X$$

and $C \cap C' = \mathrm{bd}(C) \cap \mathrm{bd}(C')$ for each $C, C' \in \mathcal{C}$ with $C \neq C'$. For regular closed partitions $\mathcal{A}$ and $\mathcal{B}$ of X, $\mathcal{A}@\mathcal{B}$ denotes the regular closed partition

$$\{\mathrm{cl}[\mathrm{int}(A) \cap \mathrm{int}(B)] \mid A \in \mathcal{A} \text{ and } B \in \mathcal{B}\}$$

of X. It is clear that $\mathrm{ord}(\mathcal{A}@\mathcal{B}) \leq \mathrm{ord}(\mathcal{A}) \cdot \mathrm{ord}(\mathcal{B})$. A collection $\{A_\lambda\}_{\lambda \in \Lambda}$ of subsets of X is called a *swelling* of a collection $\{B_\lambda\}_{\lambda \in \Lambda}$ of subsets of X provided that $B_\lambda \subset A_\lambda$ for each $\lambda \in \Lambda$, and if for any $m \in \mathbb{N}$ and $\lambda_1, \dots, \lambda_m \in \Lambda$, then

$$\bigcap_{i=1}^{m} A_{\lambda_i} \neq \emptyset \text{ if and only if } \bigcap_{i=1}^{m} B_{\lambda_i} \neq \emptyset.$$

Conversely, a family $\{B_\lambda\}_{\lambda \in \Lambda}$ of subsets of X is called a *shrinking* of a cover $\{A_\lambda\}_{\lambda \in \Lambda}$ of X if $\{B_\lambda\}_{\lambda \in \Lambda}$ is a cover of X and $B_\lambda \subset A_\lambda$ for each $\lambda \in \Lambda$.

Let X and Y be compacta. A map $f \colon X \to Y$ is *zero-dimensional* if $\dim f^{-1}(y) \leq 0$ for each $y \in Y$. A map $f \colon X \to Y$ is a *zero-dimension preserving map* if for any zero-dimensional closed subset D of X, $\dim f(D) \leq 0$. Also, a map $f \colon X \to X$ is *two-sided zero-dimensional* if f is zero-dimensional and zero-dimension preserving, i.e., for any zero-dimensional closed subset D of X, $\dim f^{-1}(D) \leq 0$ and $\dim f(D) \leq 0$. In this case, note that if Z is a zero-dimensional F_σ-subset of X, then $\dim f(Z) = 0$. A map $f \colon X \to Y$ is *semi-open* (or *quasi-open*) if for any nonempty open set U of X, $f(U)$ contains a nonempty open set of Y, i.e., $\mathrm{int} f(U) \neq \emptyset$. An onto map $p \colon X \to Y$ is *at most k-to-one* ($k \in \mathbb{N}$) if for any $y \in Y$, $|p^{-1}(y)| \leq k$.

For a map $f: X \to X$, a subset A of X is *f-invariant* if $f(A) \subset A$. We define the set

$$O(x) = \{f^p(x) | \ p \in \mathbb{Z}_+\}$$

which denotes the (positive) orbit of x. Similarly, we define the *eventual orbit* of $x \in X$:

$$EO(x) = \{z \in X | \text{ there exists } i, j \in \mathbb{Z}_+ \text{ such that } f^i(x) = f^j(z)\}$$

$$= \{z \in X | \text{ there exists } \ j \in \mathbb{Z}_+ \text{ such that } f^j(z) \in O(x)\}.$$

Note that

$$EO(x) = \bigcup_{i,j \in \mathbb{Z}_+} f^{-j}(f^i(x)),$$

the family $\{EO(x) | \ x \in X\}$ is a decomposition of X and $EO(x)$ is f-invariant, i.e., $f(EO(x)) \subset EO(x)$. Let $P(f)$ be the set of all periodic points of f;

$$P(f) = \{x \in X \mid f^j(x) = x \text{ for some } j \in \mathbb{N}\}.$$

A point $x \in X$ is *eventually periodic* if there is some $p \in \mathbb{Z}_+$ such that $f^p(x) \in P(f)$. Let $EP(f)$ be the set of all eventually periodic points of f;

$$EP(f) = \bigcup_{p=0}^{\infty} f^{-p}(P(f)).$$

Note that $P(f)$ and $EP(f)$ are F_σ-sets of X. In [15], Krupski, Omiljanowski and Ungeheuer showed that the set of maps $f: X \to X$ with zero-dimensional sets $CR(f)$ of all chain recurrent points is a dense G_δ-set of the mapping space $C(X, X)$ if X is a (compact) polyhedron. Note that a point $x \in X$ is a *chain recurrent point* of f if for any $\epsilon > 0$ there is a finite sequence $x = x_0, x_1, \cdots, x_m = x$ of points of X such that $d(f(x_i), x_{i+1}) < \epsilon$ for each $i = 0, 1, ..., m - 1$. Since $P(f) \subset CR(f)$, we see that the set of maps $f: X \to X$ with zero-dimensional sets $P(f)$ of all periodic points is residual in the mapping space $C(X, X)$ if X is a compact polyhedron. Hence, almost all maps on compact polyhedra have zero-dimensional sets of periodic points.

Let X be a compactum and $\mathcal{U}$, $\mathcal{V}$ be two covers of X. Put

$$\mathcal{U} \vee \mathcal{V} = \{U \cap V | \ U \in \mathcal{U}, V \in \mathcal{V}\}.$$

The quantity $N(\mathcal{U})$ denotes minimal cardinality of subcovers of $\mathcal{U}$. Let $f: X \to X$ be a map, and let $\mathcal{U}$ be an open cover of X. Put

$$h(f, \mathcal{U}) = \lim_{n \to \infty} \frac{\log N(\mathcal{U} \vee f^{-1}(\mathcal{U}) \vee \ldots \vee f^{-n+1}(\mathcal{U}))}{n}.$$

The *topological entropy of* f, denoted by $h(f)$, is the supremum of $h(f, \mathcal{U})$ for all open covers $\mathcal{U}$ of X. Positive topological entropy of map is one of generally accepted definitions of chaos.

3 Finite-to-one Zero-Dimensional Covers

In this section, we study finite-to-one zero-dimensional covers of some dynamical systems. We need the followings.

Lemma 3.1 (cf. [10, Lemma 3.4]) *Let* $f: X \to X$ *be a two-sided zero-dimensional map of a compactum* X *such that* $\dim X = n < \infty$ *and* $\dim P(f) \leq 0$. *Let* F *be an* F_σ*-set of* X *with* $\dim F \leq 0$. *Suppose that* $\mathcal{C} = \{C_i \mid 1 \leq i \leq M\}$ *is a finite open cover of* X *and let* $\mathcal{B} = \{B_i \mid 1 \leq i \leq M\}$ *be a closed shrinking of* $\mathcal{C}$. *Then, for each* $k = 0, 1, 2, \ldots,$ *there is an open shrinking* $\mathcal{C}'(k) = \{C'_i \mid 1 \leq i \leq M\}$ *of* $\mathcal{C}$ *such that for each* $1 \leq i \leq M$,
(1) $B_i \subset C'_i \subset C_i$,
(2) $\{f^{-p}(\mathrm{bd}(C'_i)) \mid 1 \leq i \leq M, p = 0, 1, ..., k\}$ *is in general position,*
(3) $\mathrm{bd}(C'_i) \cap (EP(f) \cup F) = \emptyset$ *for each* i.

Lemma 3.2 (cf. [10, Lemma 3.5]) *Suppose that* $f: X \to X$ *is a two-sided zero-dimensional map of a compactum* X *such that* $\dim X = n < \infty$ *and* $\dim P(f) \leq 0$. *Let* F *be an* F_σ*-set of* X *with* $\dim F \leq 0$. *Then, for each* $j \in \mathbb{N}$, *there is a finite open cover* $\mathcal{C}(j) = \{C(j)_i \mid 1 \leq i \leq m_j\}$ *of* X *such that*
(1) $\mathrm{mesh}(\mathcal{C}(j)) < 1/j$,
(2) $\mathrm{ord}(\mathcal{G}) \leq n$, *where* $\mathcal{G} = \{f^{-p}(\mathrm{bd}(C(j)_i)) \mid 1 \leq i \leq m_j,\ j \in \mathbb{N}$ *and* $p \in \mathbb{Z}_+\}$, *and*
(3) $F \cap L = \emptyset$, *where* $L = \bigcup\{\mathrm{bd}(C(j)_i) | \ 1 \leq i \leq m_j, j \in \mathbb{N}\}$.

Lemma 3.3 *Let* $f: X \to X$ *be a map of a compactum* X, *and let* H *be a subset of* X. *Suppose that for* $j \in \mathbb{N}$, $\mathcal{C}(j) = \{C(j)_i \mid 1 \leq i \leq m_j\}$ *is a finite open cover of* X *such that* $\mathrm{mesh}(\mathcal{C}(j)) < 1/j$, $H \cap \bigcup \mathcal{G} = \emptyset$ *and* $\mathrm{ord}(\mathcal{G}) \leq n$, *where*

$$\mathcal{G} = \{f^{-p}(\mathrm{bd}(C(j)_i)) \mid 1 \leq i \leq m_j,\ j \in \mathbb{N} \text{ and } p \in \mathbb{Z}_+\}.$$

Then, for $j \in \mathbb{N}$, *there is a finite regular closed partition* $\mathcal{D}(j)$ *of* X *such that the following properties hold;*
(1) $\mathrm{mesh}(\mathcal{D}(j)) \leq 1/j$,
(2) $\mathcal{D}(j+1)$ *is a refinement of* $\mathcal{D}(j)$,
(3) $\prod_{p=0}^{\infty} \mathrm{ord}_{f^p(x)} \mathcal{D}(j) \leq 2^n$ *for each* $x \in X$, *and*
(4) *if* $x \in H$, *then* $\prod_{p=0}^{\infty} \mathrm{ord}_{f^p(x)} \mathcal{D}(j) = 1$.

Lemma 3.4 *Let* $f: X \to X$ *be a map of a compactum* X, *and let* H *be a subset of* X. *Suppose that there is* $m \in \mathbb{N}$ *and a sequence of finite regular closed partitions* $\mathcal{D}(j)$ $(j \in \mathbb{N})$ *of* X *such that*

(1) $\mathrm{mesh}(\mathcal{D}(j)) \leq 1/j$,
(2) $\mathcal{D}(j+1)$ *is a refinement of* $\mathcal{D}(j)$,
(3) $\prod_{p=0}^{\infty} \mathrm{ord}_{f^p(x)} \mathcal{D}(j) \leq m$ *for each* $x \in X$, *and*
(4) $H \cap D = \emptyset$, *where* $D = \bigcup\{f^{-p}(\mathrm{bd}(d)) \mid d \in \mathcal{D}(j),\ j \in \mathbb{N},\ p \in \mathbb{Z}_+\}$, *i.e., if* $x \in H$,

$$\prod_{p=0}^{\infty} \mathrm{ord}_{f^p(x)} \mathcal{D}(j) = 1.$$

Then, there is a zero-dimensional cover $(Z, \tilde{f})$ *of* (X, f) *via an at most* m*-to-one map* $p\colon Z \to X$ *such that* $|p^{-1}(x)| = 1$ *for* $x \in H$. *Moreover, if* X *is perfect, then* Z *can be taken as a Cantor set* C.

By use of the above results, we obtain the following theorem.

Theorem 3.5 (Kato and Matsumoto [14]) *Suppose that* $f\colon X \to X$ *is a two-sided zero-dimensional map of a compactum* X *with* $\dim X = n < \infty$. *If* $\dim P(f) \leq 0$, *then there exist a dense* G_δ*-set* H *of* X *and a zero-dimensional cover* $(Z, \tilde{f})$ *of* (X, f) *via an at most* 2^n*-to-one onto map* p *such that* $P(f) \subset H$ *and* $|p^{-1}(x)| = 1$ *for* $x \in H$. *Moreover, if* X *is perfect, then* Z *can be chosen as a Cantor set. In particular,* $h(f) = h(\tilde{f})$, *where* $h(f)$ *denotes the topological entropy of* f.

We consider a generalization of Theorem 3.5. For a map $f\colon X \to X$ on a compactum X, let

$$D_0(f) = \{x \in X \mid \dim f^{-1}(x) \leq 0\}$$

and

$$D_+(f) = \{x \in X \mid \dim f^{-1}(x) \geq 1\}\ (= X - D_0(f)).$$

Note that a map $f\colon X \to X$ is a zero-dimensional map if and only if $D_+(f) = \emptyset$. The following theorem is a generalization of Theorem 3.5 which is the main theorem of this article (see [14]).

Main Theorem 3.6 (a generalization of Theorem 3.5) *Let* $f\colon X \to X$ *be a map on an* n*-dimensional compactum* X $(n < \infty)$. *Suppose that* f *is a zero-dimension preserving map,* $\dim D_+(f) \leq 0$ *and* $\dim EP(f) \leq 0$. *Then, there exist a dense* G_δ*-set* H *of* X *and a zero-dimensional cover* $(Z, \tilde{f})$ *of* (X, f) *via an at most* 2^n*-to-one onto map* p *such that* $EP(f) \subset H$ *and* $|p^{-1}(x)| = 1$ *for* $x \in H$. *Moreover, if* X *is perfect, then* Z *can be chosen as a Cantor set. In particular,* $h(f) = h(\tilde{f})$.

Also, we consider the case that $f\colon X \to X$ is a positively expansive map of a compactum X. A map $f\colon X \to X$ of a compactum X is *positively expansive* if there is $\epsilon > 0$ such that for any $x, y \in X$ with $x \neq y$, there is $k \in \mathbb{Z}_+$ such that $d(f^k(x), f^k(y)) \geq \epsilon$. Similarly, a map $f\colon X \to X$ of a compactum X is *positively continuum-wise expansive* if there is $\epsilon > 0$ such that for any nondegenarate subcontinuum A of X, there is a $k \in \mathbb{Z}_+$ such that $\mathrm{diam}(f^k(A)) \geq \epsilon$ (see [12]). Such an $\epsilon > 0$

is called an *expansive constant* for f. Note that any positively expansive map is two-sided zero-dimensional and positively continuum-wise expansive. In [12, Theorem 5.3], we know that if a compactum X admits an positively continuum-wise expansive map f on X, then $\dim X < \infty$ and every minimal set of f is zero-dimensional.

Proposition 3.7 (cf. [13, Proposition 2.5]) *Let $f: X \to X$ be a positively continuum-wise expansive map of a compact metric space X, and let*

$$I_0(f) = \bigcup\{M \mid M \text{ is a zero-dimensional } f\text{-invariant closed set of } X\}.$$

Then, $I_0(f)$ is a zero-dimensional F_σ-set of X. In particular, $\dim P(f) \le 0$.

Let $Y_k = \{1, 2, ..., k\}$ $(k \in \mathbb{N})$ be the discrete space having k-elements, and let $Y_k^{\mathbb{Z}_+} = \prod_0^\infty Y_k$ be the product space. Then, the shift map $\sigma: Y_k^{\mathbb{Z}_+} \to Y_k^{\mathbb{Z}_+}$ is defined by $\sigma(x)_i = x_{i+1}$ for $x = (x_0, x_1, x_2, \cdots) \in Y_k^{\mathbb{Z}_+}$. Note that σ is the typical positively expansive map.

Theorem 3.8 (cf. [10, Corollary 3.7] and [17, Proposition 20]) *Let $f: X \to X$ be a positively expansive map of a compactum X with* $\dim X = n < \infty$. *Then, there exist $k \in \mathbb{N}$ and a closed σ-invariant set Σ of $\sigma: Y_k^{\mathbb{Z}_+} \to Y_k^{\mathbb{Z}_+}$ such that (Σ, σ) is a zero-dimensional cover (= symbolic extension) of (X, f) via an at most 2^n-to-one map $p: \Sigma \to X$ satisfying that $|p^{-1}(x)| = 1$ for any $x \in I_0(f)$.*

Remark: For the case that $f: X \to X$ is an expansive homeomorphism of a compactum X with $\dim X = n < \infty$ (see [12] for the definition of expansive homeomorphism), there exist $k \in \mathbb{N}$ and a closed σ-invariant set Σ of $\sigma: Y_k^{\mathbb{Z}} \to Y_k^{\mathbb{Z}}$ such that (Σ, σ) is a zero-dimensional cover (= symbolic extension) of (X, f) via an at most 2^n-to-one map $p: \Sigma \to X$, where $\sigma: Y_k^{\mathbb{Z}} \to Y_k^{\mathbb{Z}}$ is the shift homeomorphism (see [10, 16]).

In the special case that X is a graph G (= compact connected one-dimensional polyhedron) and $f: G \to G$ is a piece-wise monotone map, we can omit the condition $\dim P(f) \le 0$. A map $f: G \to G$ is *piece-wise monotone* (with respect to some triangulation K) if for any edge E of K (i.e., $E \in K^1$), the restriction $f|E: E \to G$ of f to the edge E is injective. We need the following result.

Theorem 3.9 *If $f: G \to G$ is a piece-wise monotone map on a graph G, then there is a zero-dimensional cover $(C, \tilde{f})$ of (G, f) via an at most 2-to-one map, where C is a Cantor set.*

4 Zero-Dimensional Decompositions of Dynamical Systems

In dimension theory, the following decomposition theorem is well-known [6, Theorem 1.5.8]: A separable metric space X is $\dim X \le n$ $(n \in \mathbb{Z}_+)$ if and only if X

can be represented as the union of $n+1$ subspaces $Z_0, Z_1, ..., Z_n$ of X such that $\dim Z_i \leq 0$ for each $i = 0, 1, ..., n$. In this section, we study the similar dynamical decomposition theorems of two-sided zero-dimensional maps (cf. [7]). We consider bright spaces and dark spaces of maps except n times, and by use of these spaces, we prove some dynamical decomposition theorems of spaces related to given maps (see [14]).

Let $f: X \to X$ be a map. A subset Z of X is a *bright space* of f except n times ($n \in \mathbb{Z}_+$) if for any $x \in X$,

$$|\{p \in \mathbb{Z}_+ |\ f^p(x) \notin Z\}| \leq n.$$

Also, we say that $L = X - Z$ is a *dark space* of f except n times. Note that for any $x \in X$, $|O(x) \cap L| \leq n$ and $L \cap P(f) = \phi$. For each $z \in X$, put

$$t(z) = |\{p \in \mathbb{Z}_+ \ ; \ f^p(z) \in L\}|.$$

Also, we put

$$T(x) = \max\{t(z) \ ; \ z \in EO(x)\}$$

for each $x \in X$. For a dark space L of f except n times and $0 \leq j \leq n$, we put

$$A_f(L, j) = \{x \in X |\ T(x) = j\}.$$

Note that $A_f(L, j)$ is f-invariant, i.e. $f(A_f(L, j)) \subset A_f(L, j)$ and $A_f(L, i) \cap A_f(L, j) = \phi$ if $i \neq j$. Hence, we have the f-invariant decomposition related to the dark space L as follows;

$$X = A_f(L, 0) \cup A_f(L, 1) \cup \cdots \cup A_f(L, n).$$

Theorem 4.1 (cf. [7, Theorem 2.4]) *Suppose that $f: X \to X$ is a two-sided zero-dimensional map of a compactum X with* $\dim X = n < \infty$. *Then, there is a bright space Z of f except n times such that Z is a zero-dimensional dense G_δ-set of X and the dark space $L = X - Z$ of f is an $(n-1)$-dimensional F_σ-set of X if and only if* $\dim P(f) \leq 0$.

Corollary 4.2 (cf. [7, Corollary 2.5]) *Suppose that X is a compactum with* $\dim X = n\ (< \infty)$ *and $f: X \to X$ is a two-sided zero-dimensional onto map. Then, there exists a zero-dimensional G_δ-dense set Z of X such that for any $n+1$ integers $k_0 < k_1 < \cdots < k_n$ ($k_i \in \mathbb{Z}$),*

$$X = f^{k_0}(Z) \cup f^{k_1}(Z) \cup \cdots \cup f^{k_n}(Z)$$

if and only if $\dim P(f) \leq 0$.

By use of F_σ-dark spaces, we have the following decomposition theorem.

Theorem 4.3 (cf. [7, Theorem 2.6]) *Suppose that* X *is a compactum with* $\dim X = n\ (< \infty)$ *and* $f : X \to X$ *is a two-sided zero-dimensional map on* X *with* $\dim P(f) \leq 0$. *If* L *is a dark space of* f *except* n *times such that* L *is an* F_σ*-set of* X *and* $\dim\ (X - L) \leq 0$, *then* $\dim A_f(L, j) = 0$ *for each* $j = 0, 1, 2, ..., n$. *In particular, there is the* f*-invariant zero-dimensional decomposition of* X *related to the dark space* L:

$$X = A_f(L, 0) \cup A_f(L, 1) \cup \cdots \cup A_f(L, n).$$

In the case of positively expansive maps, we obtain decomposition theorem for a compact dark space L.

Theorem 4.4 (cf. [7, Theorem 2.8]) *Suppose that* X *is a compactum with* $\dim X = n\ (< \infty)$ *and* $f : X \to X$ *is a positively expansive map. Then, there exists a compact* $(n-1)$*-dimensional dark space* L *of* f *except* n *times such that* $\dim A_f(L, j) = 0$ *for each* $j = 0, 1, 2, ..., n$. *In particular, there is the* f*-invariant zero-dimensional decomposition of* X *related to the compact dark space* L:

$$X = A_f(L, 0) \cup A_f(L, 1) \cup \cdots \cup A_f(L, n).$$

References

1. R. D. Anderson, On rasing flows and mappings, Bull. AMS 69 (1963), 259–264
2. J. M. Arts, R. J. Fokkink and J. Vermeer, A dynamical decomposition theorem, Acta Math. Hung., 94(3), 2002, 191–196
3. M. Boyle, T. Downarowicz, The entropy theory of symbolic extensions. Invent. math. **156**, 119–161 (2004)
4. M. Boyle, D. Fiebig and U. Fiebig, Residual entropy, conditional entropy and subshift covers, Forum Math. 14 (2002), 713–757
5. R. Bowen, On axiom A diffeomorphisms, in *CBMS Registration Conference*, vol. 35 (American mathematical Society: Providence, RI, 1978)
6. R. Engelking, *Dimension Theory* (PWN, Warsaw, 1977)
7. M. Hiraki and H. Kato, Dynamical decomposition theorems of homeomorphisms with zero-dimensional sets of periodic points, Topol. Appl. 196 (2015), 54–59
8. W. Hurewicz, Ein Theorem der Dimensionstheorie. Ann. of Math. **31**, 176–180 (1930)
9. Y. Ikegami, H. Kato and A. Ueda, Eventual colorings of homeomorphisms, J. Math. Soc. Japan, 65 (2) (2013), 375–387
10. Y. Ikegami, H. Kato and A. Ueda, Dynamical systems of finite-dimensional metric spaces and zero-dimensional covers, Topol. Appl. 160 (2013), 564–574
11. M.V. Jakobson, On some properties of Markov Partitions. Sov. math. Dokl. **17**, 247–251 (1976)
12. H. Kato, Continuum-wise expansive homeomorphisms, Can. J. Math. 45 (1993), 576–598
13. H. Kato, *Minimal Sets and Chaos in the Sense of Devaney on Continuum-Wise Expansive Homeomorphisms*. Lecture Notes in Pure and Application Mathematics, vol. 170 (Dekker, New York, 1995)
14. H. Kato and M. Matsumoto, *Finite-to-One Zero-Dimensional Covers of Dynamical Systems*. J. Math. Soc. Japan. **72**, 819–845 (2020)
15. P. Krupski, K. Omiljanowski and K. Ungeheuer, Chain recurrent sets of generic mappings on compact spaces, Topol. Appl. 202 (2016), 251–268

16. J. Kulesza, Zero-dimensional covers of finite dimensional dynamical systems, Ergod. Th. Dynam. Sys. 15 (1995), 939–950
17. P. Kurka, *Topological and Symbolic Dynamics, Cours Spécialisés [Specialized Courses], 11* (Société Mathématique de France, Paris, 2003)
18. R. Mañé, Expansive homeomorphisms and topological dimension, Trans. Amer. Math. Soc. 252 (1979), 313–319
19. W.D. Melo, S. van Strien, *One Dimensional Dynamics* (Springer, Berlin, 1993)
20. J. van Mill, *The Infinite-Dimensional Topology of Function Spaces* (North-Holland publishing Co., Amsterdam, 2001)
21. J. Nagata, Modern Dimension Theory, North-Holland publishing Co., Amsterdam, 1965

Chaotic Continua in Chaotic Dynamical Systems

Hisao Kato

Abstract In this article, for any graph G we define a new notion of "free tracing property by free G-chains" on G-like continua and we show that a positive topological entropy homeomorphism f of a G-like continuum X admits a Cantor set Z in X such that any sequence $(z_1, z_2, ..., z_n)$ of points in Z is an IE-tuple of f, Z has the free tracing property by free G-chains and the minimal continuum H containing Z in X is indecomposable. Moreover, we show that the similar result can be obtained for positive topological entropy "monotone" maps. Also we give characterization theorems of G-like continua containing indecomposable subcontinua.

Keywords Topological entropy · Indecomposable continuum · Composants · G-like continuum · Cantor sets · Free tracing property by free G-chains · Inverse limits

1 Introduction

During the last thirty years or so, many interesting connections between dynamical systems and continuum theory have been studied by many mathematicians. Many complicated spaces frequently appear in chaotic dynamical systems. Such spaces play important roles in order to investigate behaviors of the dynamics. We are interested in the following fact that chaotic topological dynamics should imply the existence of complicated topological structures of underlying spaces. In many cases, such spaces are indecomposable continua. We know that many indecomposable continua often appear as chaotic attractors of dynamical systems. Also, in many cases, the composants of such indecomposable continua are strongly related to stable or unstable (connected) sets of the dynamics. For instance, in continuum theory and the theory of dynamical systems, the Knaster continuum (= Smale's horse shoe), the pseudo-arc, solenoids and Wada's lakes (= Plykin attractors) etc., are well-known as such

H. Kato (✉)
Institute of Mathematics, University of Tsukuba, Tsukuba, Japan
e-mail: hkato@math.tsukuba.ac.jp

R. L. Devaney et al. (eds.), *Topological Dynamics and Topological Data Analysis*, Springer Proceedings in Mathematics & Statistics 350,
https://doi.org/10.1007/978-981-16-0174-3_6

indecomposable continua. The theory of indecomposable continua is one of the most interesting branches of continuum theory in topology.

By use of ergodic theory method, Blanchard, Glasner, Kolyada and Maass proved that if a map $f: X \to X$ of a compact metric space X has positive topological entropy, then there is an uncountable δ-scrambled subset of X for some $\delta > 0$ and hence the dynamics (X, f) is Li-Yorke chaotic. Huang and Ye studied local entropy theory and they gave a characterization of positive topological entropy by use of entropy tuples. Kerr and Li developed local entropy theory and gave a new proof of the theorem of Blanchard, Glasner, Kolyada and Maass. Moreover, they proved that X contains a Cantor set Z which yields more chaotic behaviors. Barge and Diamond showed that for piecewise monotone surjections of graphs, the conditions of having positive topological entropy, containing a horse shoe and the inverse limit space containing an indecomposable subcontinuum are all equivalent. Mouron proved that if X is an arc-like continuum which admits a homeomorphism with positive topological entropy, then X contains an indecomposable subcontinuum. As an extension of the Mouron's theorem, Darji and Kato proved that if X is a G-like continuum for a graph G and X admits a homeomorphism f with positive topological entropy, then X contains an indecomposable subcontinuum. Moreover, if the graph G is a tree, then there is a pair of two distinct points x and y of X such that the pair (x, y) is an IE-pair of f and the irreducible continuum between x and y in X is an indecomposable subcontinuum.

In this article, for any graph G, we define a new notion of "free tracing property by free G-chains" on G-like continua and we prove that a positive topological entropy homeomorphism f of a G-like continuum X admits a Cantor set Z in X such that any sequence $(z_1, z_2, \ldots, z_n)$ of points in Z is an IE-tuple of f and Z has the free tracing property by free G-chains. Our main theorem is a dynamical and geometric structure theorem of positive topological entropy homeomorphism of G-like continua. Also, we show that the similar result can be obtained for positive topological entropy "monotone" maps. Also, we give characterization theorems of continua containing indecomposable subcontinua.

2 Preliminaries

In this article, we assume that all spaces are separable metric spaces and all maps are continuous. Let $\mathbb{N}$ be the set of natural numbers, $\mathbb{R}$ the real line, and $I = [0, 1]$ the unit interval. A *graph* is a compact connected 1-dimensional polyhedron. A graph T is a *tree* if T contains no simple closed curve. For a set A, $|A|$ denotes the cardinality of the set A. For a family $\mathcal{A}$ of subsets of a space, $\bigcup \mathcal{A}$ denotes the union of all elements of $\mathcal{A}$, i.e.,

$$\bigcup \mathcal{A} = \bigcup_{A \in \mathcal{A}} A \quad (= \bigcup \{A \mid A \in \mathcal{A}\}).$$

For a subset A of a space X, $\overline{A}$ denotes the closure of A in X. A subset E of X is an F_σ-set of X if E is a countable union of closed sets of X.

A *continuum* is a compact connected metric space. We say that a continuum is *nondegenerate* if it has more than one point. A continuum is *indecomposable* [24] if it is nondegenerate and it is not the union of two proper subcontinua. For any continuum H, the set $c(p)$ of all points of the continuum H, which can be joined with the point p by a proper subcontinuum of H, is said to be the *composant* of the point $p \in H$, i.e.,

$$c(p) = \bigcup\{C \mid C \text{ is a proper subcontinuum of } H \text{ containing the point } p\}.$$

Note that for an indecomposable continuum H, the following conditions are equivalent;

1. the two points p, q belong to same composant of H;
2. $c(p) \cap c(q) \neq \emptyset$;
3. $c(p) = c(q)$.

So, we know that if H is an indecomposable continuum, the family

$$\{c(p) \mid p \in H\}$$

of all composants of H is a family of uncountable mutually disjoint sets $c(p)$ which are connected and dense F_σ-sets in H. Note that a (nondegenerate) continuum X is indecomposable if and only if there are three distinct points of X such that any subcontinuum of X containing any two points of the three points coincides with X, i.e., X is irreducible between any two points of the three points [24].

Let H be an indecomposable continuum. We say that a subset Z of H is *transversal for composants of* H if no distinct two points of Z belong to the same composant of H, i.e., if x, y are any distinct points of Z and E is any subcontinuum of H containing x and y, then $E = H$. In [27], Mazurukiewicz proved that if H is an indecomposable continuum, then there is a Cantor set Z in H which is transversal for composants of H.

Let X_i ($i \in \mathbb{N}$) be a sequence of compact metric spaces and let $f_{i,i+1}: X_{i+1} \to X_i$ be a map for each $i \in \mathbb{N}$. The *inverse limit* of the inverse sequence $\{X_i, f_{i,i+1}\}_{i=1}^{\infty}$ is the space

$$\varprojlim\{X_i, f_{i,i+1}\} = \{(x_i)_{i=1}^{\infty} \mid x_i = f_{i,i+1}(x_{i+1}) \text{ for each } i \in \mathbb{N}\} \subset \prod_{i=1}^{\infty} X_i$$

which has the topology inherited as a subspace of the product space $\prod_{i=1}^{\infty} X_i$. For a map $f: X \to X$, put

$$\varprojlim(X, f) = \{(x_i)_{i=1}^{\infty} \mid x_i = f(x_{i+1}) \text{ for each } i \in \mathbb{N}\}.$$

A map g from X onto G is an ϵ-*map* ($\epsilon > 0$) if for every $y \in G$, the diameter of $g^{-1}(y)$ is less than ϵ. For any collection $\mathcal{P}$ of graphs, X is $\mathcal{P}$-like if for any $\epsilon > 0$

there exist an element $G \in \mathcal{P}$ and an ϵ-map from X onto G. A continuum$X$ is G-like if X is $\mathcal{P}$-like, where $\mathcal{P} = \{G\}$. Note that X is G-like if only if X is homeomorphic to the inverse limit of an inverse sequence of G, i.e.,

$$X = \varprojlim\{G_i, f_{i,i+1}\},$$

where $G_i = G$ and$f_{i,i+1}\colon G_{i+1} \to G_i$ is an onto map for each $i \in \mathbb{N}$. Arc-like continua (=chainable continua) are those which are G-like for $G = I$, and circle-like continua are those which are S-like, where S is a simple closed curve. Our focus in this article is on G-like continua where G is any graph.

Let $\mathcal{U}$ be a collection of subsets of X. The nerve $N(\mathcal{U})$ of $\mathcal{U}$ is the polyhedron whose vertices are elements of $\mathcal{U}$ and there is a simplex
$< U_1, U_2, ..., U_k >$ with distinct vertices $U_1, U_2, ..., U_k \in \mathcal{U}$ if

$$\bigcap_{i=1}^{k} U_i \neq \emptyset.$$

In this paper, we consider the only case that nerves are graphs.

If $\{C_1, \ldots, C_n\}$ is a subcollection of $\mathcal{U}$, we call it a *chain* if $C_i \cap C_{i+1} \neq \emptyset$ for $1 \leq i < n$ and $\overline{C_i} \cap \overline{C_j} \neq \emptyset$ implies that $|i - j| \leq 1$. We say that $\{C_1, \ldots, C_n\}$ is a *free chain in* $\mathcal{U}$ if it is a chain and, moreover, for all $1 < i < n$ we have that $C \in \mathcal{U}$ with $\overline{C} \cap \overline{C_i} \neq \emptyset$ implies that $C = C_i$, $C = C_{i-1}$ or $C = C_{i+1}$. By the *mesh* of a finite collection $\mathcal{U}$ of sets, we means the largest of diameters of elements of $\mathcal{U}$. Note that for a graph G, a continuum X is G-like if and only if for any $\epsilon > 0$, there is a finite open cover $\mathcal{U}$ of X such that $N(\mathcal{U})$ is homeomorphic to G and the mesh of $\mathcal{U}$ is less than ϵ. The Knaster continuum [21] (= Smale's horse shoe) and the pseudo-arc (= hereditarily indecomposable arc-like continuum) are arc-like continua, solenoids are circle-like continua and the Wada' lake [35] (= Plykin attractor [32]) is a $(S_1 \vee S_2 \vee S_3)$-like continuum, where $S_1 \vee S_2 \vee S_3$ denotes the one point union of 3 circles. Such spaces are typical indecomposable continua which often appear in continuum theory and chaotic dynamical systems. The reader may refer to [24, 31] for standard facts concerning continuum theory.

3 Free Tracing Property by Free G-chains

Let X be a continuum and $m \in \mathbb{N}$. Suppose that A_i $(1 \leq i \leq m)$ are nonempty m open sets in X and x_i $(1 \leq i \leq m)$ are m distinct points of X. We identify the order $A_1 \to A_2 \to \cdots \to A_m$ and the converse order $A_m \to A_{m-1} \to \cdots \to A_1$. Then we consider the equivalence class

$$[A_1 \to A_2 \to \cdots \to A_m] = \{A_1 \to A_2 \to \cdots \to A_m; A_m \to A_{m-1} \to \cdots \to A_1\}.$$

Suppose that $\mathcal{U}$ is a finite open cover of X. We say that a chain $\{C_1, \cdots, C_n\} \subseteq \mathcal{U}$ *follows from the pattern* $[A_1 \to A_2 \to \cdots \to A_m]$ [11] if there exist

$$1 \leq k_1 < k_2 < \cdots < k_m \leq n \text{ or } 1 \leq k_m < k_{m-1} < \cdots < k_1 \leq n$$

such that $C_{k_i} \subset A_i$ for each $i = 1, 2, ..., m$. In this case, more precisely we say that the chain $[C_{k_1} \to C_{k_2} \to \cdots \to C_{k_m}]$ follows from the pattern $[A_1 \to A_2 \to \cdots \to A_m]$. Similarly, we say that a chain $\{C_1, \ldots, C_n\} \subseteq \mathcal{U}$ *follows from the pattern* $[x_1 \to x_2 \to \cdots \to x_m]$ [11] if there exist

$$1 \leq k_1 < k_2 < \cdots < k_m \leq n \text{ or } 1 \leq k_m < k_{m-1} < \cdots < k_1 \leq n$$

such that $x_i \in C_{k_i}$ for each $i = 1, 2, ..., m$, where

$$[x_1 \to x_2 \to \cdots \to x_m] = \{x_1 \to x_2 \to \cdots \to x_m;\ x_m \to x_{m-1} \to \cdots \to x_1\}.$$

More precisely, we say that the chain $[C_{k_1} \to C_{k_2} \to \cdots \to C_{k_m}]$ follows from the pattern $[x_1 \to x_2 \to \cdots \to x_m]$.

Let $\mathcal{P}$ be a collection of graphs and let Z be a subset of a $\mathcal{P}$-like continuum X. We say that Z has *the free tracing property by (resp. free) $\mathcal{P}$-chains* if for any $\epsilon > 0$, any $m \in \mathbb{N}$ and any order $x_1 \to x_2 \to \cdots \to x_m$ of any m distinct points x_i $(i = 1, 2, ..., m)$ of Z, there is an open cover $\mathcal{U}$ of X such that the mesh of $\mathcal{U}$ is less than ϵ, the nerve $N(\mathcal{U})$ of $\mathcal{U}$ is homeomorphic to an element of $\mathcal{P}$ and there is a (resp. free) chain in $\mathcal{U}$ which follows from the pattern $[x_1 \to x_2 \to \cdots \to x_m]$. Especially, for a G-like continuum X, we say that a subset Z of X has *the free tracing property by (resp. free) G-chains* if Z has the free tracing property by (resp. free) $\mathcal{P}$-chains, where $\mathcal{P} = \{G\}$.

In the special the case that X itself is a graph G, for points x_i $(i = 1, 2, ..., m)$ of G, we can similarly define that an edge of G follows from the pattern $[x_1 \to x_2 \to \cdots \to x_m]$.

4 Positive Topological Entropy

Let X be a compact metric space and $\mathcal{U}, \mathcal{V}$ be two covers of X. Put

$$\mathcal{U} \vee \mathcal{V} = \{U \cap V \mid U \in \mathcal{U}, V \in \mathcal{V}\}.$$

The quantity $N(\mathcal{U})$ denotes minimal cardinality of subcovers of $\mathcal{U}$. Let $f: X \to X$ be a map and let $\mathcal{U}$ be an open cover of X. Put

$$h(f, \mathcal{U}) = \lim_{n\to\infty} \frac{\log N(\mathcal{U} \vee f^{-1}(\mathcal{U}) \vee \ldots \vee f^{-n+1}(\mathcal{U}))}{n}.$$

The *topological entropy of* f, denoted by $h(f)$, is the supremum of $h(f, \mathcal{U})$ for all open covers $\mathcal{U}$ of X. Positive topological entropy of map is one of generally accepted definitions of chaos. We say that a set $I \subseteq \mathbb{N}$ has *positive density* if

$$\liminf_{n\to\infty} \frac{|I \cap \{1, 2, ..., n\}|}{n} > 0.$$

Let X be a compact metric space and $f: X \to X$ a map. Let $\mathcal{A}$ be a collection of subsets of X. We say that a set $I \subset \mathbb{N}$ is an *independence set* for $\mathcal{A}$ if for all finite sets $J \subseteq I$, and for all $(Y_j) \in \prod_{j\in J} \mathcal{A}$, we have that

$$\bigcap_{j\in J} f^{-j}(Y_j) \neq \emptyset.$$

We now recall the definition of IE-tuple which is related to independence set in $\mathbb{N}$ and (topological) entropy (see [20]). Let $(x_1, \ldots, x_n)$ be a sequence of points in X. We say that $(x_1, \ldots, x_n)$ is an *IE-tuple of* f if whenever $A_1, \ldots, A_n$ are open sets containing $x_1, \ldots, x_n$, respectively, we have that the collection $\mathcal{A} = \{A_1, \ldots, A_n\}$ has an independence set with positive density. In the case that $n = 2$, we use the term IE-pair. We use IE_k to denote the set of all IE-tuples of length k.

Let $f: X \to X$ be a map of a compact metric space X with metric d and let $\delta > 0$. A subset S of X is a δ-*scrambled set* of f if $|S| \geq 2$ and for any $x, y \in S$ with $x \neq y$, then one has

$$\liminf_{n\to\infty} d(f^n(x), f^n(y)) = 0 \text{ and } \limsup_{n\to\infty} d(f^n(x), f^n(y)) \geq \delta.$$

We say that $f: X \to X$ is *Li-Yorke chaotic* if there is an uncountable subset S of X such that for any $x, y \in S$ with $x \neq y$, then one has

$$\liminf_{n\to\infty} d(f^n(x), f^n(y)) = 0 \text{ and } \limsup_{n\to\infty} d(f^n(x), f^n(y)) > 0.$$

In [3], by use of ergodic theory method, Blanchard, Glasner, Kolyada and Maass proved the following theorem.

Theorem 4.1 (Blanchard et al. [3]) *If a map* $f: X \to X$ *of a compact metric space* X *has positive topological entropy, then there is an uncountable* δ*-scrambled subset of* X *for some* $\delta > 0$ *and hence the dynamics* (X, f) *is Li-Yorke chaotic.*

In [20], by use of local entropy theory (IE-tuples), Kerr and Li proved the following theorem.

Theorem 4.2 (Kerr and Li [20] Theorem 3.18) *Suppose that* $f: X \to X$ *is a positive topological entropy map of a compact metric space* X, *and* $x_1, x_2, ..., x_m$ $(m \geq 2)$ *are distinct points of* X *such that the tuple* $(x_1, x_2, ..., x_m)$ *is an IE-tuple of* f. *If* A_i $(i = 1, 2, ..., m)$ *is any neighborhood of* x_i, *then there are Cantor sets* $Z_i \subset A_i$ *such that the following conditions hold;*

(1) *any sequence* $(z_1, z_, ..., z_n)$ *of points in the Cantor set* $Z = \bigcup_i Z_i$ *is an IE-tuple of* f, *and*
(2) *for all* $k \in \mathbb{N}$, k *distinct points* $y_1, y_2, ..., y_k \in Z$ *and any points* $z_1, z_2, ..., z_k \in Z$, *one has*

$$\liminf_{n\to\infty} \max\{d(f^n(y_i), z_i) | \ 1 \le i \le k\} = 0.$$

In particular, Z *is a* δ*-scrambled set of* f *for some* $\delta > 0$.

In [11], we have the following structure theorem for homeomorphisms on G-like continua.

Theorem 4.3 (Kato [11]) *In the setting of Theorem 4.2 assume additionally that* X *is a* G*-like continuum for a graph* G *and* $f: X \to X$ *is a homeomorphism. Then the Cantor sets* $Z_i \subset A_i$ $(i = 1, 2, ..., m)$ *can be chosen so as to satisfy, in addition to the above conditions* (1) *and* (2), *also the following two ones;*
(3) $Z = \bigcup_{i=1}^m Z_i$ *has the free tracing property by free* G*-chains, and*
(4) *the unique minimal subcontinuum* H *of* X *containing* Z *is indecomposable and* Z *is transversal for composants of* H.

An onto map $f: X \to Y$ of continua is *monotone* if for any $y \in Y$, $f^{-1}(y)$ is connected.

Theorem 4.4 (Kato [11]) *Let* X *be a* G*-like continuum, where* G *is a graph. If* $f: X \to X$ *is a monotone map with positive topological entropy, then there exists a Cantor set* Z *in* X *satisfying conditions (1) and (2) of Theorem 5.2 and transversal for composants of a certain indecomposable subcontinuum* H *of* X. *Moreover,* H *can be taken to be the unique minimal subcontinnum of* X *containing* Z.

5 Characterizations of Indecomposable Continua and Free Tracing Property

A continuum X is *tree-like* if X is $\mathcal{T}$-like, where $\mathcal{T}$ is the collection of all trees. For the case that X is a tree-like continuum, we have the following characterization theorem.

Theorem 5.1 ([12]) Let $\mathcal{T}$ be the collection of all trees and let X be a $\mathcal{T}$-like continuum, i.e., X is tree-like. Suppose that D is a subset of X with $|D| \ge 3$. Then, the following conditions are equivalent.
(1) For any order $x_1 \to x_2 \to x_3$ of three distinct points x_i $(i = 1, 2, 3)$ of D and any $\epsilon > 0$, there is an open cover $\mathcal{U}$ of X such that the mesh of $\mathcal{U}$ is less than ϵ, the nerve $N(\mathcal{U})$ of $\mathcal{U}$ is homeomorphic to an element of $\mathcal{T}$ and there is a chain in $\mathcal{U}$ which follows from the pattern $[x_1 \to x_2 \to x_3]$.
(2) D has the free tracing property by $\mathcal{T}$-chains.
(3) The minimal continuum H in X containing D is indecomposable and Z is transversal for composants of H.

For the special case of arc-like continua, we have the following characterization theorem.

Theorem 5.2 (Kato [12]) *Let X be an arc-like continuum. Suppose that Z is a subset of X with $|Z| \geq 3$. Then, the following conditions are equivalent.*
(1) *X is indecomposable and Z is transversal for composants of X.*
(2) *Z has the free tracing property by free I-chains and X is the minimal continuum containing Z.*

Next result is the main theorem in this section.

Theorem 5.3 (Kato [12]) *Suppose that X is any G-like continuum for a graph G and H is a subcontinuum of X. Then, the following conditions* (1), (2) *and* (3) *are equivalent.*
(1) *H is indecomposable.*
(2) *There is a Cantor set Z in H such that Z has the free tracing property by free G-chains and H is the unique minimal continuum containing Z. In particular, Z is transversal for composants of H.*
(3) *There is a dense F_σ-set Z_∞ of H such that*

$$Z_\infty = \bigcup_{i \in \mathbb{N}} Z_i$$

is the countable union of Cantor sets Z_i in H, Z_∞ has the free tracing property by free G-chains and H is the unique minimal continuum containing Z_i for each $i \in \mathbb{N}$. In particular, Z_∞ is transversal for composants of H.

Proposition 5.4 (Kato [12]) *Let X be a $\mathcal{P}$-like continuum for a collection $\mathcal{P}$ of graphs. Suppose that Z is a Cantor set in X such that Z has the free tracing property by free $\mathcal{P}$-chains and H is the unique minimal continuum H in X containing Z. Let $z \in Z$ and let $c(z, H)$ be the composant of z in the indecomposable continuum H. Then any subcontinuum A in $c(z, H)$ is an arc-like continuum.*

For hereditarily indecomposable continua, we have the following.

Corollary 5.5 (Kato [12]) *Suppose that X is any G-like continuum for a graph G and H is a subcontinuum of X. Then, the following* (1) *and* (2) *are equivalent.*
(1) *H is hereditarily indecomposable.*
(2) *For any subcontinuum K of H, there is a Cantor set Z_K in K such that Z_K has the free tracing property by free G-chains and K is the unique minimal continuum containing Z_K. In particular, Z_K is transversal for composants of K.*

The following is a characterization of pseudo-arc.

Corollary 5.6 (Kato [12]) *Suppose that X is an arc-like continuum and H is a subcontinuum of X. Then the following* (1) *and* (2) *are equivalent.*
(1) *H is the pseudo-arc.*

(2) *For any subcontinuum K of H, there is a Cantor set Z_K in K such that Z_K has the free tracing property by free 1-chains and K is the unique minimal continuum containing Z_K. In particular, Z_K is transversal for composants of K.*

In [23], Kuykendall studied irreducibility and indecomposability in inverse limits of continua. Also, we have the following.

Corollary 5.7 (Kato [12]) *Let G be a graph and let $X = \varprojlim\{G_i, f_{i,i+1}\}$ be an inverse limit with onto bonding maps $f_{i,i+1}$, where $G_i = G$ for each $i \in \mathbb{N}$. Then the followings hold.*

(1) There is an indecomposable subcontinuum in X if and only if there is a Cantor set Z in X such that for any order $z^1 \to z^2 \to \cdots \to z^m$ of any finite points $z^j = (z_i^j)_{i=1}^{\infty}$ $(j = 1, 2, ..., m)$ of Z and any $n \in \mathbb{N}$, there is $k \geq n$ and an edge of G_k which follows from the pattern

$$[z_k^1 \to z_k^2 \to \cdots \to z_k^m].$$

(2) Moreover, if G is a tree, there is an indecomposable subcontinuum in X if and only if there is a three points set Z in X such that for any order $z^1 \to z^2 \to z^3$ of Z and any $n \in \mathbb{N}$, there is $k \geq n$ and an edge of G_k which follows from the pattern $[z_k^1 \to z_k^2 \to z_k^3]$.

References

1. M. Barge and J. Martin, Chaos, periodicity, and snakelike continua, Trans. Amer. Math. Soc. 289 (1985), no. 1, 355–365
2. M. Barge and B. Diamond, The dynamics of continuous maps of finite graphs through inverse limits, Trans. Amer. Math. Soc. 344 (1994), no. 2, 773–790
3. F. Blanchard, E. Glasner, S. Kolyada and A. Maass, On Li-Yorke pairs, J. Reine Angew. Math. 547 (2002), 51–68
4. L.S. Block, W.A. Coppel, *Dynamics in One Dimension*, vol. 1513, Lecture Notes in Mathematics (Springer, 1992)
5. J. P. Boroński and P. Oprocha, On indecomposability in chaotic attractors, Proc. Amer. Math. Soc. 143 (8) (2015), 3659–3670
6. U.B. Darji, H. Kato, Chaos and indecomposability. Adv. Math. **304**, 793–808 (2017)
7. W. Dębski, E.D. Tymchatyn, Composant-like decompositions. Fund. Math. **140**, 69–78 (1991)
8. G. W. Henderson, The pseudo-arc as an inverse limit with one binding map, Duke Math. J. 31 (1964), 421–425
9. Wen Huang and Xiangdong Ye, A local variational relation and applications, Israel J. Math. 151 (2006), 237–279
10. W. T. Ingram, Periodic points for homeomorphisms of hereditarily decomposable chainable continua, Proc. Amer. Math. Soc. 107 (1989), no. 2, 549–553
11. H. Kato, Topological entropy and IE-tuples of indecomposable continua. Fund. Math. Appear **247**, 131–149 (2019)

12. H. Kato, *Characterizations of graph-like continua containing indecomposable subcontinua*, preprint
13. H. Kato, On indecomposability and composants of chaotic continua. Fund. Math. **150**, 245–253 (1996)
14. H. Kato, The nonexistence of expansive homeomorphisms of chainable continua. Fund. Math. **149**, 119–126 (1996)
15. H. Kato, Chaotic continua of (continuum-wise) expansive homeomorphisms and chaos in the sense of Li and Yorke. Fund. Math. **145**, 261–279 (1994)
16. H. Kato, Concerning continuum-wise fully expansive homeomorphisms of continua. Topology Appl. **53**, 239–258 (1993)
17. H. Kato, Continuum-wise expansive homeomorphisms. Canad. J. Math. **45**, 576–598 (1993)
18. H. Kato, *Monotone maps of G-like continua with positive topological entropy yield indecomposability*, Proc. Amer. Math. Soc. Appear **147**, 4363–4370 (2019)
19. H. Kato, C. Mouron, Hereditarily indecomposable compacta do not admit expansive homeomorphisms. Proc. Amer. Math. Soc. **136**, 3689–3696 (2008)
20. D. Kerr and H. Li, Independence in topological and C^*-dynamics, Math. Ann. 338 (2007), no. 4, 869–926
21. B. Knaster, Un continu dont tout sous-continu est indécomposable, Fund. Math. 3 (1922), 247–286
22. J. Krasinkiewicz and P. Minc, Mappings onto indecomposable continua, Bull. Acad. Polon. Sci. Math. Astronom. Phys. 25 (1977), no. 7, 675–680
23. D. P. Kuykendall, Irreducibility and indecomposability in inverse limits, Fund. Math. 80 (1973), 265–270
24. K. Kuratowski, *Topology II* (Academic Press, New York, 1968)
25. K. Kuratowski, Applications of the Baire-category method to the problem of independent sets, Fund. Math. 81 (1973), 65–72
26. J. Llibre and M. Misiurewicz, Horseshoes, entropy and periods for graph maps, Topology 32 (1993), no. 3, 649–664
27. S. Mazurkiewicz, Sur les continus indécomposables. Fund. Math. **10**, 305–310 (1927)
28. P. Minc and W. R. R. Transue, Sarkovskiĭ's theorem for hereditarily decomposable chainable continua, Trans. Amer. Math. Soc. 315 (1989), no. 1, 173–188
29. C. Mouron, Positive entropy homeomorphisms of chainable continua and indecomposable subcontinua, Proc. Amer. Math. Soc. 139 (2011), no. 8, 2783–2791
30. C. Mouron, Mixing sets, positive entropy homeomorphisms and non-Suslinean continua. Ergod. Th. Dynamical Sys. 36 (2016), 2246–2257
31. S. Nadler, *Continuum Theory. An introduction, Monographs and Textbooks in Pure and Applied Mathematics*, vol. 158 (Marcel Dekker, Inc., New York, 1992)
32. R. V. Plykin, On the geometry of hyperbolic attractors of smooth cascades, Russian Math. Survey 39 (1984), 85–131
33. S. Solecki, The space of composants of an indecomposable continuum. Advances in Math. **166**, 149–192 (2002)
34. X. Ye, Topological entropy of the induced maps of the inverse limits with bonding maps. Topology Appl. **67**(2), 113–118 (1995)
35. K. Yoneyama, Theory of continuous sets of points, Tôhok Math. J. 12 (1917), 43–158

Mandelpinski Necklaces in the Parameter Planes of Rational Maps

Robert L. Devaney and Sebastian M. Marotta

Abstract In this paper, we give a survey of some recent results involving "Mandelpinski necklaces" that occur in the family of complex rational maps of the form $z^n + \lambda/z^d$ where $\lambda \in \mathbb{C}$ and $n, d \geq 2$. A Mandelpinski necklace is a simple closed curve in the parameter plane for these maps that passes alternately through a certain number of baby Mandelbrot sets and Sierpinski holes. At the end of the paper, we describe the very special case that occurs when $n = d = 2$.

Keywords Julia set · Critical point · Critical value · McMullen domain · Mandelpinski necklace · Sierpinski holes

For the family of maps

$$F_\lambda(z) = z^n + \frac{\lambda}{z^d}$$

a "Mandelpinski necklace" is a simple closed curve in the parameter plane that passes alternately through a certain number of centers of baby Mandelbrot sets and Sierpinski holes. The center of a baby Mandelbrot set is the parameter that lies at the "center" of the main cardioid of this set and, hence, is a parameter for which one of the critical orbits is periodic. A Sierpinski hole is a disk in the parameter plane containing parameters for which the corresponding Julia sets are Sierpinski curves, i.e., sets homeomorphic to the well-known Sierpinski carpet fractal. The center of such a hole is a parameter for which the critical orbits all eventually map to ∞. The main result that we shall focus on in this paper is the following: In the parameter plane for the maps $z^n + \lambda/z^d$, there are infinitely many disjoint simple closed curves $\mathcal{S}^k$ for $k = 1, 2, 3, \ldots$ surrounding the McMullen domain, with the

2000 MSC number: Primary 37F10; Secondary 37F45.

R. L. Devaney (✉)
Department of Mathematics, Boston University, 111 Cummington Street, Boston, MA 02215, USA
e-mail: bob@bu.edu

S. M. Marotta
Instituto Hidrologico de Llanuras, Univ. Argentina de la Empresa, Lima, Argentina

R. L. Devaney et al. (eds.), *Topological Dynamics and Topological Data Analysis*, Springer Proceedings in Mathematics & Statistics 350,
https://doi.org/10.1007/978-981-16-0174-3_7

$\mathcal{S}^k$ converging down to the boundary of the McMullen domain (when n and d are not both equal to 2). The curve $\mathcal{S}^1$ passes through exactly $n-1$ centers of baby Mandelbrot sets and Sierpinski holes. The curve $\mathcal{S}^k$ for $k > 1$ passes through exactly $dn^{k-2}(n-1) - n^{k-1} + 1$ centers of baby Mandelbrot sets and Sierpinski holes.

1 Introduction

For simplicity, we shall concentrate for most of this paper on the family of complex rational maps given by

$$F_\lambda(z) = z^n + \frac{\lambda}{z^n}$$

where $\lambda \neq 0$ is a complex parameter and $n \geq 3$. The reason for this simplification is that this family has $2n$ "free" critical points. However, like the well-studied quadratic family $z^2 + c$, because of certain symmetries, there is really only one free critical orbit since all of the critical orbits behave symmetrically. Moreover, there are certain symmetries in the dynamical plane that are present when $n = d$ but not so when $n \neq d$. For complete results in the case where $n \neq d$, see [10, 11].

As another similarity with the quadratic family, the point at ∞ is a superattracting fixed point for each λ. Hence, we have an immediate basin of attraction at ∞ which we denote by B_λ. Also, 0 is a pole of order n, and so, there is an open set containing 0 that is mapped onto B_λ. If this open set is disjoint from B_λ, we call this set the "trap door" and denote it by T_λ. Note that F_λ maps both T_λ and B_λ n-to-1 over B_λ.

As usual in complex dynamics, we are interested in the Julia set for F_λ, which we denote by $J(F_\lambda)$. As in the quadratic case, the Julia set has several equivalent definitions. First, $J(F_\lambda)$ is the boundary of the set of points whose orbits tend to ∞. Second, $J(F_\lambda)$ is the closure of the set of repelling periodic points. And third, $J(F_\lambda)$ is the set on which the map F_λ is chaotic.

The following result was proved in [8].

Theorem 1 (The Escape Trichotomy) *For the family of functions*

$$F_\lambda(z) = z^n + \frac{\lambda}{z^n}$$

with $n \geq 2$

1. *If the critical values lie in* B_λ*, then the Julia set is a Cantor set.*
2. *If the critical values lie in* T_λ*, then the Julia set is a Cantor set of simple closed curves.*
3. *If the critical values lie in any other preimage of* T_λ*, then the Julia set is a Sierpinski curve.*

A *Sierpinski curve* is a planar set that is characterized by the following five properties: it is a compact, connected, locally connected, and nowhere dense set whose

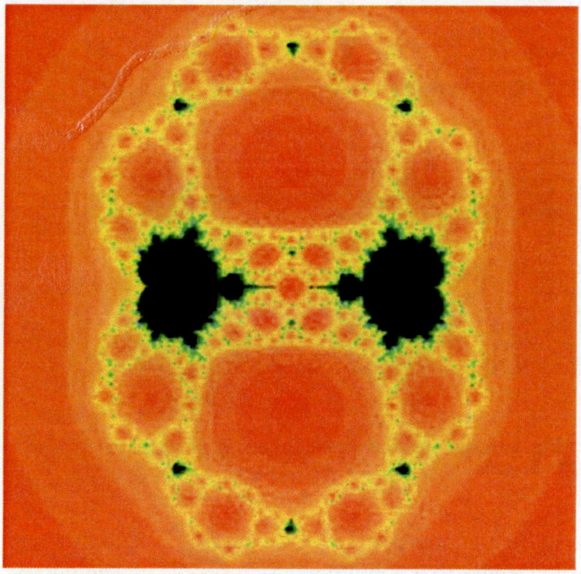

Fig. 1 The parameter plane for the family $z^3 + \lambda/z^3$

complementary domains (of which there must be at least two) are bounded by simple closed curves that are pairwise disjoint. It is known from work of Whyburn [14] that any two Sierpinski curves are homeomorphic. In fact, they are homeomorphic to the well-known Sierpinski carpet fractal. From the point of view of topology, a Sierpinski curve is a universal set in the sense that it contains a homeomorphic copy of any planar, compact, connected, and one-dimensional set. The first example of a Sierpinski curve Julia set was given by Milnor and Tan Lei [13].

Case 2 of the Escape Trichotomy was first observed by McMullen [12], who showed that this phenomenon occurs in each family provided that $n \neq 2$ and λ is sufficiently small. As we describe later, when $n = 2$, the critical values of F_λ cannot lie in T_λ.

In Fig. 1, we display the parameter plane for the family $F_\lambda(z) = z^3 + \lambda/z^3$. The external red region in this set corresponds to parameter values for which the Julia set is a Cantor set; we call this set the *Cantor set locus*. The small red region in the center is a disk surrounding the origin that contains parameter values for which the Julia set is a Cantor set of simple closed curves. We call this region the *McMullen domain*. All of the other red disks contain parameters for which the Julia set is a Sierpinski curve. These disks are called *Sierpinski holes*. In each such hole, there is a unique parameter for which the orbit of some critical point lands on 0 at some iteration and therefore on ∞ at the next iteration, say at iteration $k > 2$. We then call this parameter the center of the Sierpinski hole and k the *escape time* of the hole.

Our goal in this paper is to investigate further properties of the parameter plane for these maps and, in particular, the structure of the parameter plane in a neighborhood of the McMullen domain. It is known [1, 3, 7] that there is a unique McMullen domain

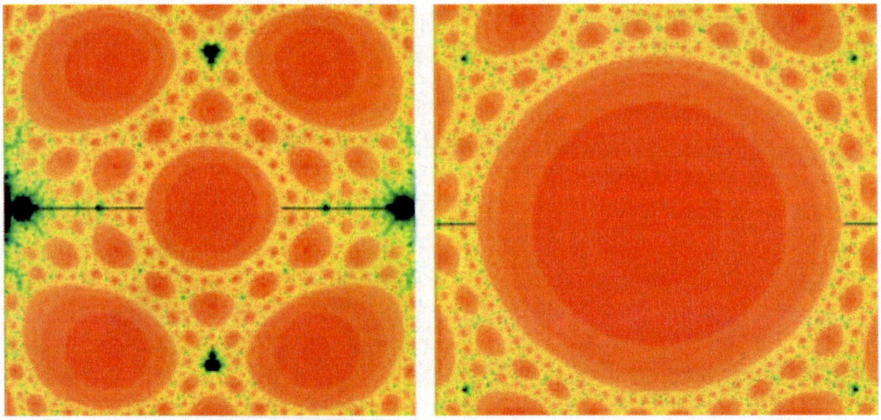

Fig. 2 Magnifications of the parameter plane for the family $z^3 + \lambda/z^3$ around the McMullen domain

in the parameter plane for each $n \geq 3$, and this region is an open disk surrounding the origin that is bounded by a simple closed curve.

In Fig. 2, we have displayed several magnifications of the region around the McMullen domain in the case $n = 3$. In the first image, note that there are four large Sierpinski holes symmetrically placed around the McMullen domain. These Sierpinski holes all have escape time 4. Between the two upper and the two lower Sierpinski holes there appear to be small copies of a Mandelbrot set; while between the two left and two right holes, we see the period two bulb of a principal Mandelbrot set and the remainder of the "tail" of this set. Indeed, one may draw a simple closed curve that encircles the McMullen domain and passes through the centers of each of these Sierpinski holes, the centers of the main cardioids of the two smaller Mandelbrot sets, and the centers of the two period two bulbs of the principal Mandelbrot sets. That is, on this simple closed curve, we find four parameter values for which the map has a a superstable periodic point and four other values for which F_λ^4 maps the critical points to ∞, and these parameter values alternate between the superstable and the centers of Sierpinski holes as the parameter winds around the closed curve.

Inside these four Sierpinski holes appear to be another simple closed curve containing ten Sierpinski holes. Each of these holes has escape time 5. Also, each pair of these holes apparently has either a small copy of a Mandelbrot set or a portion of a principal Mandelbrot set (the two largest Mandelbrot sets displayed in Fig. 1) between them. Examining the further magnification in Fig. 2, we see a smaller closed curve containing 28 Sierpinski holes with escape time 6 and, inside that curve, an even smaller curve containing 82 Sierpinski holes with escape time 7. It appears that the kth curve meets exactly $3^k + 1$ Sierpinski holes with escape time $k + 3$ as well as the same number of (portions of) Mandelbrot sets (though these are so small that they are not quite visible). These are the curves that we call Mandelpinski necklaces.

Actually, the formula in the general case is a little more complicated than that. In Fig. 3, we display the parameter plane for the case $n = 4$ as well as a magnification

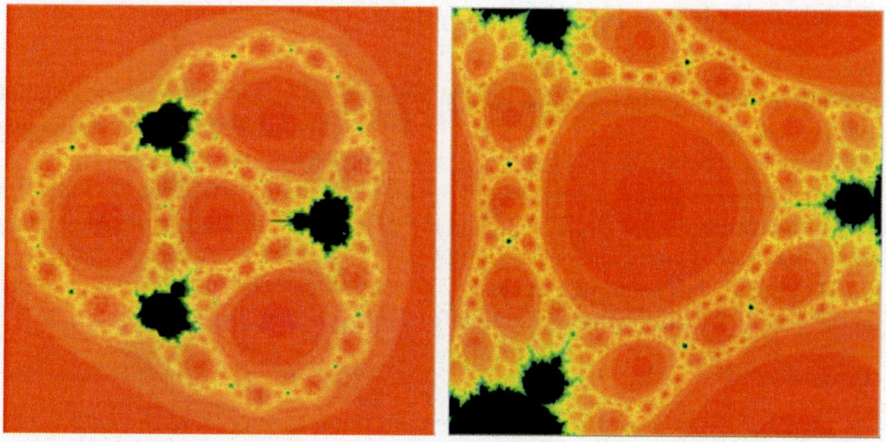

Fig. 3 The parameter plane for the family $z^4 + \lambda/z^4$ and a magnification around the McMullen domain

of the McMullen domain. Here we see three principal Mandelbrot sets arranged between three large Sierpinski holes, each of which has escape time 3. Inside these sets is a curve containing 9 Sierpinski holes, each with escape time 4, and inside another curve containing 33 holes of escape time 5. Further magnification shows that there are $2 \cdot 4^{k-1} + 1$ holes with escape time $k + 2$ in case $n = 4$.

Our main goal in this paper is to make these observations rigorous. We shall prove:

Theorem 2 (Mandelpinski Necklace Theorem) *For each $n \geq 3$, the McMullen domain for the family $z^n + \lambda/z^n$ is surrounded by infinitely many simple closed curves (or rings) $\mathcal{S}^k$ for $k = 1, 2, \ldots$ having the property that:*

1. *Each ring $\mathcal{S}^k$ surrounds the McMullen domain as well as $\mathcal{S}^{k+1}$, and the $\mathcal{S}^k$ accumulate on the boundary of the McMullen domain as $k \to \infty$;*
2. *The ring $\mathcal{S}^k$ meets the centers of τ_k^n Sierpinski holes, each with escape time $k + 2$ where*
$$\tau_k^n = (n - 2)n^{k-1} + 1.$$
3. *The ring $\mathcal{S}^k$ also passes through τ_k^n superstable parameter values where a critical point is periodic of period k or $2k$.*

Using techniques from complex dynamics, it has been shown [4] that these superstable parameter values each lie at the center of the main cardioid of a Mandelbrot set when $k \neq 2$, while the Sierpinski holes surrounding the centers are all simply connected sets. When $k = 2$, $\mathcal{S}^2$ passes through exactly $n - 1$ centers of period 2 bulbs of the largest Mandelbrot sets and also the centers of $\tau_2^n - (n - 1)$ centers of smaller baby Mandelbrot sets. As a remark, the case where $n = 2$ is very different and quite special. We shall describe the result in this case at the end of this paper.

2 Elementary Mapping Properties

Besides 0 and ∞, F_λ has $2n$ other critical points given by $\lambda^{1/2n}$. We call these points the *free critical points* for F_λ. There are, however, only two critical values, and these are given by $\pm 2\sqrt{\lambda}$. We denote a free critical point by c_λ and a critical value by v_λ. The map also has $2n$ prepoles given by $(-\lambda)^{1/2n}$. Note that all of the critical points and prepoles lie on the circle of radius $|\lambda|^{1/2n}$ centered at the origin. We call this circle the *critical circle* and denote it by C_λ.

The map F_λ has some very special properties when restricted to circles centered at the origin. The following is a straightforward computation (see [3]):

Proposition 1

1. *F_λ takes the critical circle $2n$-to-one onto the line interval connecting the two critical values $\pm 2\sqrt{\lambda}$;*
2. *F_λ takes any other circle centered at the origin to an ellipse whose foci are the critical values.*

We call the image of the critical circle the *critical segment*. We call the straight line connecting the origin to ∞ and passing through one of the critical points (resp., prepoles) a *critical point ray* (resp., *prepole ray*). Similar straightforward computations show that each of the critical point rays is mapped in two-to-one fashion onto one of the two straight line segments of the form tv_λ, where $t \geq 1$ and v_λ is the image of the critical point on this ray. So the image of a critical point ray is a straight ray connecting either v_λ or $-v_\lambda$ to ∞. Thus, the critical segment together with these two rays forms a straight line through the origin.

Similarly, each of the $2n$ prepole rays is mapped in one-to-one fashion onto the straight line given by $it\sqrt{\lambda}$, where t is now any real number. Note that the image of the prepole rays is the line that is perpendicular to the line tv_λ for $t \in \mathbb{R}$, i.e., the line that contains the critical segment and the images of all of the critical point rays.

Let U_λ be a sector bounded by two prepole rays corresponding to adjacent prepoles on C_λ, i.e., U_λ is a sector in the plane with angle $2\pi/2n$. We call U_λ a *critical point sector* since it contains at its "center" a unique critical point of F_λ. Similarly, let V_λ be the sector bounded by two critical point rays corresponding to adjacent critical points on C_λ. We call V_λ a *prepole sector*. The next result follows immediately from the above:

Proposition 2 (Mapping Properties of F_λ)

1. *F_λ maps the interior of each critical point sector in two-to-one fashion onto the open half plane that is bounded by the image of the prepole rays and contains the critical value that is the image of the unique critical point in the sector;*
2. *F_λ maps the interior of each prepole sector in one-to-one fashion onto the entire plane minus the two half lines $\pm tv_\lambda$ where $t \geq 1$;*
3. *F_λ maps the region in either the interior or the exterior of the critical circle onto the complement of the critical segment as an n-to-one covering map of $\mathbb{C}$.*

We now turn to the symmetry properties of F_λ in both the dynamical and parameter planes. Let ν be the primitive $2n^{\text{th}}$ root of unity given by $\exp(\pi i/n)$. Then, for each j, we have $F_\lambda(\nu^j z) = (-1)^j F_\lambda(z)$. Hence, if n is even, we have $F_\lambda^2(\nu^j z) = F_\lambda^2(z)$ for each j. Therefore, the points z and $\nu^j z$ land on the same orbit after two iterations, and so, their orbits have the same eventual behavior for each j. If n is odd, the orbits of $F_\lambda(z)$ and $F_\lambda(\nu^j z)$ are either the same or else they are the negatives of each other after the first iteration. In either case, it follows that the orbits of $\nu^j z$ behave symmetrically under $z \mapsto -z$ for each j. Hence, the Julia set of F_λ is always symmetric under $z \mapsto \nu z$. In particular, each of the free critical points eventually maps onto the same orbit (in case n is even) or onto one of two symmetric orbits (in case n is odd). Thus, these orbits all have the same behavior, and so the λ-plane is a natural parameter plane for each of these families. Note also that, if n is even and the orbit of some critical point eventually lands on some other critical point at iteration $j \geq 1$, then in fact one of the critical points of F_λ must be periodic of period j. If n is odd, then there are two possibilities: either one of the critical points has period j or else it has period $2j$.

Let $H_\lambda(z)$ be one of the n involutions given by $H_\lambda(z) = \lambda^{1/n}/z$. Then we have $F_\lambda(H_\lambda(z)) = F_\lambda(z)$, so that the Julia set is also preserved by each of these involutions. Note that each H_λ maps the critical circle to itself and also fixes a pair of critical points of the form $\pm\sqrt{\lambda^{1/n}}$. H_λ also maps circles centered at the origin outside the critical circle to similar circles inside the critical circle and vice versa. It follows that two such circles, one inside and one outside the critical circle, are mapped onto the same ellipse by F_λ.

The parameter plane (see Figs. 1 and 3) for F_λ also possesses several symmetries. First of all, we have

$$\overline{F_\lambda(z)} = F_{\overline{\lambda}}(\overline{z})$$

so that F_λ and $F_{\overline{\lambda}}$ are conjugate via the map $z \mapsto \overline{z}$. Therefore, the parameter plane is symmetric under the map $\lambda \mapsto \overline{\lambda}$.

We also have $(n-1)$-fold symmetry in the parameter plane for F_λ. To see this, let ω be the primitive $(n-1)^{\text{st}}$ root of unity given by $\exp(2\pi i/(n-1))$. Then, if n is even, we compute that

$$F_{\lambda\omega}(\omega^{n/2} z) = \omega^{n/2}(F_\lambda(z)).$$

As a consequence, for each $\lambda \in \mathbb{C}$, the maps F_λ and $F_{\lambda\omega}$ are conjugate under the linear map $z \mapsto \omega^{n/2} z$. In particular, since, when λ is real, the real line is preserved by F_λ, and it follows that the straight line passing through 0 and $\omega^{n/2}$ is preserved by $F_{\lambda\omega}$.

When n is odd, the situation is a little different. We now have

$$F_{\lambda\omega}(\omega^{n/2} z) = -\omega^{n/2}(F_\lambda(z)).$$

Since $F_\lambda(-z) = -F_\lambda(z)$ when n is odd, we therefore have that $F_{\lambda\omega}^2$ is conjugate to F_λ^2 via the map $z \mapsto \omega^{n/2} z$. This means that the dynamics of F_λ and $F_{\lambda\omega}$ are "essentially" the same, though subtly different. For example, if F_λ has a fixed point, then under the conjugacy, this fixed point and its negative are mapped to a 2-cycle for $F_{\lambda\omega}$. Since the real line is invariant when λ is real, it follows that the straight lines passing through the origin and $\pm\omega^{n/2}$ are interchanged by $F_{\lambda\omega}$ and hence invariant under $F^2_{\lambda\omega}$.

To summarize the symmetry properties of F_λ, we have:

Proposition 3 (Symmetries in the dynamical and parameter plane) *The dynamical plane for F_λ is symmetric under the map $z \mapsto \nu z$ where $\nu = \exp(\pi i/n)$. The parameter plane is symmetric under both $z \mapsto \overline{z}$ and $z \mapsto \omega z$ where $\omega = \exp(2\pi i/(n-1))$.*

The following result shows that the McMullen domain, and all of the Sierpinski holes are located inside the unit circle in parameter space.

Proposition 4 (Location of the Cantor set locus) *Suppose $|\lambda| \geq 1$. Then v_λ lies in B_λ so that λ lies in the Cantor set locus.*

Proof Suppose $|z| \geq 2|\lambda|^{1/2} \geq 2$. Then, since $|z| \geq |\lambda|^{1/2}$, we have

$$|F_\lambda(z)| \geq |z|^n - \frac{|\lambda|}{|z|^n} \geq |z|^n - |\lambda|^{1-\frac{n}{2}} \geq |z|^n - 1 \geq |z|^{n-1} > |z|$$

since $n > 2$. Hence $|F^j_\lambda(z)| \to \infty$ so the region $|z| \geq 2|\lambda|^{1/2}$ lies in B_λ. In particular, $v_\lambda \in B_\lambda$.

For each n, let $\lambda^* = \lambda^*_n$ be the unique real solution to the equation

$$|v_\lambda| = 2|\sqrt{\lambda}| = |\lambda|^{1/2n} = |c_\lambda|.$$

Using this equation, we compute easily that

$$\lambda^* = \left(\frac{1}{4}\right)^{\frac{n}{n-1}}.$$

The circle of radius λ^* plays an important role in the parameter plane; if λ lies on this circle, it follows that both of the critical values lie on the critical circle for F_λ. If λ lies inside this circle, then F_λ maps the critical circle strictly inside itself. We call the circle of radius λ^* in parameter plane the *dividing circle*. We denote by $\mathcal{O} = \mathcal{O}_n$ the open set of parameters inside the dividing circle. We will be primarily concerned in later sections with values of the parameter that lie in $\mathcal{O}$. In particular, we shall show that all of the rings around the McMullen domain $\mathcal{S}^k$ with $k > 1$ lie in this region, while the ring $\mathcal{S}^1$ is the dividing circle itself.

3 Some Special Cases

In this section, we discuss the dynamics of several special cases of F_λ that will help define the rings around the McMullen domain later.

First suppose that λ lies on the dividing circle, i.e., $|\lambda| = \lambda^*$. In this case, all of the critical points, critical values, and prepoles of F_λ lie on the same circle (the critical circle) in dynamical plane, namely the circle

$$|z| = \left(\frac{1}{2}\right)^{\frac{1}{n-1}}.$$

As λ winds once around the dividing circle in the counterclockwise direction beginning on the real axis, the critical points and prepoles of F_λ wind $1/2n$ of a turn around the critical circle, while the critical values wind one-half of a turn around the critical circle, all in the counterclockwise direction. Hence, there are exactly $n-1$ special parameter values on the dividing circle for which a critical point of the corresponding map equals a critical value, so for these special λ-values, we have a superattracting fixed or period two point for F_λ. Equivalently, one computes that these $n-1$ parameters are given by

$$\lambda = \left(\frac{1}{4}\right)^{\frac{n}{n-1}}.$$

There are $n-1$ other parameters on this circle for which the critical value is a prepole, and these are given by

$$\lambda = \left(\frac{-1}{4}\right)^{\frac{n}{n-1}}.$$

This proves the case $k = 1$ of the Mandelpinski Necklace Theorem.

Theorem 3 *The ring $\mathcal{S}^1$ is the dividing circle in parameter plane. It contains $n-1$ superstable parameters and the same number of centers of Sierpinski holes.*

See Fig. 4.

We next restrict attention to values of λ lying in $\mathbb{R}^+$. The graph of F_λ shows that, in this case, F_λ maps $\mathbb{R}^+$ to itself and that there is a unique critical point lying in $\mathbb{R}^+$. We denote this critical point by $c_0 = c_0(\lambda)$. See Fig. 5.

It is known [2] that there is a Mandelbrot set (a principal Mandelbrot set) whose central spine lies along an interval $[\lambda_-, \lambda_+]$ contained in $\mathbb{R}^+$. Moreover, if $\lambda > \lambda_+$, then λ lies in the Cantor set locus, whereas if $0 < \lambda < \lambda_-$, then λ lies in the McMullen domain. The graph of $F_\lambda \mid \mathbb{R}^+$ shows that F_λ undergoes a saddle-node bifurcation at λ_+ and that the critical point c_λ maps onto the repelling fixed point in $\partial B_\lambda \cap \mathbb{R}^+$ after two iterations when $\lambda = \lambda_-$. Since each F_λ is conjugate on the real line to a real quadratic polynomial of the form $Q_c(x) = x^2 + c$, standard facts from quadratic dynamics yield the following:

Proposition 5 (Superstable parameters for $\lambda \in \mathbb{R}^+$) *There is a decreasing sequence of parameters in $\mathbb{R}^+$ $\lambda_1 > \lambda_2 \ldots$ converging to λ_- such that, for $\lambda = \lambda_k$, the critical*

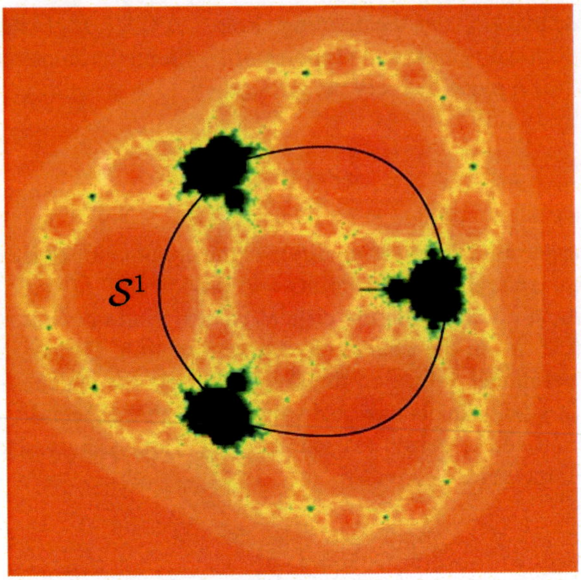

Fig. 4 The ring $\mathcal{S}^1$ in the parameter plane for $n = 4$

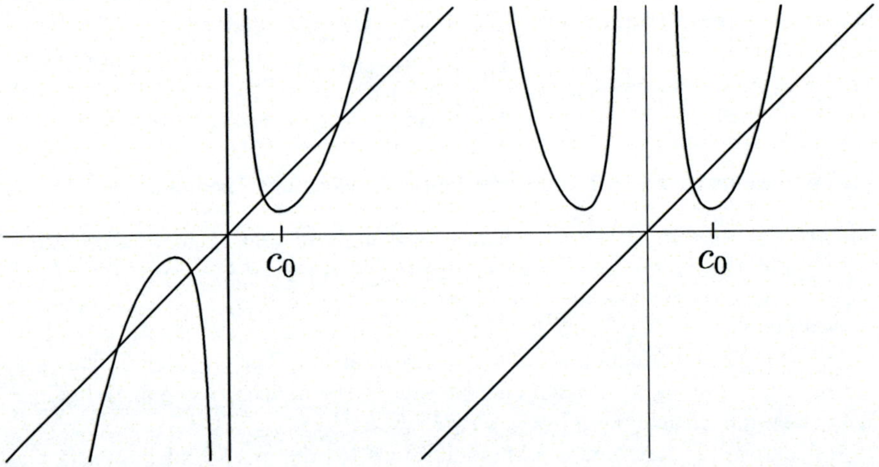

Fig. 5 The graphs of $x^3 + 0.01/x^3$ and $x^4 + 0.01/x^4$

point c_0 is periodic with period k and the critical orbit in $\mathbb{R}^+$ has the special form when $k \geq 2$:

$$0 < v_\lambda = F_\lambda(c_0) < c_0 = F_\lambda^k(c_0) < F_\lambda^{k-1}(c_0) < \ldots < F_\lambda^3(c_0) < F_\lambda^2(c_0).$$

In particular, λ_k is a superstable parameter value of period k, and the orbit of $F_{\lambda_k}^2(c_0)$ is monotonically decreasing for $k - 1$ iterations along $\mathbb{R}^+$.

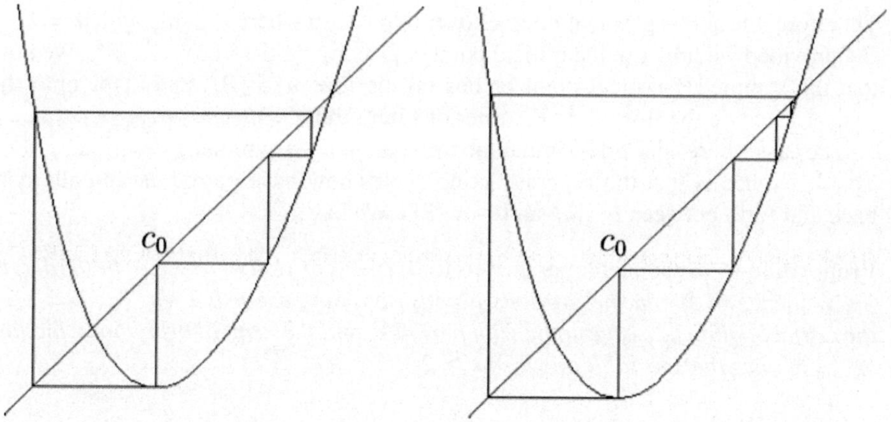

Fig. 6 The graphs of F_λ for $\lambda = \lambda_4$ and $\lambda = \lambda_5$ when $n = 4$

Portions of the graphs of F_{λ_k} for $k = 4$ and $k = 5$ when $n = 4$ are displayed in Fig. 6. Note that the parameter λ_1 necessarily lies on the dividing circle $\mathcal{S}^1$. We shall show below that each λ_k lies on $\mathcal{S}^k$.

Because of the $(n-1)$-fold symmetry in the parameter plane, we have a similar sequence of superstable parameter values along the ray $\lambda = \omega \cdot \mathbb{R}^+$ in parameter plane. To be more precise, first suppose that n is even. Suppose that $\lambda = a\omega$ with $a > 0$ and, as before, $\omega = \exp(2\pi i/(n-1))$. Then, using the results in Sect. 2, we have that, if $t > 0$,

$$F_\lambda(\omega^{\frac{n}{2}} t) = \omega^{\frac{n}{2}} F_a(t)$$

so that F_λ on the line $\omega^{n/2} \cdot \mathbb{R}^+$ is conjugate to F_a on $\mathbb{R}^+$.

Now F_λ has critical points at

$$\begin{aligned} c_0 &= (a\omega)^{\frac{1}{2n}} \\ c_1 &= \nu(a\omega)^{\frac{1}{2n}} \\ c_{n+1} &= \nu^{n+1}(a\omega)^{\frac{1}{2n}} = -\nu(a\omega)^{\frac{1}{2n}} = -c_1. \end{aligned}$$

Note that the critical point c_{n+1} lies on the line $\omega^{n/2} \cdot \mathbb{R}^+$. This follows since

$$\begin{aligned} -\nu(a\omega)^{\frac{1}{2n}} &= -(a)^{\frac{1}{2n}} \left(\exp\left(\frac{\pi i}{n}\right) \exp\left(\frac{\pi i}{n(n-1)}\right)\right) \\ &= -(a)^{\frac{1}{2n}} \exp\left(\frac{\pi i}{n-1}\right) \\ &= -a^{\frac{1}{2n}} \omega^{\frac{1}{2}} = a^{\frac{1}{2n}} \omega^{\frac{n}{2}}. \end{aligned}$$

Therefore, the above proposition goes over to the case where $\lambda = a\omega$ with $a = \lambda_k \in \mathbb{R}^+$ provided we now use the critical point c_{n+1} lying on the line $\omega^{n/2} \cdot \mathbb{R}^+$. We note that the symmetric critical point c_1 lies on the line $\omega^{1/2} \cdot \mathbb{R}^+$ and maps onto the critical value on the line $\omega^{n/2} \cdot \mathbb{R}^+$ after one iteration.

The case where n is odd is similar modulo the $z \mapsto -z$ symmetry described earlier. The difference is that the superattracting cycles now have period $2k$ and alternate back and forth between $\omega \cdot \mathbb{R}^+$ and $-\omega \cdot \mathbb{R}^+$. We have:

Proposition 6 (Superstable parameters for $\lambda \in \omega \cdot \mathbb{R}^+$) *Let $\lambda_1 > \lambda_2 \ldots$ be the decreasing sequence in $\mathbb{R}^+$ in the previous proposition. Suppose n is even. For $\lambda = \lambda_k\omega$, the critical point c_{n+1} is periodic with period k, and the critical orbit along the line $\omega^{n/2} \cdot \mathbb{R}^+$ has the special form when $k \geq 2$*

$$F_\lambda(c_{n+1}) < c_{n+1} = F_\lambda^k(c_{n+1}) < F_\lambda^{k-1}(c_{n+1}) < \ldots < F_\lambda^3(c_{n+1}) < F_\lambda^2(c_{n+1}).$$

In particular, $\lambda = \lambda_k\omega$ is a superstable parameter value of period k, and the orbit of $F_\lambda^2(c_{n+1})$ is monotonically decreasing for $k-1$ iterations along $\omega^{n/2} \cdot \mathbb{R}^+$. When n is odd, replace F_λ with F_λ^2. The cycle corresponding to $\lambda = \lambda_k\omega$ now has period $2k$.

4 Rings in Dynamical Plane

In this section, we prove the existence of infinitely many rings γ_λ^k for $k = 0, 1, \ldots$ in the dynamical plane. Each ring γ_λ^k is a smooth, simple closed curve that is mapped n^k-to-1 onto the critical circle by F_λ^k. We shall use these rings in the next section to construct the rings S^k in the parameter plane.

We begin by defining γ_λ^0 to be the critical circle. Recall that, if $\lambda \in \mathcal{O}$, then F_λ maps γ_λ^0 strictly inside itself. Since all of the critical points of F_λ lie on γ_λ^0, it follows that F_λ takes the exterior of γ_λ^0 as an n-to-1 covering onto the plane minus the critical segment and hence over the entire exterior of γ_λ^0. Thus, there is a preimage γ_λ^1 lying outside of γ_λ^0 and mapped n-to-1 onto γ_λ^0 by F_λ. Since F_λ is a covering map, it follows that γ_λ^1 must be a single simple closed curve. Then F_λ maps the exterior of γ_λ^1 as an n-to-1 covering onto the exterior of γ_λ^0, so there is a preimage of γ_λ^1 lying in this region and mapped n-to-1 to γ_λ^1. Call this simple closed curve γ_λ^2. Continuing inductively, we find a collection of simple closed curves γ_λ^k for $k \geq 1$ having the properties that:

1. γ_λ^{k+1} lies in the exterior of γ_λ^k;
2. F_λ takes γ_λ^{k+1} as an n-to-1 covering onto γ_λ^k;
3. so F_λ takes γ_λ^{k+1} as an n^{k+1}-to-1 covering of the critical circle;
4. the γ_λ^{k+1} converge outward to the boundary of B_λ as $k \to \infty$.

We now construct a parameterization of each of the γ_λ^k. In order for this parametrization to be well-defined, we need to restrict attention to parameters in the region $\mathcal{O}' = \mathcal{O} - (-\lambda^*, 0]$, so that $-\pi < \operatorname{Arg} \lambda < \pi$. We therefore assume that λ lies in $\mathcal{O}'$ for the remainder of this paper.

For $\lambda \in \mathcal{O}'$, there is a unique critical point of F_λ lying in the region $|\operatorname{Arg} z| < \pi/2n$. Call this critical point $c_0 = c_0(\lambda)$. Note that $c_0 \in \mathbb{R}^+$ if $\lambda \in \mathbb{R}^+$. We index the remaining critical points by c_j with the argument of c_j increasing as j increases.

To parametrize the critical circle γ_λ^0, we set $\gamma_\lambda^0(0) = c_0(\lambda)$. By the mapping properties proposition, for each $\theta \in \mathbb{R}$, we then let $\gamma_\lambda^0(\theta)$ be the natural continuation of this parametrization of the circle in the counterclockwise direction. So $\gamma_\lambda^0(\theta)$ is 2π-periodic in θ and depends analytically on λ for $\lambda \in \mathcal{O}'$.

To parametrize $\gamma_\lambda^1(\theta)$, consider the portion of the critical point sector containing $c_0(\lambda)$ that lies outside the critical circle. There is a unique point in this region mapped to c_0 by F_λ; call this point $\gamma_\lambda^1(0)$. Then define $\gamma_\lambda^1(\theta)$ by requiring that

$$F_\lambda(\gamma_\lambda^1(\theta)) = \gamma_\lambda^0(\theta)$$

and that $\gamma_\lambda^1(\theta)$ varies continuously with θ. Note that $\gamma_\lambda^1(\theta)$ is $2n\pi$ periodic since F_λ is n-to-1 on γ_λ^1. We then proceed inductively to define $\gamma_\lambda^k(\theta)$ by first specifying that, in the outside portion of the critical point sector containing c_0, $\gamma_\lambda^k(0)$ is the unique point that is mapped by F_λ to $\gamma_\lambda^{k-1}(0)$ and then using F_λ to complete this parameterization. As above, for each k, $\gamma_\lambda^k(\theta)$ is $2n^k\pi$ periodic in θ and depends analytically on λ.

To prove the existence of the rings in the parameter plane, we need to be more specific about the location of the rings in the dynamical plane. Let V_+ be the portion of the prepole sector lying on and outside the critical circle and also between the two critical point rays through c_0 and c_1. That is,

$$V_+ = \left\{ z \,\middle|\, |z| \geq |\lambda|^{1/2n},\ \frac{\operatorname{Arg} \lambda}{2n} \leq \operatorname{Arg} z \leq \frac{\operatorname{Arg} \lambda}{2n} + \frac{\pi}{n} \right\}.$$

Let $V_- = \nu^{-1} \cdot V_+$. So V_- is the portion of the prepole sector bounded by the critical lines through c_0 and c_{-1} and lying on or outside the critical circle. Let $V_\lambda = V_+ \cup V_-$. See Fig. 7.

Since $|\operatorname{Arg} \lambda| < \pi$ and $n \geq 3$, we have for $z \in V_\lambda$

$$|\operatorname{Arg} z| \leq \left| \frac{\operatorname{Arg} \lambda}{2n} \right| + \frac{\pi}{n} < \frac{3\pi}{2n} \leq \frac{\pi}{2}.$$

So for each $\lambda \in \mathcal{O}'$, the region V_λ is contained in the half plane $\operatorname{Re} z > 0$.

Now F_λ maps the portion of boundary of V_+ lying along the critical circle one-to-one to the critical segment since the endpoints of this arc are adjacent critical points along C_λ that are mapped to distinct critical values. Also, F_λ maps the portion of the critical point line containing c_0 lying on the boundary of V_+ one-to-one onto the ray $tv_\lambda = 2t\sqrt{\lambda}$ with $t \geq 1$ and $\operatorname{Arg} \sqrt{\lambda} > 0$, while F_λ maps the other boundary ray containing c_1 to the negative of this ray. Hence, the boundary of V_+ is mapped onto the entire straight line passing through $\pm v_\lambda$ and the origin. Therefore, F_λ maps V_+

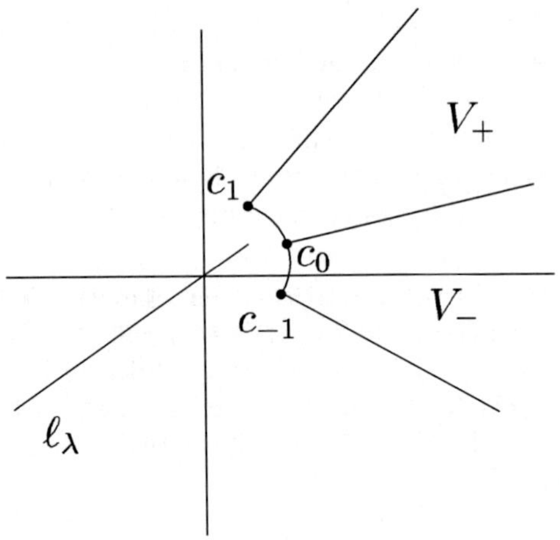

Fig. 7 The region $V_\lambda = V_+ \cup V_-$

univalently onto one of the half planes bounded by this line. Similarly, F_λ maps V_- univalently onto the opposite half plane.

Let ℓ_λ be the straight line given by $2t\sqrt{\lambda}$ where $t \in (-\infty, 1]$. So ℓ_λ is the straight line that starts at $2\sqrt{\lambda}$ at $t = 1$ and passes through the origin and $-2\sqrt{\lambda}$ enroute to ∞ as $t \to \infty$. Note that the boundary of V_λ is mapped two-to-one onto ℓ_λ by F_λ. Hence, F_λ maps the interior of V_λ univalently onto $\mathbb{C} - \ell_\lambda$. Now, for each $\lambda \in \mathcal{O}'$, the critical segment lies outside V_λ since neither V_+ nor V_- meets the interior of the critical circle. Also, the portion of ℓ_λ extending from $-2\sqrt{\lambda}$ to ∞ lies in the left half plane, so the entire line ℓ_λ does not intersect V_λ. So we have:

Proposition 7 *For each $\lambda \in \mathcal{O}'$, F_λ maps the interior of V_λ univalently onto $\mathbb{C} - \ell_\lambda$ and so the image of V_λ contains V_λ.*

Recall that the kth ring in the dynamical plane is parametrized by $\gamma_\lambda^k(\theta)$ and is periodic with period $2n^k\pi$.

Proposition 8 *For each $k \geq 1$, the portion of the ring $\gamma_\lambda^k(\theta)$ with $|\theta| \leq n^{k-1}\pi$ lies in the region*

$$-\frac{3\pi}{2n} < \operatorname{Arg} z < \frac{3\pi}{2n}.$$

Proof We deal first with the case $0 \leq \theta \leq n^{k-1}\pi$; the other case is handled by applying the $z \mapsto \nu^{-1}z$ symmetry, as we describe below.

We claim that the portion of the ring $\gamma_\lambda^k(\theta)$ with $0 \leq \theta \leq n^{k-1}\pi$ actually lies in the smaller region

$$-\frac{\pi}{2n} < \operatorname{Arg} z < \frac{3\pi}{2n}.$$

To see this, we first consider the simplest case where $\lambda \in \mathbb{R}^+$. In this case, V_+ is bounded by $\mathbb{R}^+$ and $\nu \cdot \mathbb{R}^+$ and F_λ maps V_+ univalently onto $\operatorname{Im} z \geq 0$. Recall that

$\gamma_\lambda^0(\theta)$ lies in the region $\text{Im}\, z \geq 0$ if $\theta \in [0, \pi]$. Hence, there is a continuous preimage of $\gamma_\lambda^0(\theta)$ lying in V_+. This preimage is, by definition, $\gamma_\lambda^1(\theta)$ for $\theta \in [0, \pi]$. So $\gamma_\lambda^1(\theta)$ lies in the region $0 \leq \text{Arg}\, z \leq \pi/n$, and thus, the result is true when $k = 1$.

Next note that $\gamma_\lambda^1(\pi)$ lies on the line $\nu \cdot \mathbb{R}^+$ and is given by $\nu\gamma_\lambda^1(0)$. So we can use the symmetry in the dynamical plane to extend the definition of $\gamma_\lambda^1(\theta)$ to a continuous curve defined for $\theta \in [0, n\pi]$ as follows: if $\theta \in [j\pi, (j+1)\pi]$, let $\gamma_\lambda^1(\theta) = \nu^j \gamma_\lambda^1(\theta - j\pi)$ for $j = 1, \ldots, n-1$. So $\gamma_\lambda^1(\theta)$ lies in $\text{Im}\, z \geq 0$ for $\theta \in [0, n\pi]$. Then the sector V_+ is again mapped over $\gamma_\lambda^1(\theta)$ for these θ-values, so we have a continuous preimage $\gamma_\lambda^2(\theta)$ lying in V_+, mapped onto $\gamma_\lambda^1(\theta)$, and defined for $\theta \in [0, n\pi]$.

Then we extend the definition of $\gamma_\lambda^2(\theta)$ to $[0, n^2\pi]$ as above using the symmetry in the dynamical plane. So we have that $\gamma_\lambda^3(\theta)$ lies in V_+ for all $\theta \in [0, n^2\pi]$. Continuing in this fashion proves the stronger result that $\gamma_\lambda^k(\theta)$ in fact lies in V_+ for $\theta \in [0, n^{k-1}\pi]$ for all k as long as $\lambda \in \mathbb{R}^+$.

Now suppose that $0 < \text{Arg}\, \lambda < \pi$. We no longer have the fact that V_+ is mapped over $\gamma_\lambda^0(\theta)$ for $0 \leq \theta \leq \pi$. Indeed, the point $\gamma_\lambda^1(0)$ now lies in V_-. This follows from the fact that the critical point ray through c_0 is mapped to a line whose argument is strictly larger than that of c_0, so the preimage of c_0 must lie below this critical point line. By the previous proposition, we have that F_λ maps the interior of the entire region V_λ univalently onto $\mathbb{C} - \ell_\lambda$. Let ℓ'_λ denote the portion of ℓ_λ lying in the lower half plane. Then

$$\pi < \frac{\text{Arg}\, \lambda}{2} + \pi = \text{Arg}\, \ell'_\lambda < \frac{3\pi}{2}.$$

Since, for $\theta \in [0, \pi]$, we have

$$0 < \text{Arg}\, c_0 \leq \text{Arg}\, \gamma_\lambda^0(\theta) \leq \text{Arg}\, c_0 + \pi < \frac{\text{Arg}\, \lambda}{2} + \pi = \text{Arg}\, \ell'_\lambda,$$

it follows that the entire line ℓ_λ never meets $\gamma_\lambda^0(\theta)$ for these θ-values. Hence, there is a continuous preimage of $\gamma_\lambda^0(\theta)$ in $V_+ \cup V_-$ for each $\theta \in [0, \pi]$. This defines $\gamma_\lambda^1(\theta)$ over this interval. Note that $\gamma_\lambda^1(\pi) = \nu\gamma_\lambda^1(0)$ must lie in V_+. In fact, we can say more:

$$-\frac{\pi}{2n} < \frac{\text{Arg}\, \lambda}{2n} - \frac{\pi}{2n} \leq \text{Arg}\, \gamma_\lambda^1(\theta)$$

for $0 \leq \theta \leq \pi$. This follows since F_λ maps the prepole line in V_- to a line perpendicular to ℓ_λ in $-\pi/2 < \text{Arg}\, z < 0$. This line does not intersect the curve $\gamma_\lambda^0(\theta)$ for $\theta \in [0, \pi]$. So $\gamma_\lambda^1(\theta)$ does not meet the prepole line in V_-. We therefore have

$$-\frac{\pi}{2n} < \text{Arg}\, \gamma_\lambda^1(\theta) < \frac{3\pi}{2n}$$

for $\theta \in [0, \pi]$, so this proves the case $k = 1$ when $0 < \text{Arg}\, \lambda < \pi$.

Now we extend the definition of $\gamma_\lambda^1(\theta)$ to $\theta \in [0, n\pi]$ as in the previous case using symmetry. Then we have, for $0 \leq \theta \leq n\pi$,

$$-\frac{\pi}{2n} < \text{Arg}\, \gamma_\lambda^1(\theta) \le \text{Arg}\, c_0 + \pi.$$

But $\text{Arg}\, c_0 + \pi < \text{Arg}\, \lambda/2 + \pi = \text{Arg}\, \ell'_\lambda$. So again ℓ_λ does not meet the extension of $\gamma_\lambda^1(\theta)$. So we have that $\gamma_\lambda^2(\theta)$ lies in the interior of $V_+ \cup V_-$ for $0 \le \theta \le n\pi$ and so $\text{Arg}\, \gamma_\lambda^2(\theta) < 3\pi/2n$. As above, we in fact also have $-\pi/2n \le \text{Arg}\, \gamma_\lambda^2(\theta)$, so this proves the case $k = 2$. Continuing inductively proves the result for all k-values when $0 < \text{Arg}\, \lambda < \pi$ and $0 \le \theta \le n^{k-1}\pi$.

The case of negative values of θ is handled by symmetry as follows. We again assume that $0 < \text{Arg}\, \lambda < \pi$. For each k, we have, since $\gamma_\lambda^k(\theta)$ is $2n^k\pi$ periodic,

$$\begin{aligned} F_\lambda(\nu^{-1}\gamma_\lambda^k(\theta)) &= -F_\lambda(\gamma_\lambda^k(\theta)) \\ &= -\gamma_\lambda^{k-1}(\theta) \\ &= \gamma_\lambda^{k-1}(\theta - n^{k-1}\pi) \\ &= F_\lambda(\gamma_\lambda^k(\theta - n^{k-1}\pi)). \end{aligned}$$

Therefore

$$\nu^{-1}\gamma_\lambda^k(\theta) = \gamma_\lambda^k(\theta - n^{k-1}\pi)$$

follows since $\gamma_\lambda^k(\theta)$ is continuous in θ. Therefore, we have that when $\theta \in [-n^{k-1}\pi, 0]$, $\gamma_\lambda^k(\theta)$ lies in the region

$$-\frac{3\pi}{2n} < \text{Arg}\, z < \frac{\pi}{2n}.$$

So altogether the curve $\gamma_\lambda^k(\theta)$ lies in the region $|\text{Arg}\, z| < 3\pi/2n$ for all $|\theta| \le n^{k-1}\pi$. This concludes the proof when $0 \le \text{Arg}\, \lambda < \pi$.

If $-\pi < \text{Arg}\, \lambda < 0$, we invoke the $z \mapsto \overline{z}$ symmetry in the parameter plane. Since F_λ is conjugate to $F_{\overline{\lambda}}$ via $z \mapsto \overline{z}$, it follows that the curves $\gamma_\lambda^k(\theta)$ are mapped to $\gamma_{\overline{\lambda}}^k(-\theta)$ by the conjugacy. Hence, these curves lie in the same region when $-\pi < \text{Arg}\, \lambda < 0$. This concludes the proof.

5 Rings in Parameter Plane

Before turning to the proof of the existence of the Mandelpinski necklaces in the parameter plane, we need to examine more carefully the parametrizations of the rings in the dynamical plane in two of the special cases discussed earlier, namely when $\lambda \in \mathbb{R}^+$ and $\lambda \in \omega \cdot \mathbb{R}^+$.

First suppose that $\lambda \in \mathbb{R}^+$. For the special parameters λ_k among the superstable parameters in $\mathbb{R}^+$, we have seen that $F_{\lambda_k}(c_0)$ always lies in $\mathbb{R}^+$ and satisfies

$$0 < F_{\lambda_k}(c_0) < c_0 = F_{\lambda_k}^k(c_0) < F_{\lambda_k}^{k-1}(c_0) < \ldots < F_{\lambda_k}^2(c_0).$$

Hence, $F_{\lambda_k}^2(c_0)$ lies on $\gamma_{\lambda_k}^{k-2} \cap \mathbb{R}^+$ and $F_{\lambda_k}^j(c_0)$ lies on $\gamma_{\lambda_k}^{k-j} \cap \mathbb{R}^+$ for $j = 2, \ldots, k$.

In particular, since the definition of the parametrization requires that $F_\lambda(\gamma_\lambda^j(0)) = \gamma_\lambda^{j-1}(0)$, it follows that, for the special parameter value λ_k, we have

$$\begin{aligned} \gamma_{\lambda_k}^0(0) &= c_0 \\ \gamma_{\lambda_k}^{k-2}(0) &= F_{\lambda_k}^2(c_0) \\ \gamma_{\lambda_k}^{k-3}(0) &= F_{\lambda_k}^3(c_0) \\ &\vdots \\ \gamma_{\lambda_k}^1(0) &= F_{\lambda_k}^{k-1}(c_0) \end{aligned}$$

Next we turn attention to the special parameter values $\lambda_k\omega$ lying along the line $\omega \cdot \mathbb{R}^+$ in the parameter plane. Here the situation is somewhat more complicated. For simplicity of notation, we fix a value of k and set $\mu = \lambda_k\omega$.

As we showed earlier, the line $\omega^{n/2} \cdot \mathbb{R}^+$ contains the critical point c_{n+1} and is either invariant under F_μ (if n is even) or interchanged with the symmetric line $-\omega^{n/2} \cdot \mathbb{R}^+$ by F_μ (if n is odd). In either case, the symmetric line $-\omega^{n/2} \cdot \mathbb{R}^+$ is mapped to this line by F_μ and contains the critical point $c_1 = -c_{n+1}$. Also, the critical point line through c_0 is mapped to $-\omega^{n/2} \cdot \mathbb{R}^+$ by F_μ and then to $\omega^{n/2} \cdot \mathbb{R}^+$ by F_μ^2.

We have, by definition, $\gamma_\mu^0(0) = c_0$. Since $c_1 = \nu c_0$ where, as usual, $\nu = \exp(\pi i/n)$, we also have

$$\begin{aligned} c_1 &= \gamma_\mu^0\left(\frac{\pi}{n}\right) \\ c_{n+1} &= \gamma_\mu^0\left(\frac{\pi}{n} + \pi\right). \end{aligned}$$

Consider the portion of the critical point sector containing c_0 and lying on or outside C_λ. $\gamma_\mu^1(0)$ is the unique point in this region that is mapped to c_0 by F_μ. Since F_μ takes the critical point line through c_0 to the critical point line through c_1, it follows that $\gamma_\mu^1(0)$ lies below this line and that $\gamma_\mu^1(\pi/n)$, the preimage of c_1, lies on the critical point line through c_0. By symmetry, $\gamma_\mu^1((\pi/n) + \pi)$ then lies on the critical point line through c_1 and, since γ_μ^1 is $2n\pi$-periodic, the point

$$\gamma_\mu^1\left(\frac{\pi}{n} + \pi + n\pi\right)$$

lies on the line $\omega^{n/2} \cdot \mathbb{R}^+$ containing c_{n+1}.

Continuing, we have that $\gamma_\mu^2((\pi/n) + \pi)$ lies on the critical point line through c_0 and is mapped by F_μ to $\gamma_\mu^1((\pi/n) + \pi)$. The point

$$\gamma_\mu^2\left(\frac{\pi}{n} + \pi + n\pi\right)$$

then lies on the critical point line through c_1 and is mapped to

$$\gamma_\mu^1\left(\frac{\pi}{n}+\pi+n\pi\right)$$

on $\omega^{n/2}\cdot\mathbb{R}^+$.

Continuing inductively, we see that the critical point line through c_0 contains the points

$$\begin{aligned}
&c_0=\gamma_\mu^0(0)\\
&\gamma_\mu^1\left(\frac{\pi}{n}\right)\\
&\gamma_\mu^2\left(\frac{\pi}{n}+\pi\right)\\
&\vdots\\
&\gamma_\mu^j\left(\frac{\pi}{n}+\pi+n\pi+\ldots+n^{j-2}\pi\right)=\gamma_\mu^j\left(\frac{\pi}{n}\left(1+n+\ldots+n^{j-1}\right)\right).
\end{aligned}$$

and the critical point line through c_1 contains the points

$$\begin{aligned}
&c_1=\gamma_\mu^0\left(\frac{\pi}{n}\right)\\
&\gamma_\mu^1\left(\frac{\pi}{n}+\pi\right)\\
&\gamma_\mu^2\left(\frac{\pi}{n}+\pi+n\pi\right)\\
&\vdots\\
&\gamma_\mu^j\left(\frac{\pi}{n}+\pi+n\pi+\ldots+n^{j-1}\pi\right)=\gamma_\mu^j\left(\frac{\pi}{n}\left(1+n+\ldots+n^{j}\right)\right).
\end{aligned}$$

Equivalently, $\gamma_\mu^j(\theta)$ lies on the critical point line through c_1 for

$$\theta=\frac{\pi}{n}\left(\frac{n^{j+1}-1}{n-1}\right).$$

Now consider the corresponding points on the critical point line through c_{-1}. Since the parametrization corresponding to points on this line and γ_μ^j is obtained by subtracting $n^{j-1}\pi$ from the corresponding critical point line through c_0, we find the following points on this critical point line:

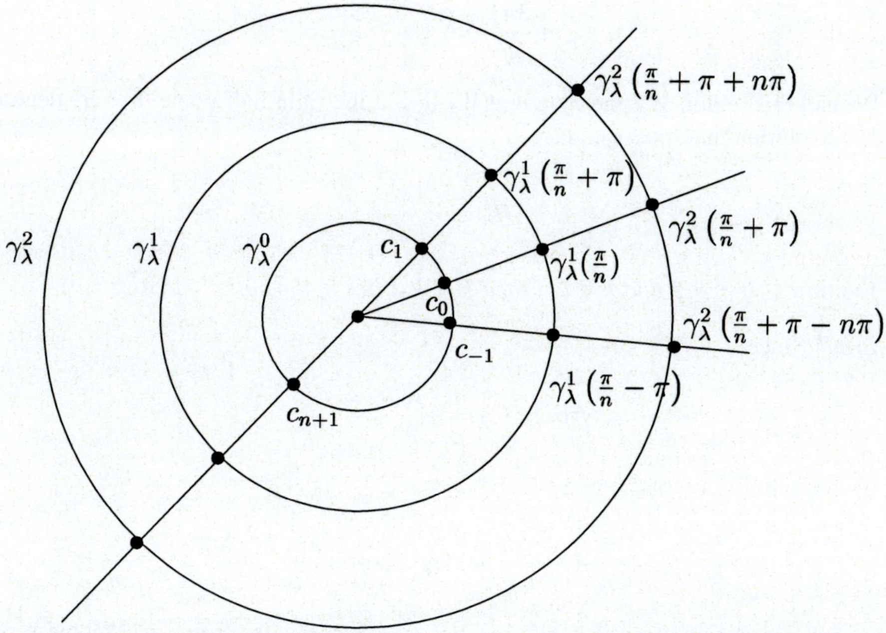

Fig. 8 Parametrization of $\gamma_\lambda(\theta)$ when $\lambda = \lambda_k\omega$

$$\begin{aligned}
c_{-1} &= \gamma_\mu^0\left(-\frac{\pi}{n}\right)\\
&\gamma_\mu^1\left(\frac{\pi}{n}-\pi\right)\\
&\gamma_\mu^2\left(\frac{\pi}{n}+\pi-n\pi\right)\\
&\vdots\\
&\gamma_\mu^j\left(\frac{\pi}{n}+\pi+n\pi+\ldots+n^{j-2}\pi-n^{j-1}\pi\right).
\end{aligned}$$

Equivalently, $\gamma_\mu^j(\theta)$ lies on the critical point line through c_{-1} for

$$\theta = \frac{\pi}{n}\left(1+n+n^2+\ldots+n^{j-1}-n^j\right) = \frac{\pi}{n}\left(\frac{n^j-1}{n-1}\right) - n^{j-1}\pi.$$

For later use, this value of θ is called $\theta_{n,j}$. See Fig. 8.

We now turn to the proof of the existence of the rings $\mathcal{S}^k$ in parameter plane for $k > 1$. For simplicity, we consider only the case when $n \geq 5$ in this section; the special cases $n = 3, 4$ are described in [9].

Recall that, from the results of the previous section, we have that, when $k \geq 1$, the portion of the curve $\gamma_\lambda^k(\theta)$ for $|\theta| \leq n^{k-1}\pi$ lies in the region

$$-\frac{3\pi}{2n} < \operatorname{Arg} z < \frac{3\pi}{2n}.$$

We call this region W_n and note that W_n lies in the right half plane. Let H_λ denote the involution that fixes c_0, i.e.,

$$H_\lambda(z) = \frac{\lambda^{1/n}}{z}.$$

Lemma 1 *If $n \geq 5$ and $\lambda \in \mathcal{O}'$, then $H_\lambda(W_n)$ lies in the half plane* $\operatorname{Re} z > 0$.

Proof Since

$$\operatorname{Arg} H_\lambda(z) = \frac{\operatorname{Arg} \lambda}{n} - \operatorname{Arg} z,$$

we have, if $z \in W_n$ and $n \geq 5$,

$$-\frac{\pi}{2} \leq -\frac{5\pi}{2n} \leq -\frac{3\pi}{2n} + \frac{\operatorname{Arg} \lambda}{n} < \operatorname{Arg} H_\lambda(z) < \frac{3\pi}{2n} + \frac{\operatorname{Arg} \lambda}{n} \leq \frac{5\pi}{2n} \leq \frac{\pi}{2}.$$

We remark that this result is false when $n = 3, 4$; that is the reason why these are special cases.

Now consider the curves

$$\xi_\lambda^k(\theta) = H_\lambda(\gamma_\lambda^k(\theta)).$$

Since the involution H_λ interchanges the inside and outside of C_λ, each of the curves ξ_λ^k is a simple closed curve lying inside the critical circle. We have

$$F_\lambda(\xi_\lambda^k(\theta)) = \gamma_\lambda^{k-1}(\theta)$$

since $F_\lambda(H_\lambda(z)) = F_\lambda(z)$. By the Lemma, we also have that $\xi_\lambda^k(\theta)$ lies in $\operatorname{Re} z > 0$ for $|\theta| \leq n^{k-1}\pi$, at least if $n \geq 5$.

Theorem 4 *For each $k \geq 1$ and any θ satisfying $|\theta| \leq n^{k-1}\pi$, there exists a unique parameter $\lambda = \lambda_{\theta,k}$ such that*

$$v_\lambda = 2\sqrt{\lambda} = \xi_\lambda^k(\theta).$$

Proof The function $G(\lambda) = v_\lambda = 2\sqrt{\lambda}$ takes the subset $\mathcal{O}'$ of the parameter plane univalently onto an open subset of $\operatorname{Re} z > 0$. For each $\lambda \in \mathcal{O}'$, $G(\lambda)$ lies inside C_λ, but for λ on the dividing circle (which is the circular boundary of $\mathcal{O}'$), $G(\lambda)$ lies on the critical circle. Hence, G maps $\mathcal{O}'$ univalently onto the interior of a half disk in the right half plane that contains the region inside C_λ in $\operatorname{Re} z > 0$ for each $\lambda \in \mathcal{O}'$. Call this half disk D.

Also, for fixed θ, the function $\lambda \mapsto \xi_\lambda^k(\theta)$ is analytic on $\mathcal{O}'$ and takes this set strictly inside the portion of the critical circle bounded by the rays $|\text{Arg}\, z| = 3\pi/2n$. Hence, for each θ, the set of points $\xi_\lambda^k(\theta)$ lies inside a compact sector in D. That is, this set of points can possibly accumulate on the boundary of D only at the origin. Hence, we may consider the composition $Q(\lambda) = G^{-1}(\xi_\lambda^k(\theta))$. As a function of λ, Q is analytic and maps the simply connected region $\mathcal{O}'$ inside itself. By the Schwarz Lemma, Q has a unique fixed point in this set or on its boundary. But the fixed point cannot lie at $\lambda = 0$ since 0 is surrounded by the McMullen domain so that the curves ξ_λ^k are bounded away from the origin. Hence, there must be a unique fixed point in the interior of D. This fixed point is $\lambda_{\theta,k}$.

Note that the fixed points $\lambda_{\theta,k}$ vary continuously with θ, so $\theta \mapsto \lambda_{\theta,k}$ is a curve in the parameter plane.

The following proposition identifies the specific values of $\lambda_{\theta,k}$ corresponding to the special cases considered earlier.

Proposition 9 *When $\theta = 0$ and $k \geq 1$, the parameter values $\lambda_{0,k}$ are given by the parameters $\lambda_{k+1} \in \mathbb{R}^+$. When $\theta = \theta_{n,k}$, $\lambda(\theta, k)$ is given by $\omega\lambda_{k+1}$ on the symmetry line $\omega \cdot \mathbb{R}^+$.*

Proof When $\lambda \in \mathbb{R}^+$, the points $\gamma_\lambda^j(0)$ also lie in $\mathbb{R}^+$ for each j. Since, as shown earlier, the parameter λ_{k+1} has the property that $v_{\lambda_{k+1}} \in \xi_{\lambda_{k+1}}^k$, $F_{\lambda_{k+1}}^2(c_0) \in \gamma_{\lambda_{k+1}}^{k-1} \cap \mathbb{R}^+$ and the forward orbit of this point decreases along $\mathbb{R}^+$ until meeting c_0, it follows from the uniqueness of the parameter $\lambda_{0,j}$ that we must have $\lambda_{0,k} = \lambda_{k+1}$ for each $k \geq 1$.

When $\lambda = \lambda_{k+1}\omega$ and $\theta = \theta_{n,k}$, we know that the point $\gamma_\lambda^k(\theta_{n,k})$ lies on the critical point line through c_{-1}. Hence, $H_\lambda(\gamma_\lambda^k(\theta_{n,k}))$ lies on the critical point line through c_1 and is given by $\xi_\lambda^k(\theta_{n,k})$. This point is then mapped by F_λ to the point on $\omega^{n/2} \cdot \mathbb{R}^+$ whose orbit meets c_{n+1} after $k-1$ iterations of F_λ or F_λ^2, depending upon whether n is even or odd. Hence, $\lambda_{\theta_{n,k},k} = \lambda_{k+1}\omega$ as claimed.

Now the parameters in the previous proposition are the unique parameters on the corresponding lines in parameter space for which the orbit of the second iterate of the appropriate critical point monotonically decreases along the corresponding line(s) for $k-1$ iterations before returning to itself and becoming periodic. So the curve $\theta \mapsto \lambda_{\theta,k}$ meets each of these two symmetry lines only once. Hence, the portion of this curve defined for $0 \leq \theta \leq \theta_{n,k}$ either lies outside the sector

$$0 \leq \text{Arg}\, \lambda \leq \frac{2\pi}{n-1}$$

for all values of θ or else this entire curve lies inside the sector. But the former cannot occur since this would imply that some $\lambda_{\theta,k}$ would lie in $\mathbb{R}^-$, contradicting the fact that each $\lambda_{\theta,k}$ lies in $\mathcal{O}'$. Hence, the portion of the curve $\lambda_{\theta,k}$ defined for $0 \leq \theta \leq \theta_{n,k}$ is a continuous arc connecting $\theta = 0$ and $\theta = 2\pi/(n-1)$. It then follows by the $(n-1)$-fold symmetry that, for each $k \geq 1$, $\lambda_{\theta,k}$ is a simple closed curve in parameter space which is periodic of period

$$\begin{aligned}(n-1)\theta_{n,k} &= (n-1)\left(\frac{\pi}{n}\left(\frac{n^k-1}{n-1}\right)-n^{k-1}\pi\right)\\ &= \frac{\pi}{n}\left(-n^{k+1}+2n^k-1\right).\end{aligned}$$

We therefore define the ring $\mathcal{S}^{k+1}$ to be the simple closed curve $\theta \mapsto \lambda_{\theta,k}$. That is, $\mathcal{S}^{k+1}$ consists of parameter values for which the critical orbit has the following behavior:

1. both critical values lie inside the critical circle;
2. $F_\lambda^2(c_\lambda)$ lies on γ_λ^{k-1};
3. subsequent iterates decrease through the γ_λ^j until, at the kth iterate, the critical orbit lands back on the critical circle.

We have shown:

Theorem 5 *When $n \geq 5$, the ring $\mathcal{S}^{k+1}$ in parameter space is a simple closed curve that is parameterized by $\theta \mapsto \lambda_{\theta,k}$ and is periodic of period*

$$\frac{\pi}{n}\left(n^{k+1}-2n^k+1\right) = \frac{\pi}{n}\left((n-2)n^k+1\right).$$

In particular, since the critical points (resp., prepoles) of F_λ are located on $\gamma_\lambda^0(\theta)$ at $\theta = \pi j/n$ (resp., $(2j+1)\pi/2n$) for $0 \leq j < 2n$, we have the following count of superstable parameters and centers of Sierpinski holes along $\mathcal{S}^{k+1}$:

Corollary 1 *There are precisely $(n-2)n^k+1$ parameters along $\mathcal{S}^{k+1}$ that are superstable parameters. There are the same number of parameters that are centers of Sierpinski holes. These parameters alternate between these two types as the parameter winds around $\mathcal{S}^{k+1}$.*

This proves the existence of the Mandelpinski necklaces when $n \geq 5$.

6 The Special Case $n = 2$

In this section, we give three examples of how the case $n = 2$ is so much different from the cases where $n > 2$. The first example of this difference is the fact that there is no McMullen domain when $n = 2$. The reason for this is as follows. Recall that the critical values of F_λ are given by $v_\lambda = \pm 2\sqrt{\lambda}$. By McMullen's result [12], the critical values must lie in the trap door if the Julia set is a Cantor set of simple closed curves. But, in the case $n = 2$, we have

$$F_\lambda(v_\lambda) = 4\lambda + \frac{1}{4}.$$

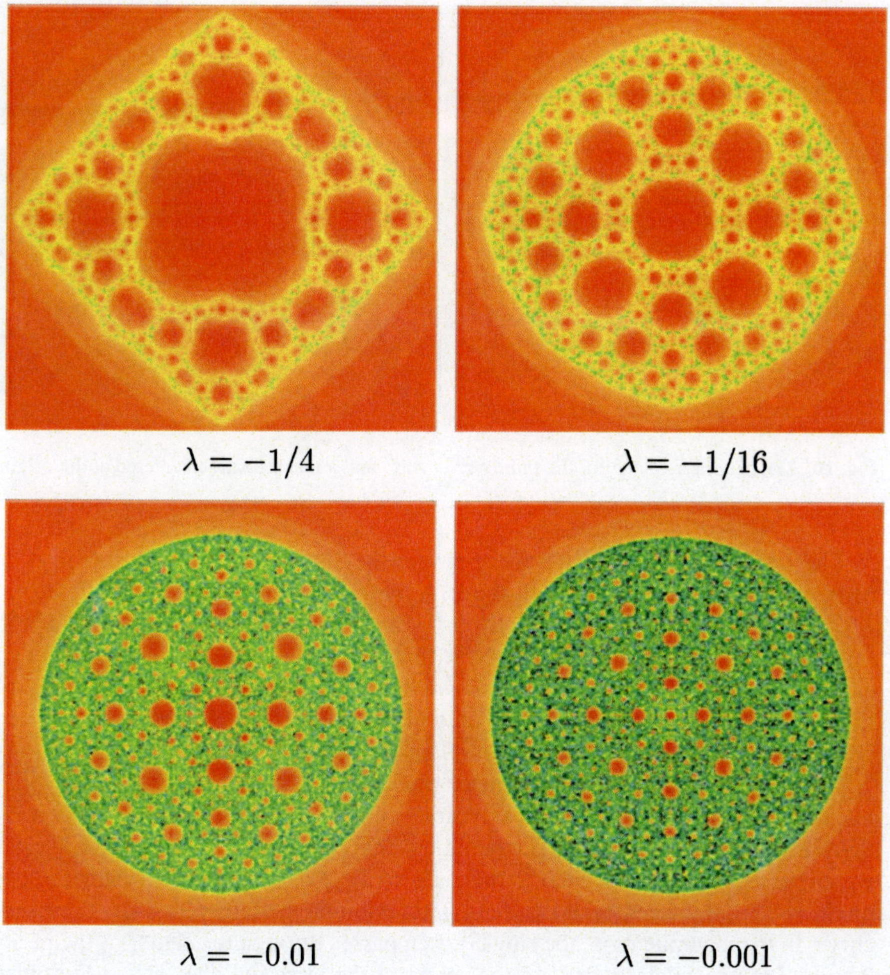

Fig. 9 Sierpinski curve Julia sets for various negative values of λ in the case $n = 2$

So, as $\lambda \to 0$, $F_\lambda(v_\lambda) \to 1/4$, which is nowhere near B_λ since, when $|\lambda|$ is small, the boundary of B_λ is close to the unit circle.

A second reason why the case $n = 2$ is different involves the Julia sets of the maps F_λ when $|\lambda|$ is small. When $n > 2$, these Julia sets are always Cantor sets of simple closed curves surrounding the origin. It is known [6] that there is a round annulus of some given width lying inside the unit circle and separating two of these curves when $|\lambda|$ is small. Hence, these Julia sets never converge to the unit disk as $\lambda \to 0$. However, when $n = 2$, it is also shown in [6] that the Julia sets for F_λ do converge to the closed unit disk as $\lambda \to 0$. In Fig. 9 we display four Julia sets with λ small and $n = 2$. All of these Julia sets are in fact Sierpinski curves. But notice how the preimages of T_λ get smaller and smaller as $|\lambda|$ decreases.

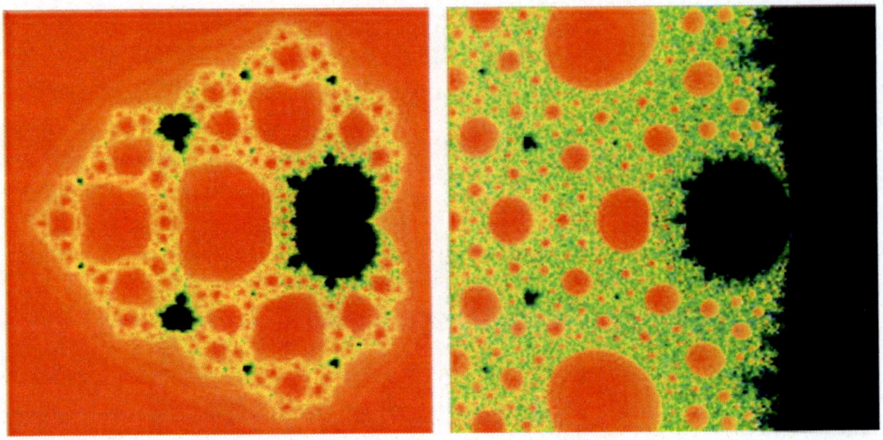

Fig. 10 The parameter plane for the family $z^2 + \lambda/z^2$ and a magnification centered at the origin

The final example of the difference between the cases $n = 2$ and $n > 2$ involves the Mandelpinski necklaces described above. As we showed earlier, when $n > 2$, the ring $\mathcal{S}^k$ passes alternately through exactly $(n-2)n^{k-1} + 1$ centers of baby Mandelbrot sets and centers of Sierpinski holes. Note that, when $n = 2$, this formula yields 1 for each k. And that, in fact, is true. As shown in [5], we do have these special rings $\mathcal{S}^k$ in this case. The single center of the only Mandelbrot set in $\mathcal{S}^k$ now lies along $\mathbb{R}^+$, while the single center of the corresponding Sierpinski hole lies in $\mathbb{R}^-$.

In Fig. 10 we display the parameter plane for the case $n = 2$ together with a magnification. The large red central region is not a McMullen domain; rather it is a Sierpinski hole and it does not contain the origin. The ring $\mathcal{S}^1$ is the dividing circle which passes through the center of the main cardioid of the principal Mandelbrot set on the right and the center of that large red region on the left, which is a Sierpinski curve. In the magnification, the ring $\mathcal{S}^2$ then passes through the center of the period 2 bulb of the Mandelbrot set and the center of the large red disk, also a Sierpinski hole, that lies to the left of the origin.

References

1. P. Blanchard, R.L. Devaney, D.M. Look, P. Seal, Y. Shapiro, Sierpinski Curve Julia sets and singular perturbations of complex polynomials. Ergodic Theory Dynam. Systems **25**, 1047–1055 (2005)
2. R.L. Devaney, Baby Mandelbrot sets adorned with Halos. Complex dynamics: twenty five years after the appearance of the Mandelbrot set. Contemporary Math. AMS **396**, 37–50 (2006)
3. R.L. Devaney, Structure of the McMullen domain in the parameter space of rational maps. Fundam. Math. **185**, 267–285 (2005)
4. R.L. Devaney, The McMullen domain: satellite Mandelbrot Sets and Sierpinski Holes. Conform. Geom. Dyn. **11**, 164–190 (2007)

5. R.L. Devaney, D. Cuzzocreo, Simple Mandelpinski necklaces for $z^2 + C/z^2$, in *Difference Equations, Discrete Dynamical Systems and Applications.*(Springer-Verlag, 2016), pp. 63–72
6. R.L. Devaney, A. Garijo, Julia sets converging to the unit disk. Proc. AMS **136**, 981–988 (2008)
7. R.L. Devaney, D.M. Look, A criterion for Sierpinski curve Julia sets. Topology Proceedings **30**, 163–179 (2006)
8. R.L. Devaney, D.M. Look, D. Uminsky, The escape trichotomy for singularly perturbed rational maps. Indiana U. Math. J **54**, 1621–1634 (2005)
9. R.L. Devaney, S. Marotta, The McMullen domain: rings around the boundary. Trans. AMS. **359**, 3251–3273 (2007)
10. A. Garijo, H. Jang, S. Marotta, Generalized rings around the mcmullen domain. *Qualitative Theory of Dynamical Systems*
11. H. Jang, S. Marotta, S. So, Generalized baby Mandelbrot sets adorned with halos in families of rational maps. J. Differ. Equations Appl. **23**, 503–520 (2017)
12. C. McMullen, Automorphisms of rational maps. *Holomorphic Functions and Moduli*, vol. 1 (Springer, New York, 1988); Math. Sci. Res. Inst. Publ. **10** (1988)
13. J. Milnor, A. Tan Lei, "Sierpinski Carpet" as Julia Set. Appendix F in geometry and dynamics of quadratic rational maps. Experiment. Math.**2**, 37–83 (1993)
14. G.T. Whyburn, Topological characterization of the Sierpinski curve. Fund. Math. **45**, 320–324 (1958)

Some Examples of Hypercyclic Operators and Universal Sequences of Operators

Kit C. Chan

Abstract Many examples of hypercyclicity take place in analytic function spaces, such as spaces of entire functions, Hardy spaces, Bergman spaces, and Dirichlet spaces. Using unique features of these analytic function spaces, we explore properties of some hypercyclic operators, such as spectral properties, orbital properties, as well as their hypercyclicity with respect to different topologies of the spaces.

Keywords Hypercyclic operators · Universal sequence of operators · Analytic function spaces

2010 Mathematics Subject Classification. Primary: 47A16 · Secondary: 47B37, 46B45

1 Introduction

Let X be an separable, infinite dimensional Fréchet space over $\mathbb{C}$. A continuous linear operator T is *hypercyclic* if there is a vector x for which the orbit $\text{orb}(T, x) = \{x, Tx, T^2x, T^3x, \ldots\}$ is dense in X. Such a vector x is called a *hypercyclic vector*. When we generalize the sequence of powers $\{T, T^2, T^3, \ldots\}$ in the orbit to a sequence of operators $\{T_1, T_2, T_3, \ldots\}$, we have the notion of universality. To be precise, we say that a sequence of continuous linear operators $T_n : X \to X$ is *universal* if there is a vector x for which $\{x, T_1x, T_2x, T_3x, \ldots\}$ is dense in X. Such a vector x is called a *universal vector*. In this teminology, the sequence $\{T^n\}$ of powers of T is universal if and only if T is *hypercyclic*.

Some early examples of hypercyclicity take place in the Fréchet space $H(\Omega) = \{f : \Omega \to \mathbb{C} | f \text{ is analytic}\}$, where Ω is a region in $\mathbb{C}$ and $H(\Omega)$ carries the compact-open topology. That is, a sequence $f_n \to f$ in $H(\Omega)$ if and only if $f_n \to f$ uniformly

K. C. Chan (✉)
Department of Mathematics and Statistics, Bowling Green State University, Bowling Green, OH 43403, USA
e-mail: kchan@bgsu.edu

R. L. Devaney et al. (eds.), *Topological Dynamics and Topological Data Analysis*, Springer Proceedings in Mathematics & Statistics 350,
https://doi.org/10.1007/978-981-16-0174-3_8

on compact subsets of Ω. Two well-known examples of hypercyclicity are the following results on the Fréchet space $H(\mathbb{C})$ of all entire functions.

Theorem 1.1 (Birkhoff [3]) *There is a function $f \in H(\mathbb{C})$ so that $\{f(z), f(z+1), f(z+2), \ldots\}$ is dense in $H(\mathbb{C})$. In other words, the translation operator $T : H(\mathbb{C}) \to H(\mathbb{C})$ defined by $Tf(z) = f(z+1)$ is hypercyclic.*

Theorem 1.2 (MacLane [13]) *There is a function $f \in H(\mathbb{C})$ so that the set of successive derivatives $\{f(z), f'(z), f''(z), \ldots\}$ is dense in $H(\mathbb{C})$. In other words, the differentiation operator $D : H(\mathbb{C}) \to H(\mathbb{C})$ defined by $Df(z) = f'(z)$ is hypercyclic.*

One early example of universality was given on the Fréchet space $H(\mathbb{D})$ for the open unit disk $\mathbb{D}$.

Theorem 1.3 (Seidel and Walsh [14]) *Suppose $\{a_n\} \subset \mathbb{D}$ with $a_n \to 1$, and*

$$\varphi_n(z) = \frac{a_n - z}{1 - \overline{a_n} z}.$$

Then there exists a function $f \in H(\mathbb{D})$ for which the set of non-Euclidean translates $\{f \circ \varphi_n\}$ is dense in $H(\mathbb{D})$. In other words, the sequence of composition operators $C_n : H(\mathbb{D}) \to H(\mathbb{D})$ given by $C_n f(z) = f \circ \varphi_n(z)$ is universal.

All of the above theorems concern analytic functions, which provide the setting for our discussion of hypercyclicity and universality.

2 Entire Functions

Putting the three examples of hypercyclicity and universality in the previous section into one setting, Gethner and Shapiro ([9]) obtained the following sufficient condition for a sequence of continuous linear operators on a Fréchet space X to be universal. This sufficient condition is now known as the *Universality Criterion.*

Theorem 2.1 (Gethner and Shapiro [9]) *For each integer $n \geq 1$, let $T_n : X \to X$ be a continuous linear operator on a separable, infinite dimensional Fréchet space X. The sequence $\{T_n\}$ is universal if there are dense subsets D_1 and D_2 of X and maps $S_n : X \to X$ such that*
(1) $T_n S_n =$ identity,
(2) $T_n x \to 0$ for each $x \in D_1$, and
(3) $S_n x \to 0$ for each $x \in D_2$.

Using Theorem 2.1, Gethner and Shapiro reproved the results of Birkhoff, MacLane, and Seidel and Walsh. We remark that if there is an operator T such that each T_n in Theorem 2.1 satisfies $T_n = T^n$, then the Universailty Criterion can be used to show that T is hypercyclic.

Continuing with Birkhoff and MacLane's results, Godefroy and Shapiro ([10]) showed that a continuous linear operator $L : H(\mathbb{C}) \to H(\mathbb{C})$ commutes with the differentiation $D : H(\mathbb{C}) \to H(\mathbb{C})$ if and only if it commutes with every translation $T_a : H(\mathbb{C}) \to H(\mathbb{C})$ given by $T_a g(z) = g(z + a)$. Furthermore, they provided the following result.

Theorem 2.2 (Godefroy and Shapiro [10]) *If $L : H(\mathbb{C}) \to H(\mathbb{C})$ is a continuous, linear, nonscalar operator that communtes with D, then L is hypercyclic.*

The set of all entire functions $H(\mathbb{C})$ cannot be a Hilbert space in a meaningful way. However, it is possible to give a dense linear manifold M of $H(\mathbb{C})$ a topology that makes M a Hilbert space of entire functions. If M is invariant under the translation operator T, then we can study the hypercyclicity of the translation operator on M. To proceed with this idea, we introduce the following definition given by Chan and Shapiro ([8]).

Let $\gamma = \{\gamma_n > 0 \mid n \geq 0\}$ be a sequence of positive numbers with $\frac{\gamma_{n+1}}{\gamma_n} \downarrow 0$. Let

$$E^2(\gamma) = \left\{ f(z) = \sum_0^\infty \hat{f}(n) z^n \;\middle|\; \|f\|_\gamma^2 = \sum_0^\infty \frac{|\hat{f}(n)|^2}{\gamma_n^2} < \infty \right\}$$

be a Hilbert space of entire functions. One can easily check that the norm topology of $E^2(\gamma)$ is stronger than the compact-open topology inherited from $H(\mathbb{C})$, and also that the differentiation operator $D : E^2(\gamma) \to E^2(\gamma)$ given by $Df = f'(z)$ is bounded if and only if $\{n\gamma_n/\gamma_{n-1}\}$ is a bounded sequence. In that case, the translation operator $T_a : E^2(\gamma) \to E^2(\gamma)$ given by $T_a(f) = f(z + a)$ is bounded. This follows from the observation that

$$e^{aD}(f)(z) = \sum_{n=0}^\infty \frac{a^n}{n!} D^n f(z) = \sum_{n=0}^\infty \frac{f^{(n)}(z)}{n!} ((a + z) - z)^n = f(z + a),$$

and hence,

$$T_a = e^{aD} = \sum_0^\infty a^n \frac{D^n}{n!} = I + D\left(\sum_1^\infty a^n \frac{D^{n-1}}{n!}\right).$$

One can check that $D : E^2(\gamma) \to E^2(\gamma)$ is compact if and only if $n\gamma_n/\gamma_{n-1} \to 0$, and in that case $T_a = I + K$, where K is a compact operator.

Theorem 2.3 (Chan and Shapiro [8]) *If the sequence $\{n\gamma_n/\gamma_{n-1}\}$ is monotonically decreasing and if $a \neq 0$, then $T_a : E^2(\gamma) \to E^2(\gamma)$ is hypercyclic.*

In the case that D is compact, T_a is the first natural example of a hypercyclic operator that is of the form $I + K$ where K is a compact operator, with the singleton spectrum $\sigma(T_a) = \{1\}$.

3 Hilbert Spaces of Analytic Functions

Let H be a Hilbert space of analytic functions on a region $\Omega \subset \mathbb{C}$ satisfying (a) $H \neq \{0\}$, and (b) For each $\omega \in \Omega$, the point evaluation functional $k_\omega : f \mapsto f(\omega)$ is continuous. An analytic function $\varphi : \Omega \to \mathbb{C}$ is a *multiplier* for H if $\varphi \cdot H \subset H$. Using the Closed Graph Theorem, one can show that the *multiplication operator* $M_\varphi : H \to H$ given by $M_\varphi(f) = \varphi f$ is a bounded linear operator.

Let $H^\infty(\Omega)$ be the algebra of all bounded analytic functions on Ω. We claim that every multiplier φ is in $H^\infty(\Omega)$. To prove that, we see that $|\varphi(\omega)| \cdot |\langle f, k_\omega\rangle| = |\varphi(\omega) f(\omega)| = |\langle \varphi f, k_\omega\rangle| \leq \|M_\varphi\| \|f\| \|k_\omega\|$. Putting $f = k_\omega / \|k_\omega\|$, we have $|\varphi(\omega)| \leq \|M_\varphi\|$ for all $\omega \in \Omega$. This proves our claim.

In fact, it is quite easy for the adjoint multiplication operator M_φ^* to be hypercyclic.

Theorem 3.1 (Godefroy and Shapiro [10]) *If φ is nonconstant and $\varphi(\Omega)$ intersects the unit circle, then $M_\varphi^* : H \to H$ is hypercyclic.*

The next result shows that hypercyclicity of the adjoint multiplication operator on a Hilbert space H of analytic functions can be completely determined under additional hypotheses.

Theorem 3.2 (Godefroy and Shapiro [10]) *Suppose every function φ in $H^\infty(\Omega)$ is a multiplier for H with $\|M_\varphi\| = \|\varphi\|_\infty$. If φ is a nonconstant multiplier, then M_φ^* is hypercyclic if and only if $\varphi(\Omega)$ intersects the unit circle.*

To illustrate the theorem, let

$$L_a^2(\Omega) = \left\{ f : \Omega \to \mathbb{C} \,\middle|\, f \text{ is analytic and } \|f\|^2 = \int_\Omega |f|^2 \, dA < \infty \right\}$$

be the *Bergman space*. Clearly every $\varphi \in H^\infty(\Omega)$ is a multiplier for $L_a^2(\Omega)$ with $\|M_\varphi\| = \|\varphi\|_\infty$. Thus it follows from Theorem 3.2 that the adjoint multiplication operator $M_\varphi^* : L_a^2(\Omega) \to L_a^2(\Omega)$ is hypercyclic if and only if $\varphi(\Omega)$ intersects the unit circle.

However, in the case that not every bounded analytic function is a multiplier, then the conclusion of Theorem 3.2 may not be true. For example, we consider the *Dirichlet space* for the open unit disk $\mathbb{D}$, which is given by

$$Dir(\mathbb{D}) = \left\{ f : \mathbb{D} \to \mathbb{C} \,\middle|\, f \text{ is analytic and } \|f\|^2 = |f(0)|^2 + \int_\mathbb{D} |f'|^2 \, dA < \infty \right\}.$$

Example 3.3 *(Chan and Seceleanu* [7]*, 2012)* Let $\varphi(z) = z$. The operator $M_\varphi^* : Dir(\mathbb{D}) \to Dir(\mathbb{D})$ is hypercyclic, but $\varphi(\mathbb{D}) = \mathbb{D}$, which does not intersect the unit circle.

Before we return to the Bergman space, we take a look at a hypercyclicity result for a general operator on a Banach space.

Theorem 3.4 (Bourdon and Feldman [4]) *For any bounded linear operator* $T : X \to X$ *on a separable, infinite dimensional Banach space* X*, an orbit* $orb(T, x)$ *is somewhere dense if and only if* $orb(T, x)$ *is everywhere dense.*

However, for the adjoint multiplication operator M_φ^* on the Bergman space $L_a^2(\Omega)$ to be hypercyclic, the equivalent condition of having a somewhere dense orbit in Theorem 3.4 can be relaxed.

Theorem 3.5 (Chan and Seceleanu [6]) *Let* φ *be a nonconstant function in* $H^\infty(\Omega)$. *For the adjoint multiplication operator* $M_\varphi^* : L_a^2(\Omega) \to L_a^2(\Omega)$*, the following statements are equivalent.*

(A) M_φ^* *is hypercyclic.*
(B) M_φ^* *has an orbit with a nonzero limit point.*
(C) M_φ^* *has an orbit* $orb(M_\varphi^*, f)$ *with infinitely many members* $M_\varphi^{*n} f$ *contained in an open ball whose closure avoids the origin.*

Theorem 3.5 does not hold true for all classes of operators on a Hilbert space H of analytic functions. To explain that, let $\mathbb{D}$ be the open unit disk, and let

$$H^2 = \left\{ f : \mathbb{D} \to \mathbb{D} \,\middle|\, f(z) = \sum_0^\infty a_n z^n \text{ analytic and } \sum_0^\infty |a_n|^2 < \infty \right\}$$

be the *Hardy space*. Let $\varphi : \mathbb{D} \to \mathbb{D}$ be an analytic function, and define the composition operator $C_\varphi : H^2 \to H^2$ by $C_\varphi f = f \circ \varphi$.

Example 3.6 *(Chan and Seceleanu* [6], *2012)* If α is an irrational number and $\varphi(z) = e^{2\pi i \alpha} z$,, then C_φ has an orbit with the identity function $\psi(z) \equiv z$ as a nonzero limit point, but C_φ is not hypercyclic.

Now we turn our attention to the weak topology of a separable, infinite dimensional Hilbert space H. The linear span of an orbit orb(T, x) is a convex set. So, it is dense with the weak topology if and only if it is dense with the norm topology. What about the orbit itself, without taking the linear span? In other words, must a weakly dense orbit be norm dense? To provide an answer for that question, we introduce the following definition. If there is a vector $h \in H$ such that its orbit orb(T, h) is dense with the weak topology of the Hilbert space H, then T is said to be *weakly hypercyclic*. To provide an example of weak hypercyclicity, let $A = \{1 < |z| < 2\}$ be the annulus with radii 1 and 2, centered at 0. The corresponding *Hardy space* $H^2(A)$ is given by $H^2(A) = \{f(z) = \sum_{-\infty}^\infty a_n z^n \mid \sum_{-\infty}^0 |a_n|^2 + \sum_1^\infty 2^{2n} |a_n|^2 < \infty\}$.

Theorem 3.7 (Chan and Sanders [5]) *The adjoint multiplication operator* $M_z^* : H^2(A) \to H^2(A)$ *is weakly hypercyclic but not hypercyclic.*

As an immediate corollary, we have the following unexpected result: *There is a norm increasing, and yet weakly dense sequence in a separable, infinite dimensional Hilbert space!*

To conclude this paper, we remark that universality can take place in a nonlinear setting. For example, $\overline{\text{Ball}}(H^\infty(\mathbb{D})$ carries no linear structure. Heins ([12]) showed that if $\{a_n : n \geq 1\} \subset \mathbb{D}$ with $a_n \to 1$, and

$$\varphi_n(z) = \frac{a_n - z}{1 - \overline{a_n} z},$$

then there exists a Blaschke product B such that the set

$$\{B \circ \varphi_1,\ B \circ \varphi_2,\ B \circ \varphi_3, \ldots\}$$

is dense in $\overline{\text{Ball}}(H^\infty(\mathbb{D}))$, with the compact-open topology. In other words, if T_n : $\overline{\text{Ball}}(H^\infty(\mathbb{D}) \to \overline{\text{Ball}}(H^\infty(\mathbb{D})$ is given by $T_n f = f \circ \varphi_n$. Then the sequence $\{T_n\}$ is *universal*. The Blaschke product B is a *universal element*.

Recently, many authors have obtained results related to Heins' results. For example, Aron and Gorkin ([1]), Bayart, Gorkin, Grivaux, and Mortini ([2]), and also Gorkin and Mortini ([11]).

References

1. R. Aron, P. Gorkin, An infinite dimensional vector space of universal functions for H^∞ of the ball. Canad. Math. Bull. **50**(2), 172–181 (2007)
2. F. Bayart, P. Gorkin, S. Grivaux, R. Mortini, Bounded universal functions for sequences of holomorphic self-maps of the disk. Ark. Mat. **47**(2), 205–229 (2009)
3. G.D. Birkhoff, Démonstration d'un théorème elementaire sur les fonctions entieres. C. R. Acad. Sci. Paris **189**, 473–475 (1929)
4. P.S. Bourdon, N.S. Feldman, Somewhere dense orbits are everywhere dense. Indiana Univ. Math. J. **52**, 811–819 (2003)
5. K.C. Chan, R. Sanders, A weakly hypercyclic operator that is not norm hypercyclic. J. Oper. Theory **52**(1), 39–59 (2004)
6. K.C. Chan, I. Seceleanu, Hypercyclicity of shifts as a zero-one law of orbital limit points. J. Oper. Theory **67**, 257–277 (2012)
7. K.C. Chan, I. Seceleanu, Orbital limit points and hypercyclicity of operators on analytic function spaces. Math. Proc. R. Irish Acad. **110A**(1), 99–109 (2010)
8. K.C. Chan, J.H. Shapiro, The cyclic behavior of translation operators on Hilbert spaces of entire functions. Indiana Univ. Math. J. **40**(4), 1421–1449 (1991)
9. R.M. Gethner, J.H. Shapiro, Universal vectors for operators on spaces of holomorphic functions. Proc. Amer. Math. Soc. **100**(2), 281–288 (1987)
10. G. Godefroy, J.H. Shapiro, Operators with dense, invariant, cyclic vector manifolds. J. Funct. Anal. **98**, 229–269 (1991)
11. P. Gorkin, R. Mortini, Universal Blaschke products. Math. Proc. Cambridge Philos. Soc. **136**(1), 175–184 (2004)
12. M. Heins, A universal Blaschke product. Arch. Math. (Basel) **6**, 41–44 (1954)
13. G.R. MacLane, Sequences of derivatives and normal families. J. Analyse Math. **2**, 72–87 (1952)
14. W.P. Seidel, J.L. Walsh, On approximation by Euclidean and non-Euclidean translates of an analytic function. Bull. Am. Math. Soc. **47**, 916–920 (1941)

Some Basic Properties of Hypercyclic Operators

Kit C. Chan

Abstract Using a few classical examples and the invariant subspace problem, we motivate the definition of a hypercyclic operator on a Banach space. We state a sufficient condition for an uncountable family of operators to have a dense G_δ set of common hypercyclic vectors. Then we exhibit a few examples of such uncountable families. Finally, we switch our focus to some results on extending of an operator defined on a Hilbert subspace to a hypercyclic operator on the whole Hilbert space.

Keywords A path of hypercyclic operators · Common hypercyclic vector · Hypercyclic extension

2010 Mathematics Subject Classification. Primary: 47A16 · Secondary: 47A58, 47B37, 47A05

1 Introduction

A continuous linear operator $T : X \to X$ on a separable, infinite-dimensional Fréchet space is said to be *hypercyclic* if there is a vector x in X for which the orbit $\text{orb}(T, x) = \{x, Tx, T^2x, T^3x, \ldots\}$ is dense in X. Such a vector x is called a hypercyclic vector. Two classical examples of hypercyclicity take place in the Fréchet space of all entire functions $H(\mathbb{C})$, which carries the compact-open topology. Thus, a sequence $\{f_n\}$ in $H(\mathbb{C})$ converges to a function f in $H(\mathbb{C})$ if and only if $f_n \to f$ uniformly on compact subsets of $\mathbb{C}$.

One of the two examples is due to Birkhoff ([4]) who showed that the translation operator $T : H(\mathbb{C}) \to H(\mathbb{C})$ given by $Tf(z) = f(z+1)$ on the Fréchet space $H(\mathbb{C})$ of all entire functions is hypercyclic. The other example is due to MacLane ([20]) who showed that the differentiation operator $D : H(\mathbb{C}) \to H(C)$ defined by $Df = f'$ is

K. C. Chan (✉)
Department of Mathematics and Statistics, Bowling Green State University,
Bowling Green, OH 43403, USA
e-mail: kchan@bgsu.edu

R. L. Devaney et al. (eds.), *Topological Dynamics and Topological Data Analysis*,
Springer Proceedings in Mathematics & Statistics 350,
https://doi.org/10.1007/978-981-16-0174-3_9

hypercyclic. The first Banach space example of hypercyclicity is due to Rolewicz ([21]) who showed that if $B : \ell^p \to \ell^p$, where $1 \le p < \infty$, is the unilateral backward shift defined by

$$B(a_0, a_1, a_2, \ldots) = (a_1, a_2, a_3, \ldots)$$

and if $t > 1$, then tB is hypercyclic.

Another motivation for studying hypercyclic operators comes from the well-known open problem, the Invariant Subspace Problem which has been around since 1900. The problem asks whether every bounded linear operator $T : H \to H$ on a separable, infinite-dimensional Hilbert space H has a nontrivial invariant closed subspace. The problem will be solved in the negative, if one shows that there is a bounded linear operator $T : H \to H$ for which every nonzero vector is a hypercyclic vector. In that case, T does not have a nontrivial invariant closed subspace, and indeed it does not even have a nontrivial invariant closed subset.

Of course not every operator is hypercyclic. Some examples of non-hypercyclic operators include normal operators, compact operators, and those operators of the form $T = I + F$ where F is a finite rank operator.

Kitai [19] offered a sufficient condition for a bounded linear operator $T : X \to X$ on a Banach space X to be hypercyclic. The condition was rediscovered in much greater generality by Gethner and Shapiro ([16]). The condition is now known as the Hypercyclicity Criterion.

Theorem 1.1 (Kitai [19], Gethner and Shapiro [16]) *A continuous linear operator $T : X \to X$ on a separable, infinite-dimensional Fréchet space X is hypercyclic if there is a dense subset D of X and if T has a right inverse S so that $T^n x \to 0$ and $S^n x \to 0$ for each vector $x \in D$.*

Using the Criterion, Gethner and Shapiro ([16]) reproved the aforementioned results of Birkhoff, MacLane, and Rolewicz. Since then the Hypercyclicity Criterion has been a basic tool for showing an operator is hypercyclic.

2 Common Hypercyclic Vectors

For a given countable dense subset $\{x_j : j \ge 1\}$ of a separable, infinite-dimensional Banach space X, one can easily check that the set of hypercyclic vectors $\mathcal{HC}(T)$ of a bounded linear operator $T : X \to X$ is given by

$$\mathcal{HC}(T) = \bigcap_{j,k=1}^{\infty} \bigcup_{n=1}^{\infty} T^{-n} B\left(x_j, \frac{1}{k}\right).$$

Since T is bounded, the union

$$\bigcup_{n=1}^{\infty} T^{-n} B\left(x_j, \frac{1}{k}\right)$$

is an open set. Observe that if the orbit $\{x, Tx, T^2x, \ldots\}$ is dense then every member $T^n x$ in the orbit is a hypercyclic vector. Thus, $\mathcal{HC}(T)$ is dense in X, and hence, $\mathcal{HC}(T)$ is a dense G_δ subset of X. Furthermore by the Baire Category Theorem, if $\{T_n : n \geq 1\}$ is a countable collection of hypercyclic operators, then

$$\text{the set of common hypercyclic vectors} = \bigcap_{n=1}^{\infty} \mathcal{HC}(T_n)$$

is again a dense G_δ set. What about an uncountable family of hypercyclic operators? Can their set of common hypercyclic vectors be a dense G_δ set? One such example in relation to Rolewicz' result that we have mentioned in Sect. 1 was given as follows.

Theorem 2.1 (Abakumov and Gordon [1]) *If B is the unilateral backward shift, then the set of common hypercyclic vectors* $\bigcap_{t>1} \mathcal{HC}(tB)$ *is a dense G_δ set.*

The above theorem was given a simpler proof, by introducing the concept of a path of operators. To give a definition, let $B(X) = \{T : X \to X | T \text{ is bounded and linear}\}$ be the operator algebra of a Banach space X, and let I be an interval of real numbers. The collection $\{T_t \in B(X) | t \in I\}$ is a *path of operators* if the map $t \mapsto T_t$ is continuous with the usual topology of $\mathbb{R}$ and the operator norm topology of $B(X)$. For example, if B is the unilateral backward shift, then tB with $t \in (1, \infty)$ is a path of operators.

Theorem 2.2 (Chan and Sanders [12]) *Suppose $\{T_t : X \to X | t \in [a, b]\}$ is a path of operators on a separable, infinite-dimensional Banach space X. Then*

$$\bigcap_{t \in [a,b]} \mathcal{HC}(T_t) \text{ is a dense } G_\delta \text{ set}$$

if and only if for each pair of nonempty open subsets U_1, U_2 of X, there exist a partition $P = \{a = t_0 < t_1 < t_2 < \cdots < t_k = b\}$ of $[a, b]$, positive integers $n_1, n_2, \ldots, n_k$, and a nonempty open set V such that $V \subset U_1$ and

$$T_t^{n_i}(V) \subset U_2, \text{ whenever } 1 \leq i \leq k \text{ and } t \in [t_{i-1}, t_i].$$

Applying Theorem 2.2, Chan and Sanders ([12]) reproved Theorem 2.1. Furthermore, they also use the concept of path of operators to obtain other results on shift operators. To explain that, let $1 \leq p < \infty$, A bounded linear operator $T : \ell^p \to \ell^p$ is said to be a *unilateral weighted backward shift*, if there is a bounded positive weight sequence $\{w_j : j \geq 1\}$ such that

$$T(a_0, a_1, a_2, \ldots) = (w_1 a_1, w_2 a_2, w_3 a_3, \ldots).$$

Note that if B is the unilateral backward shift, then the operators tB in Theorem 2.1 is a unilateral weighted backward shift with constant weight sequence $w_n = t$.

A bounded linear operator $T : \ell^p \to \ell^p$ is said to be a *bilateral weighted backward shift*, if there is a bounded positive weight sequence $\{w_j : -\infty < j < \infty\}$ such that

$$T(\ldots, a_{-1}, \overbrace{a_0}^{\text{zeroth}}, a_1, \ldots) = (\ldots, w_{-1}a_{-1}, w_0a_0, \overbrace{w_1a_1}^{\text{zeroth}}, w_2a_2, w_3a_3, \ldots).$$

Continuing with the concept of a path of operators, we have the following result for the shift operators in the operator algebra $B(\ell^p)$ of the Banach sequence space ℓ^p.

Theorem 2.3 (Chan and Sanders [12]) *Let $1 \leq p < \infty$. Between any two hypercyclic unilateral weighted backward shifts in $B(\ell^p)$, there is a path of such operators in $B(\ell^p)$ with a dense G_δ set of common hypercyclic vectors. Also, there is another path of such operators in $B(\ell^p)$ with no common hypercyclic vector.*

Immediately from Theorem 2.3, we see that the hypercyclic unilateral weighted backward shifts form a path-connected subset in the operator algebra $B(\ell^p)$.

Theorem 2.3 continues to hold true if we replace the unilateral weighted backward shifts by bilateral weighted shifts. These results lead to a natural question: Can we have "a lot" of operators in a path and yet they still have a dense G_δ set of common hypercyclic vectors? What do we mean by "a lot?" Before we look at the question, let us first quote the following result showing the existence of a hypercyclic operator.

Theorem 2.4 (Ansari [2], Bernal [3]) *For every separable, infinite-dimensional Banach space X, there is a hypercyclic operator T in $B(X)$.*

However, the Banach space X may not admit an operator of a more restrictive class. To explain that, we need the following definitions.

Definition 1 A vector $x \in X$ is said to be a *periodic point* of an operator T in $B(X)$ if there is a positive integer n such that $T^n x = x$. An operator T in $B(X)$ is said to be *chaotic* if it is hypercyclic and has a dense set of periodic points.

Unlike the result in Theorem 2.4, we cannot assume that we can always have a chaotic operator on a Banach space.

Theorem 2.5 (Bonet et al. [5]) *There is a separable, infinite-dimensional Banach space which admits no chaotic operator.*

In relation to Theorems 2.4 and 2.5, we have the following results about the density of hypercyclic and chaotic operators. Here, we use "SOT" to denote the strong operator topology of the operator algebra $B(X)$.

Theorem 2.6 (Chan [7]) *For a separable, infinite-dimensional Hilbert space H, the hypercyclic operators on H are* SOT-*dense in $B(H)$.*

Instead of a Hilbert space H in the above theorem, we have the following result for the case of a Banach space X.

Theorem 2.7 (Bès and Chan [6]) *The set of chaotic operators on a separable, infinite-dimensional Banach space X is either empty or* SOT*-dense in $B(X)$.*

Indeed, if $T \in B(X)$ is hypercyclic, then its similarity orbit $\{A^{-1}TA : A$ invertible on $X\}$ is SOT-dense in $B(X)$. In the case that X is a Hilbert space H over $\mathbb{C}$, Theorem 2.7 states that the chaotic operators are SOT-dense in $B(H)$. Using these results, we can continue our discussion on common hypercyclic vectors.

Theorem 2.8 (Chan and Sanders [13]) *There is a path of chaotic operators in $B(H)$ that is* SOT*-dense in $B(H)$, and each operator of the path has the exact same set $\mathcal{G}$ of hypercyclic vectors.*

It is further shown in [13] that there is such a path of which each operator satisfies the Hypercyclicity Criterion. From Theorem 2.8, we immediately have the following result.

Corollary 2.9 *The hypercyclic operators in $B(H)$ are* SOT*-connected. The chaotic operators in $B(H)$ are* SOT*-connected.*

From Theorem 2.8 we also have the following fact about the set $\mathcal{G}$ of hypercyclic vectors in the statement of the theorem: *Hypercyclic operators T in $B(H)$ with a set of common hypercyclic vectors $\mathcal{G}$ are* SOT*-connected.* In light of Theorem 2.5, Theorem 2.8 cannot hold true for any separable, infinite-dimensional Banach space X. However, in that case, we can offer the following result for the similarity orbit $\mathcal{S}(T) = \{A^{-1}TA \,|\, A : X \to X \text{ is invertible}\}$ of a hypercyclic operator $T : X \to X$.

Theorem 2.10 (Chan and Sanders [13]) *Let $T : X \to X$ be a hypercyclic operator on a separable, infinite-dimensional Banach space X. The similarity orbit $\mathcal{S}(T)$ contains a path $\mathcal{P}$ of operators which is* SOT*-dense in $B(X)$ and the set of common hypercyclic vectors $\bigcap_{T\in\mathcal{P}} \mathcal{HC}(T)$ for $\mathcal{P}$ is a dense G_δ set.*

It is easy to see from the definition of $\mathcal{S}(T)$ that we have the following remarks.

(1) If $\mathcal{HC}(T) = X \setminus \{0\}$, the set of common hypercyclic vectors for $\mathcal{S}(T)$ is also $X \setminus \{0\}$.
(2) If $\mathcal{HC}(T) \neq X \setminus \{0\}$, the set of common hypercyclic vectors for $\mathcal{S}(T)$ is empty.

Since we do not know whether there is a bounded linear operator $T : H \to H$ on a separable, infinite-dimensional Hilbert space H such that $\mathcal{HC}(T) = H \setminus \{0\}$, Remark (1) above may or may not make sense in the Hilbert space case.

For an operator T on H, we let $\mathcal{U}(T) = \{U^{-1}TU | U : H \to H \text{ is unitary}\}$ be the unitary orbit of T. Since the set of all unitary operators on H is a path-connected subset of $B(H)$, the unitary orbit $\mathcal{U}(T)$ is path-connected. Every operator in $\mathcal{U}(T)$ has the same norm as T, and so $\mathcal{U}(T)$ does not contain a path that is SOT-dense in $B(H)$. However, we can offer the following result for their common hypercyclic vectors.

Theorem 2.11 (Chan and Sanders [15]) *If $T \in B(H)$ is hypercyclic, then $\mathcal{U}(T)$ contains a path $\mathcal{P}$ of operators so that $\overline{\mathcal{P}}^{\text{SOT}}$ contains $\mathcal{U}(T)$ and the set of common hypercyclic vectors $\bigcap_{T \in \mathcal{P}} \mathcal{HC}(T)$ for $\mathcal{P}$ is a dense G_δ set.*

Observe that between any two unit vectors in H, there is a unitary U that takes one vector to the other. Thus, if $\mathcal{HC}(T) \neq H \setminus \{0\}$, then the set of common hypercyclic vectors for $\mathcal{U}(T)$ is empty, same as Remark (2) above.

3 Hypercyclic Extension

We begin this section with an observation that if M is a closed subspace of a separable, infinite-dimensional Hilbert space H with $\dim H/M < \infty$, then no bounded linear operator $A : M \to M$ can have an extension $T : H \to H$ that is hypercyclic. To prove that by way of contradiction, suppose $T \in B(H)$ is a hypercyclic extension of A. Let $\pi : H \to H/M$ be the quotient map; that is,

$$\pi(f) = [f] = f + M.$$

If h is a hypercyclic vector for T, then the set

$$\pi\{h, Th, T^2h, \ldots\} = \{[h], [Th], [T^2h], \ldots\} \text{ is dense in } H/M.$$

If $S : H/M \to H/M$ is the linear map defined by $S[x] = [Tx]$, then S is a hypercyclic operator on a finite dimensional space H/M, but that is impossible. However, if M is a closed subspace with infinite codimension then we have the following hypercyclic extension result.

Theorem 3.1 (Grivaux [17]) *If $\dim H/M = \infty$, then every operator $A \in B(M)$ has a chaotic extension $T \in B(H)$; that is, a chaotic operator $T : H \to H$ for which $T|_M = A$.*

If A is one-one and has closed range, then the chaotic extension T in Theorem 3.1 can be chosen to be one-one. However when A is not one-one, such an extension can be chosen to preserve the kernel of A. Indeed we have the following result.

Theorem 3.2 (Chan and Kadel [10]) *If $\dim H/M = \infty$, and A in $B(M)$ has closed range, then A has a right invertible chaotic extension T in $B(H)$ with* $\ker A = \ker T$.

To explain what the extension T in the above theorem looks like, we write $M = \operatorname{ran} A \oplus \operatorname{ran} A^\perp$. Let $M_0, M_1, M_2, \ldots$ be orthogonal subspaces of $M^\perp$, each of which is isomorphic to M. Identify M_0 with the original subspace M. Furthermore, let $M_{-1}, M_{-2}, \ldots$ be orthogonal subspaces of $M^\perp$, each of which is isomorphic to $\operatorname{ran} A$ so that

$$H = \cdots \oplus M_{-2} \oplus M_{-1} \oplus M_0 \oplus M_1 \oplus M_2 \oplus \cdots.$$

Since the restriction $A|_{\ker A^\perp} : \ker A^\perp \to \operatorname{ran} A$ is invertible, there is a bounded linear operator $B : \operatorname{ran} A \to \ker A^\perp$ such that $AB = I$ on $\operatorname{ran} A$.

An chaotic extension T is given by

$$Th = \left(\cdots, \frac{1}{\alpha} h_{-2}, \frac{1}{\alpha} h_{-1}, \alpha h_1', \overbrace{Ah_0 + \alpha h_1}^{\text{zeroth position}}, \alpha h_2, \alpha h_3, \cdots \right),$$

where $\alpha > 1$ and h_1' is the orthogonal component of h_1 in $\operatorname{ran} A$. Then a right inverse S of T is given by

$$Sh = \left(\cdots, \alpha h_{-3}, \alpha h_{-2}, \overbrace{B(h_0' - h_{-1})}^{\text{zeroth position}}, \frac{1}{\alpha}(h_{-1} \oplus h_0''), \frac{1}{\alpha} h_1, \frac{1}{\alpha} h_2, \cdots \right),$$

where h_0' and h_0'' are orthogonal components of h_0 in $\operatorname{ran} A$ and $\operatorname{ran} A^\perp$, respectively.

After explaining what the extension T looks like, we review the statement of Theorem 3.2 and obtain the following two corollaries.

Corollary 3.3 *Suppose* $\dim H/M = \infty$. *An operator* $A \in B(M)$ *has an invertible chaotic extension* $T \in B(H)$ *if and only if* A *is bounded below.*

The property of invertibility in the above corollary naturally raises the question about Fredholm operators.

Corollary 3.4 *An operator* $A \in B(M)$ *has a chaotic Fredholm extension* $T \in B(H)$ *if and only if* A *is left semi-Fredholm. Moreover,* $\operatorname{ind} T \geq \operatorname{ind} A$.

Another property of the hypercyclic extension we study is dual hypercyclicity. For the definition, a bounded linear operator $T : H \to H$ is said to be *dual hypercyclic*, if both T and T^* are hypercyclic. Herrero ([18]) asked whether an operator can be dual hypercyclic. The question was answered in the positive by Salas ([22]). In this direction, we can offer the following result.

Theorem 3.5 (Chan [8]) *Let* M *be a closed subspace of* H *with* $\dim H/M = \infty$, *and* $P : H \to H$ *be the orthogonal projection onto* M. *For any operator* $A \in B(M)$, *there exists an operator* $T \in B(H)$ *such that*

(1) T *is dual hypercyclic,*
(2) $PTP|_M = A$,
(3) $PT^*P|_M = A^*$.

In general, we can only get A to be the compression $PTP|_M$ of a dual hypercyclic operator T as stated in the above result, but not a restriction $T|_M$ of a dual hypercyclic operator T. However, such an extension exists when A^* is hypercyclic.

Theorem 3.6 (Chan and Kadel [9]) *Suppose* $\dim H/M = \infty$. *An operator* $A \in B(M)$ *has a dual hypercyclic extension* $T \in B(H)$ *if and only if* A^* *is hypercyclic.*

To explain the "only if" part of the theorem, suppose h is a hypercyclic vector for T^*. Write $h = f + g$, where $f \in M$ and $g \in M^\perp$. Since $TM \subset M$, we have $T^*M^\perp \subset M^\perp$. Also $A^{*n} = PT^{*n}|_M$, where $P : H \to H$ is the orthogonal projection onto M. Thus, $T^{*n}h = T^{*n}f + T^{*n}g = A^{*n}f + g_n$, where $g_n \in M^\perp$. Hence, $f \in M$ is a hypercyclic vector for A^*.

The above theorems and corollaries show how an operator $A : M \to M$ on a Hilbert subspace M can be extended to a hypercyclic operator $T : H \to H$. What about operators $A : M \to H$? Can we extend A to hypercyclic operator on H? This does not seem to be possible, particularly when A is onto H. Nevertheless, we have the following counterintuitive result.

Theorem 3.7 (Chan and Pinheiro [11]) *Suppose* $\dim H/M = \infty$. *Every bounded linear operator* $A : M \to H$ *has a chaotic extension* $T : H \to H$.

For a countable collection of operators $\{A_n : n \geq 1\}$ in $B(M, H)$, we can take the point of view that $A_n : H \to H$ with $A_n = 0$ on $M^\perp$. Does there exist one operator $V : M^\perp \to H$ such that each operator $A_n + V : H \to H$ is chaotic, taking the point of view that $V = 0$ on M? It was proved in [11] that such a bounded linear operator $V : M^\perp \to H$ exists provided that $\{A_n\}$ is uniformly bounded.

References

1. E. Abakumov, J. Gordon, Common hypercyclic vectors for multiples of backward shift. J. Funct. Anal. **200**, 494–504 (2003)
2. S.I. Ansari, Existence of hypercyclic operators on topological vector spaces. J. Funct. Anal. **148**, 384–390 (1997)
3. L. Bernal, On hypercyclic operators on Banach spaces. Proc. Am. Math. Soc. **127**, 1003–1010 (1999)
4. G.D. Birkhoff, Démonstration d'un théorème elementaire sur les fonctions entieres. C. R. Acad. Sci. Paris **189**, 473–475 (1929)
5. J. Bonet, F. Martiínez-Giménez, A. Peris, A Banach space which admits no chaotic operators. Bull. London Math. Soc. **33**, 196–198 (2001)
6. J. Bès, K.C. Chan, Denseness of hypercyclic operators on a Fréchet space. Houston J. Math. **29**, 195–206 (2003)
7. K.C. Chan, The density of hypercyclic operators on a Hilbert space. J. Operator Theory **47**, 131–143 (2002)
8. K.C. Chan, Prescribed compressions of dual hypercyclic operators. Proc. Am. Math. Soc. **140**, 3133–3143 (2012)
9. K.C. Chan, G.R. Kadel, Dual hypercyclic extension for an operator on a Hilbert subspace. Houston J. Math. **41**, 1221–1256 (2015)
10. K.C. Chan, G.R. Kadel, Restrictions of an invertible chaotic operator to its invariant subspace. J. Math Anal. Appl. **409**, 996–1004 (2014)
11. K.C. Chan, L. Pinheiro, Simultaneous chaotic extensions for general operators on a Hilbert subspace (preprint, 2016)
12. K.C. Chan, R. Sanders, Two criteria for a path of operators to have common hypercyclic vectors. J. Operator Theory **61**, 191–223 (2009)
13. K.C. Chan, R. Sanders, An SOT-dense path of chaotic operators with same hypercyclic vectors. J. Operator Theory **66**, 107–124 (2011)

14. K.C. Chan, R. Sanders, Common hypercyclic vectors for the conjugate class of a hypercyclic operator. J. Math. Anal. Appl. **375**, 139–148 (2011)
15. K.C. Chan, R. Sanders, Common hypercyclic vectors for the unitary orbit of a hypercyclic operator. J. Math. Anal. Appl. **387**, 17–23 (2012)
16. R.M. Gethner, J.H. Shapiro, Universal vectors for operators on spaces of holomorphic functions. Proc. Am. Math. Soc. **100**, 281–288 (1987)
17. S. Grivaux, Topologically transitive extensions of bounded operators. Math. Z. **249**, 85–96 (2005)
18. D.A. Herrero, Limits of hypercyclic and supercyclic operators. J. Funct. Anal. **99**, 179–190 (1991)
19. C. Kitai, Invariant closed sets for linear operators. Ph. D. Thesis, University of Toronto (1982)
20. G.R. MacLane, Sequences of derivatives and normal families. J. Analyse Math. **2**, 72–87 (1952)
21. S. Rolewicz, On orbits of elements. Studia Math. **32**, 17–22 (1969)
22. H. Salas, A hypercyclic operator whose adjoint is also hypercyclic. Proc. Am. Math. Soc. **112**, 765–770 (1991)

The Testing Ground of Weighted Shift Operators for Hypercyclicity

Kit C. Chan

Abstract We explore the hypercyclicity of unilateral weighted backward shifts and bilateral weighted shifts on ℓ^p, where $1 \leq p \leq \infty$, with the weak or weak-star topologies. Then, we turn our attention to see how a nonzero limit point of an orbit of such an operator determines the hypercyclicity of the operator. Lastly, we explore a recent result that a unilateral weighted backward shift can be factored as the product of two hypercyclic shifts.

Keywords Unilateral weighted backward shift · Bilateral shifts · Hypercyclic vector · Weak topology · Weak-star topology

2010 Mathematics Subject Classification. Primary: 47A16 · Secondary: 47B37, 46A45, 46A03, 46B26

1 Introduction

Let X be a separable, infinite dimensional Banach space over $\mathbb{C}$, and $T : X \to X$ be a bounded linear operator. The *orbit* of T with respect to a vector x is $\text{orb}(T, x) = \{x, Tx, T^2x, \ldots\}$. The operator T is *hypercyclic* if there is an orbit $\text{orb}(T, x)$ that is dense in X. Such a vector x is called a *hypercyclic vector*. One property of an operator T that is weaker than hypercyclicity is called supercyclicity. An operator T is *supercyclic* if there is a vector x such that $\mathbb{C} \cdot \text{orb}(T, x) = \{\alpha T^n x : n \geq 0, \alpha \in \mathbb{C}\}$ is dense in X. Such a vector x is called a *supercyclic vector*. Another property weaker than supercyclicity is called cyclicity. The operator T is *cyclic* if there is a vector x such that the linear span of its orbit

$$\text{span}\,\text{orb}(T, x) = \text{span}\{x, Tx, T^2x, \ldots\} = \{p(T)x : p \text{ polynomial}\}$$

K. C. Chan (✉)
Department of Mathematics and Statistics, Bowling Green State University, Bowling Green, OH 43403, USA
e-mail: kchan@bgsu.edu

R. L. Devaney et al. (eds.), *Topological Dynamics and Topological Data Analysis*, Springer Proceedings in Mathematics & Statistics 350,
https://doi.org/10.1007/978-981-16-0174-3_10

is dense in X. Such a vector x is called a *cyclic vector*.

In this paper, we focus on the shift operators on ℓ^p, where $1 \le p \le +\infty$. An operator $T : \ell^p \to \ell^p$, is a *unilateral weighted backward shift* if there is a bounded positive weight sequence $\{w_j : j \ge 1\}$ such that $T(a_0, a_1, a_2, \ldots) = (w_1a_1, w_2a_2, w_3a_3, \ldots)$.

If we represent every ℓ^p sequence as a two-sided sequence, then a bounded linear operator $T : \ell^p \to \ell^p$ is a *bilateral weighted backward shift* if there is a bounded two-sided positive weight sequence $\{w_j : -\infty < j < \infty\}$ such that

$$T(\ldots, a_{-1}, \overbrace{a_0}^{\text{zeroth}}, a_1, \ldots) = (\ldots, w_{-1}a_{-1}, w_0a_0, \overbrace{w_1a_1}^{\text{zeroth}}, w_2a_2, \ldots).$$

For the above two shift operators to be hypercyclic, we have the following equivalent conditions in terms of their weight sequences.

Theorem 1.1 (Salas [12]) *Suppose* $1 \le p < \infty$. *We have the following necessary and sufficient conditions for hypercyclicity.*
(1) A unilateral weighted backward shift T on ℓ^p *is hypercyclic iff* $\sup\{w_1w_2\ldots w_n : n \ge 1\} = \infty$.
(2) A bilateral weighted backward shift T on ℓ^p *is hypercyclic iff for any* $\epsilon > 0$, *and* $q \in \mathbb{N}$, *there is an arbitrarily large n such that whenever* $|k| \le q$,

$$\prod_{j=1}^{n} w_{k+j} > \frac{1}{\epsilon} \quad \textit{and} \quad \prod_{j=0}^{n-1} w_{k-j} < \epsilon.$$

To explain why Statement (1) in the above theorem holds true, suppose $x = (a_0, a_1, a_2, \ldots)$ is a hypercyclic vector for T. Then, there is sequence of positive integers $\{n_k\}$ such that $T^{n_k}x \to (1, 0, 0, \ldots)$. Thus, $w_1w_2\ldots w_{n_k}a_{n_k} \to 1$. Since x is a $p-$summable sequence, we have $a_{n_k} \to 0$. Thus, $w_1w_2\ldots w_{n_k} \to \infty$.

Conversely, suppose there is a sequence $\{n_k\}$ of positive integers such that $w_1w_2\ldots w_{n_k} \to \infty$. We now see how to construct a vector x so that there is subsequence $\{n_{k_i}\}$ with $T^{n_{k_i}}x \to (1, 1, 0, 0, 0, \ldots)$.

Since $\{w_n\}$ is bounded, we have $w_1w_2\ldots w_{-1+n_k} \to \infty$, and so we can select a subsequence $\{n_{k_i}\}$, and construct $\{a_{n_{k_i}}\}$ and $\{a_{-1+n_{k_i}}\}$ so that $w_1w_2\ldots w_{-1+n_{k_i}} a_{-1+n_{k_i}} \to 1$ and $w_2\ldots w_{n_{k_i}}a_{n_{k_i}} \to 1$. Hence, we can construct a vector x so that $T^{-1+n_{k_i}}x \to (1, 1, 0, 0, 0, \ldots)$. Instead of $(1, 1, 0, 0, 0, \ldots)$ one use the above argument on any ℓ^p sequence with finite number of nonzero rational entries, and then, carefully construct a hypercyclic vector x in ℓ^p.

The above theorem provides us with a way to see whether a unilateral weighted backward shift or a bilateral weighted backward shift is hypercyclic. In the rest of the paper, we explore other aspects of hypercyclicity of these two types of shift operators.

2 Weak Topologies

Every Banach space naturally carries the weak topology, which is weaker than the norm topology as its name suggests. Thus, an orbit $\operatorname{orb}(T, x)$ may be dense with the weak topology, without being dense with the norm topology. This leads to the following definitions: If there is an orbit $\operatorname{orb}(T, x)$ that is dense with the weak topology of X, then T is said to be *weakly hypercyclic*. Similarly, if there is an orbit $\operatorname{orb}(T, x)$ that is dense with the weak-star topology of X, then T is said to be *weak-star hypercyclic*. A subset E of X is *weakly sequentially dense* (*resp. weak-star sequentially dense*) if for each vector y in X, there is a sequence in E converging to y with the weak topology (resp. weak-star topology). If there is an orbit $\operatorname{orb}(T, x)$ that is sequentially dense with the weak topology (resp. weak-star topology) of X, then T is said to be *weakly sequentially hypercyclic* (resp. weak-star sequentially hypercyclic).

Obviously, if an orbit $\operatorname{orb}(T, x)$ is dense with the norm topology, then it is dense with the weak topology. Thus, every hypercyclic operator is weakly hypercyclic. However, a convex subset of a Banach space is closed with the weak topology if and only if it is closed with the norm topology. Thus, the linear span of an orbit is weakly dense if and only if it is norm dense. In other words, an operator T is cyclic if and only if T is weakly cyclic. This naturally leads to the question whether T is hypercyclic if T is weakly hypercyclic.

Theorem 2.1 (Chan and Sanders [5]) *The following bilateral weighted backward shift $T : \ell^2 \to \ell^2$ is weakly hypercyclic but not hypercyclic.*

$$T(\ldots, a_{-2}, a_{-1}, \overbrace{a_0}^{zeroth}, a_1, a_2 \ldots) = (\ldots, a_{-2}, a_{-1}, a_0, \overbrace{2a_1}^{zeroth}, 2a_2, 2a_3, \ldots).$$

If $x = (\ldots, a_{-2}, a_{-1}, \overbrace{a_0}^{\text{zeroth}}, a_1, a_2 \ldots)$ is a weakly hypercyclic vector for T in Theorem 2.1, then there are infinitely many entries a_n with positive n such that $a_n \neq 0$. Thus, we have the following corollary.

Corollary 2.2 *There is a strictly norm-increasing, and yet weakly dense orbit in ℓ^2!*

It was proved by Kitai ([11]) that if $T : X \to X$ on a Banach space X is hypercyclic and invertible, then its inverse T^{-1} is also hypercyclic. However, the example of a weakly hypercyclic operator T in Theorem 2.1 shows that Kitai's result does not extend to weak hypercyclicity, because $\|T^{-1}\| = 1$.

Corollary 2.3 *There exists a bounded linear operator T that is weakly hypercylic and invertible, but its inverse T^{-1} is not weakly hypercylic.*

If a sequence in a Banach space is weakly convergent, then the sequence is bounded. Thus, the following corollary follows immediately from Corollary 2.2.

Corollary 2.4 *There exists a weakly hypercyclic operator $T : \ell^2 \to \ell^2$ that is not weakly sequentially hypercyclic.*

Before we turn our attention away from Theorem 2.1, we remark that in the case of ℓ^2, its weak topology coincides with its weak-star topology. Thus Theorem 2.1 and Corollaries 2.2, 2.3, 2.4 continue to hold true if we replace the weak topology in their statements by the weak-star topology.

It is no accident that the operator T in Theorem 2.1 is a bilateral shift, because the property stated in Theorem 2.1 is not shared by a unilateral weighted backward shfit.

Theorem 2.5 (Chan and Sanders [5]) *A unilateral weighted backward shift $T : \ell^2 \to \ell^2$ is weakly hypercyclic if and only if it is hypercyclic.*

To explain why the theorem holds true, take $e_0 = (1, 0, 0, 0, \ldots)$ and suppose $x \in \ell^2$ with $|\langle T^{n_k}x - e_0, e_0\rangle| < 1/k$. Then, the weight sequence $\{w_n\}$ of T satisfies $w_1 w_2 \cdots w_{n_k} \to \infty$. Thus, T is hypercyclic, by Theorem 1.1.

The next question about weak topology is whether every weakly supercyclic operator is indeed supercyclic. Before we look into the question, we quote one supercyclicity result for a unilateral weighted backward shift.

Theorem 2.6 (Hilden and Wallen [10]) *Let $1 \le p < \infty$. Every unilateral weighted backward shift $T : \ell^p \to \ell^p$ is supercyclic.*

To explain why the theorem holds true, let $D = \{(a_0, a_1, \ldots, a_n, 0, 0, 0, \ldots) : n \ge 0$ and $a_0, a_1, \ldots a_n \in \mathbb{Q}[i]\}$ be a countable dense subset of ℓ^p. Enumerate D as $D = \{d_1, d_2, d_3, \ldots\}$. Let $F : D \to D$ be the unilateral weighted forward shift defined by $F(a_0, a_1, a_2, \ldots) = (0, w_1^{-1}a_0, w_2^{-1}a_1, w_3^{-1}a_2, \ldots)$. Thus, $TF = \text{Id}$ on D. Using this, we can create a supercyclic vector x of the form $x = \sum \alpha_k F^{n_k} d_k$, by choosing large enough integers n_k and small enough positive α_k.

Even though Theorem 2.6 shows that supercyclicity is automatic for every unilateral weighted backward shift. We have the following result for weak supercyclicity.

Theorem 2.7 (Sanders [13]) *There exists a bounded linear operator that is weakly supercyclic, but not supercyclic.*

For a given formula $T(a_0, a_1, a_2, \ldots) = (w_1 a_1, w_2 a_2, \ldots)$ of a unilateral weighted backward shift T with a bounded positive weight sequence $\{w_j\}$, we can consider T as a bounded linear operator on any ℓ^p with $1 \le p < \infty$ or $p = \infty$. In addition, T also defines a bounded linear operator on a closed subspace c_0 of ℓ^∞, where $c_0 = \{(a_0, a_1, a_2, \ldots) : a_n \to 0\}$. Since ℓ^∞ is not separable with its norm topology, we can only study hypercyclicity of T on ℓ^∞ with the separable weak-star topology.

Theorem 2.8 (Bès, Chan and Sanders [2]) *Let $1 \le p < \infty$ and $\{w_j : j \ge 1\}$ be a bounded sequence of positive weights. Suppose T is a unilateral weighted backward shift defined by $T(a_0, a_1, a_2, \ldots) = (w_1 a_1, w_2 a_2, \ldots)$. Then, the following statements are equivalent.*

- *T is weak-star hypercyclic on ℓ^∞.*
- *T is weak-star sequentially hypercyclic on ℓ^∞.*

- *T is hypercyclic on ℓ^p.*
- *T is weakly sequentially hypercyclic on ℓ^p.*
- *T is weakly hypercyclic on ℓ^p.*
- *T is weak-star sequentially hypercyclic on ℓ^p.*
- *T is hypercyclic on c_0.*
- *T is weakly sequentially hypercyclic on c_0.*
- *T is weakly hypercyclic on c_0.*
- $\sup\{w_1 w_2 \dots w_n : n \geq 1\} = \infty$ *(Salas [12]).*

When we switch our focus to bilateral shifts, we have the following two results.

Theorem 2.9 (Bès et al. [1]) *For a bilateral weighted backward shift T on ℓ^p with $1 \leq p < \infty$, we have the following two statements.*

(1) T is weakly sequentially hypercyclic iff T is hypercyclic.
(2) T is weakly sequentially supercyclic iff T is supercyclic.

The above theorem cannot hold true for ℓ^∞ because it is not separable with the norm topology. Nevertheless we have the following result for ℓ^∞.

Proposition 2.10 (Bès et al. [1]) *There exists a weak-star hypercyclic bilateral weighted backward shift $T : \ell^\infty \to \ell^\infty$ that is not weak-star sequentially hypercyclic.*

We return to case when $1 \leq p < \infty$. Let $T : \ell^p \to \ell^p$ be a unilateral weighted backward shift or a bilateral weighted backward shift. Comparing to the hypercyclicity of T with the norm topology of ℓ^p, we conclude from the above results that hypercyclicity of T with the weak or weak-star topology can be totally different in some ways, but can also be the same in some other ways.

3 Orbital Limit Points

The definition for a hypercyclic operator $T : X \to X$ on a Banach space X requires it to have a dense orbit $\text{orb}(T, x)$. However, if we know that the closure of an orbit $\text{orb}(T, x)$ contains an open set, then the closure is actually the whole space X.

Theorem 3.1 (Bourdon and Feldman [3]) *If an orbit $\text{orb}(T, x)$ is somewhere dense in a Banach space X then the orbit $\text{orb}(T, x)$ is everywhere dense.*

However, if we know that T is a unilateral weighted backward shift or a bilateral weighted backward shift, then the condition for hypercyclicity in Theorem 3.1 can be further relaxed.

Theorem 3.2 (Chan and Seceleanu [7]) *Let $1 \leq p < \infty$ and $T : \ell^p \to \ell^p$ be a unilateral weighted backward shift. The following statements are equivalent:*

(A) T is hypercyclic.
(B) There is a vector x whose orbit orb(T, x) has a nonzero limit point
(C) There is a vector x whose orbit orb(T, x) has a nonzero weak limit point.
(D) There is a vector x whose orbit orb(T, x) has infinitely many members $T^n x$ contained in an open ball whose closure avoids the origin.

For the case of a bilateral weighted backward shift $T : \ell^p \to \ell^p$, conditions (A), (B), (D) are equivalent.

When the operator T in Theorem 3.2 is a contraction satisfying $\|T\| < 1$, then T cannot be hypercyclic and for any vector $x \in X$, we have $T^n x \to 0$. That explains why we need the limit point in Conditions (B) and (C) of Theorem 3.2 to be nonzero. If condition (B) does not hold true, then the orbit's closure is the same as the orbit except for the zero vector. Hence, we have the following corollary: *The operator T in Theorem 3.2 is not hypercyclic if and only if every set of the form orb $(T, x) \cup \{0\}$ is closed.*

If orb(T, x) has a nonzero limit point, we can only conclude T is hypercyclic as stated in Theorem 3.2, but we cannot conclude that the vector x is a hypercyclic vector, and in fact not even a cyclic vector.

Theorem 3.3 (Chan and Seceleanu [8]) *Let $1 \leq p < \infty$. Suppose $T : \ell^p \to \ell^p$ is a unilateral weighted backward shift and orb(T, x) has a nonzero limit point. The vector x is a cyclic vector for T, if*

(1) the weight sequence $\{w_j : j \geq 1\}$ of T is bounded below, and
(2) orb(T, x) has a nonzero limit point $f = (a_0, a_1, \ldots, a_n, 0, 0, 0, \ldots)$ with finite number of nonzero entries.

While Conditions (1) and (2) in Theorem 3.3 do not appear to be necessary, there are examples in [8] showing that the vector x is not a cyclic vector if either Condition (1) or Condition (2) is not satisfied.

4 Hypercyclic Factorization

In this section, we take a look at sums and products of cyclic and hypercyclic operators. We first focus on the case when the underlying space is a separable, infinite dimensional Hilbert space H. The following two results concern the sum of operators.

Theorem 4.1 (Wu [14]) *For any bouned linear operator $T : H \longrightarrow H$, there exist cyclic operators T_1, T_2 for which $T = T_1 + T_2$.*

Indeed there is a hypercycllc improvement of Theorem 4.1.

Theorem 4.2 (Grivaux [9]) *For any bounded linear operator $T : H \longrightarrow H$, there exist hypercyclic operators T_1, T_2 for which $T = T_1 + T_2$.*

The sum in Theorems 4.1 and 4.2 work for any bounded linear operator $T : H \longrightarrow H$. However, not every operator $T : H \longrightarrow H$ can be written as the product of two cyclic operators, for the reason that the range ran T of a cyclic operator T has co-dimension at most 1 by the definition of a cyclic operator.

Theorem 4.3 (Wu [14]) *If $2 \leq k < \infty$, and if the operator $T : H \longrightarrow H$ satisfies* dim (ran $T)^{\perp} \leq k$*, then the operator T can be written as the product of at most $k + 2$ cyclic operators.*

In the rest of the paper, we discuss the factorization of a unilateral weighted backward shift T as the product of two hypercyclic operators. To facilitate our discussion, we rewrite the definition of T in terms of a canonical basis.

Suppose the canonical basis of ℓ^p is denoted by a one-sided sequence $\{e_n : n \geq 0\}$, where $e_0 = (1, 0, 0, 0, \ldots)$, and $e_1 = (0, 1, 0, 0, \ldots)$, and $e_2 = (0, 0, 1, 0, 0, \ldots)$ etc. A unilateral weighted backward shift

$$T(a_0, a_1, a_2, \ldots) = (w_1 a_1, w_2 a_2, w_3 a_3, \ldots)$$

can be rewritten as

$$T\left(\sum_{i=0}^{\infty} a_i e_i\right) = \sum_{i=1}^{\infty} w_i a_i e_{i-1}.$$

Suppose the canonical basis of ℓ^p is denoted by a two-sided sequence $\{f_n : -\infty < n < \infty\}$ where $f_{-1} = (\ldots, 0, 0, 1, \overbrace{0,}^{zeroth} 0, \ldots)$, $f_0 = (\ldots, 0, 0, \overbrace{1,}^{zeroth} 0, \ldots)$, and $f_1 = (\ldots, 0, \overbrace{0,}^{zeroth} 1, 0, 0, \ldots)$ etc.. A bilateral weighted backward shift

$$T(\ldots, a_{-1}, \overbrace{a_0,}^{\text{zeroth}} a_1, \ldots) = (\ldots, w_{-1}a_{-1}, w_0 a_0, \overbrace{w_1 a_1,}^{\text{zeroth}} w_2 a_2, \ldots)$$

can be rewritten as

$$T\left(\sum_{i=-\infty}^{\infty} a_i f_i\right) = \sum_{i=-\infty}^{\infty} w_i a_i f_{i-1}.$$

Let us use $\{e_0, e_1, e_2, \ldots\}$ as the canonical basis of ℓ^p, where $1 \leq p < \infty$. We can generalize the definition of a unilateral weighted backward shift on ℓ^p by allowing a permutation of the basis vector. An operator $T : \ell^p \longrightarrow \ell^p$ is a *unilateral weighted backward shift* if there exist (1) a bijection $\sigma : \mathbb{Z}^+ \to \mathbb{Z}^+$ to reorder the canonical basis as $\{e_{\sigma(0)}, e_{\sigma(1)}, e_{\sigma(2)}, \ldots\}$, and (2) a bounded, positive weight sequence $\{w_i : i \geq 1\}$ for which

$$T\left(\sum_{i=0}^{\infty} a_i e_{\sigma(i)}\right) = \sum_{i=1}^{\infty} w_i a_i e_{\sigma(i-1)}.$$

An operator $T : \ell^p \longrightarrow \ell^p$ is a *bilateral weighted (backward) shift* if there exist (1) a bijection $\rho : \mathbb{Z} \to \mathbb{Z}^+$ to reorder the canonical basis as a 2-sided sequence $\{\dots, e_{\rho(-1)}, e_{\rho(0)}, e_{\rho(1)}, \dots\}$, and (2) a bounded, positive 2-sided weight sequence $\{w_i : -\infty < i < \infty\}$ for which

$$T\left(\sum_{i=-\infty}^{\infty} a_i e_{\rho(i)}\right) = \sum_{i=-\infty}^{\infty} w_i a_i e_{\rho(i-1)}.$$

Theorem 4.4 (Chan and Sanders [6]) *Every unilateral weighted backward shift $T : \ell^p \longrightarrow \ell^p$, with $1 \le p < \infty$, can be factored as*

$$T = U_1 B_1 = B_2 U_2,$$

where U_1, U_2 are hypercyclic unilateral weighted backward shifts and B_1, B_2 are hypercyclic bilateral weighted shifts.

In the following, we show how to define U_1 and B_1 in [6] so that $T = U_1 B_1$. Without loss of generality, assume T is given by

$$T\left(\sum_{i=0}^{\infty} a_i e_i\right) = \sum_{i=1}^{\infty} w_i a_i e_{i-1}.$$

For $\epsilon > 0$, let $b = (1+\epsilon)\max\{1, \|T\|\}$ and select a scalar a satisfying $b^{-1} < a < 1$. Select a sequence $\{j_k : k \ge 0\}$ of positive, even integers very carefully.

Let $\rho : \mathbb{Z} \longrightarrow \mathbb{Z}^+$ be the bijection that reorders the canonical basis as

$$\dots e_{2+j_2}, \overbrace{e_{1+j_2}, e_{-1+j_2}}^{e_{j_2}\, missing}, e_{-2+j_2}, \dots, e_{2+j_1}, \overbrace{e_{1+j_1}, e_{-1+j_1}}^{e_{j_1}\, missing}, e_{-2+j_1}, \dots, e_2, e_1, e_0,$$
$$e_{j_1}, e_{j_2}, e_{j_3}, \dots$$

Define the bilateral weighted shift $B_1 : \ell^p \longrightarrow \ell^p$ by

$$B_1 e_{\rho(i)} = \begin{cases} a^{-1} e_{\rho(i-1)}, & \text{if } i \ge 1, \\ e_{\rho(-1)}, & \text{if } i = 0, \\ b^{-1} w_{\rho(i)} e_{\rho(i-1)}, & \text{if } i \le -1. \end{cases}$$

Use another bijection $\sigma : \mathbb{Z}^+ \longrightarrow \mathbb{Z}^+$ to reorder the canonical basis as

$$\underbrace{\overbrace{e_1, e_3, \dots, e_{-1+j_1}}^{odd\ indices}, \overbrace{e_0, e_2, \dots, e_{-2+j_1}}^{even\ indices}}_{indices\ 0,1,\dots,-1+j_1}, \underbrace{\overbrace{e_{1+j_1}, \dots, e_{-1+j_2}}^{odd\ indices}, \overbrace{e_{j_1}, \dots, e_{-2+j_2}}^{even\ indices}}_{indices\ j_1,\dots,-1+j_2}, \dots$$

Define the unilateral weighted shift $U_1 : \ell^p \longrightarrow \ell^p$ by

$$U_1 e_{\sigma(i)} = \begin{cases} b e_{\sigma(i-1)}, & \text{if } \sigma(i) \neq j_k \text{ for any } k, \\ a w_{j_{k+1}} e_{\sigma(i-1)}, & \text{if } \sigma(i) = j_k \text{ for some k}, \\ 0, & \text{if } i = 0. \end{cases}$$

To conclude our paper, we remark Chan and Sanders ([6]) proved that if the weight sequence of the unilateral weighted backward shift T in Theorem 4.4 is bounded below, then the two factors U_1 and B_1 can be chosen to have their weight sequence bounded below also, and in particular B_1 is invertible.

References

1. J. Bès, K.C. Chan, R. Sanders, Every weakly sequentially hypercyclic shift is norm hypercyclic. Math. Proc. R. Irish Acad. **105A**, 79–85 (2005)
2. J. Bès, K. Chan, R. Sanders, Weak * hypercyclicity and supercyclicity of shifts on ℓ^∞. Integr. Equ. Oper. Theory **55**, 363–376 (2006)
3. P.S. Bourdon, N.S. Feldman, Somewhere dense orbits are everywhere dense. Indiana Univ. Math. J. **52**, 811–819 (2003)
4. K.C. Chan, The density of hypercyclic operators on a Hilbert space. J. Oper. Theory **47**, 131–143 (2002)
5. K.C. Chan, R. Sanders, A weakly hypercyclic operator that is not norm hypercyclic. J. Oper. Theory **52**(1), 39–59 (2004)
6. K.C. Chan, R. Sanders, Hypercyclic shift factorizations for unilateral weighted backward shift operators. J. Oper. Theory **80**(2), 349–374 (2018)
7. K.C. Chan, I. Seceleanu, Hypercyclicity of shifts as a zero-one law of orbital limit points. J. Op. Theory **67**, 257–277 (2012)
8. K.C. Chan, I. Seceleanu, Cyclicity of vectors with orbital limit points for backward shifts. Integr. Equ. Oper. Theory **78**, 225–232 (2014)
9. S. Grivaux, Sums of hypercyclic operators. J. Funct. Anal. **202**, 486–503 (2003)
10. H.M. Hilden, L.J. Wallen, Some cyclic and non-cyclic vectors of certain operators. Indiana Univ. Math. J. **23**(7), 557–565 (1974)
11. C. Kitai, Invariant closed sets for linear operators. Ph. D. Thesis, University of Toronto (1982)
12. H. Salas, Hypercyclic weighted shifts. Trans. Am. Math. Soc. **347**(3), 993–1004 (1995)
13. R. Sanders, Weakly supercyclic operators. J. Math. Anal. Appl. **292**, 148–159 (2004)
14. P.Y. Wu, Sums and products of cyclic operators. Proc. Am. Math. Soc. **112**, 1053–1063 (1994)

On δ-deformations of Polygonal Dendrites

Dmitry Drozdov, Mary Samuel, and Andrei Tetenov

Abstract We find the conditions under which the attractor $K(\mathcal{S}')$ of a deformation $\mathcal{S}'$ of a contractible P-polygonal system $\mathcal{S}$ in $\mathbb{R}^2$ is a dendrite. The most important one is the parameter matching condition at the points where the images of the vertices of the polygon P meet.

Keywords Self-similar dendrite · Generalized polygonal system · Attractor · Post-critically finite set · Parameter matching theorem · Zipper

2010 Mathematics Subject Classification Primary: 28A80

Though the study of topological properties of dendrites from the viewpoint of general topology proceed for more than three quarters of a century [3, 11, 12], the attempts to study the geometrical properties of self-similar dendrites are rather fragmentary.

In 1985 Hata [8] studied the connectedness properties of self-similar sets and proved that if a dendrite is an attractor of a system of weak contractions in a complete metric space, then the set of its endpoints is infinite. In 1990 Bandt showed in his unpublished paper [2] that the Jordan arcs connecting pairs of points of a post-critically finite self-similar dendrite are self-similar, the set of possible values for

D. Drozdov
Novosibirsk State University, Novosibirsk 630090, Russia
e-mail: d.drozdov1@g.nsu.ru

M. Samuel
Indian Institute of Information Technology Lucknow, Lucknow, India
e-mail: marysamuel@iiitl.ac.in

A. Tetenov (✉)
Sobolev Institute of Mathematics, Siberian Branch, Russian Academy of Sciences, 630090 Novosibirsk, Russia

Novosibirsk State University, 630090 Novosibirsk, Russia

Gorno-Altaisk State University, 649000 Gorno-Altaisk, Russia
e-mail: a.tetenov@g.nsu.ru

R. L. Devaney et al. (eds.), *Topological Dynamics and Topological Data Analysis*, Springer Proceedings in Mathematics & Statistics 350,
https://doi.org/10.1007/978-981-16-0174-3_11

dimensions of such arcs is finite. Kigami in his work [10] applied the methods of harmonic calculus on fractals to dendrites; on a way to this, he developed effective approaches to the study of structure of self-similar dendrites. Croydon in his thesis [7] obtained heat kernel estimates for continuum random tree and for certain family of p.c.f. random dendrites on the plane. A special kind of dendrites, which appear as a particular case of fractal squares, was studied in papers of Christea and Steinsky [4–6].

A systematic approach to the study of self-similar dendrites required to find the answers to the following questions: What kind of topological restrictions characterize the class of dendrites generated by systems of similarities in $\mathbb{R}^d$? What are the explicit construction algorithms for self-similar dendrites? What are the metric and analytic properties of morphisms of self-similar structures on dendrites?

To approach these questions, we started from simplest and most obvious settings, which were used by many authors [2, 14]. In [13, 16, 17], we considered systems $\mathcal{S}$ of contraction similarities in $\mathbb{R}^d$ defined by some polyhedron $P \subset \mathbb{R}^d$, which we called contractible P-polyhedral systems.

We proved that the attractor of such system $\mathcal{S}$ is a dendrite K in $\mathbb{R}^d$, and there is a dense subset of K such that punctured neighbourhoods of its points split to a finite disjoint union of subsets of solid angles Ω_l, equal to the solid angles of P (Theorem 4); we showed that the orders of points $x \in K$ have an upper bound, depending only on P and that Hausdorff dimension of the set $CP(K)$ of the cut points of K is strictly smaller than the dimension of the set $EP(K)$ of its end points unless K is a Jordan arc.

This is a very convenient though rather restrictive way to define post-critically finite self-similar dendrites in the plane using contractible P-polygonal systems. Nevertheless, if we move slightly the vertices of the main polygon P and of polygons P_i, defining the polygonal system $\mathcal{S}$, and change the system $\mathcal{S}$ accordingly, we often obtain a system $\mathcal{S}'$ of a more general type whose attractor K' is a dendrite too. We call such systems generalized polygonal systems (Definition 8) and in the case when polygons P_i' differ from the polygons P_i less than by δ, we call such systems δ-deformations (Definition 12) of the polygonal system $\mathcal{S}$. In this paper, we begin initial study of generalized polygonal systems and δ-deformations of contractible polygonal systems.

In Theorem 9, we formulate sufficient conditions under which the attractor K of a generalized polygonal system $\mathcal{S}$ is a dendrite. These conditions are expressed in terms of intersections $K_i \cap K_j$ of the pieces of the attractor K. In Theorem 14, we show that a δ-deformation $\mathcal{S}'$ of a contractible polygonal system $\mathcal{S}$ defines a continuous map $\hat{f} : K \to K'$ of respective attractors of these systems which agrees with the action of $\mathcal{S}$ and $\mathcal{S}'$ and give conditions under which $\hat{f}$ is a homeomorphism. In Theorem 20, we show that parameter matching condition is a necessary condition for a generalized polygonal system to generate a dendrite. In Theorem 27, we show that if δ is sufficiently small and the system $\mathcal{S}'$ is δ-deformation of a contractible P-polygonal system $\mathcal{S}$, which satisfies parameter matching condition, then the attractor $K(\mathcal{S}')$ is a dendrite, homeomorphic to $K(\mathcal{S})$.

1 Preliminaries

1.1 Self-similar Sets

Definition 1 Let $\mathcal{S} = \{S_1, S_2, \ldots, S_m\}$ be a system of (injective) contraction maps on the complete metric space (X, d). A non-empty compact set $K \subset X$ is called the attractor of the system $\mathcal{S}$, if $K = \bigcup_{i=1}^m S_i(K)$.

The system $\mathcal{S}$ defines its Hutchinson operator T by $T(A) = \bigcup_{i=1}^m S_i(A)$. By Hutchinson's Theorem, the attractor K is unique for $\mathcal{S}$ and for any compact set $A \subset X$ the sequence $T^n(A)$ converges to K. We also call the subset $K \subset X$ self-similar with respect to $\mathcal{S}$.
Throughout the whole paper, the maps $S_i \in \mathcal{S}$ are supposed to be similarities and the set X to be $\mathbb{R}^2$. We will use complex notation for the point on the plane, so each similarity will be written as $S_j(z) = q_j e^{i\alpha_j}(z - z_j) + z_j$, where $q_j = \text{Lip} S_j$ and $z_j = \text{fix}(S_j)$. For a system $\mathcal{S}$, let $q_{min} = \min\{q_j, j \in I\}$ and $q_{max} = \max\{q_j, j \in I\}$. Here, $I = \{1, 2, ..., m\}$ is the set of indices, while $I^* = \bigcup\limits_{n=1}^{\infty} I^n$ is the set of all finite I-tuples, or multiindices $\mathbf{j} = j_1 j_2 \ldots j_n$. The length n of the multiindex $\mathbf{j} = j_1 \ldots j_n$ is denoted by $|\mathbf{j}|$ and $\mathbf{ij}$ denote the concatenation of the corresponding multiindices. We say $\mathbf{i} \sqsubset \mathbf{j}$, if $\mathbf{j} = \mathbf{il}$ for some $\mathbf{l} \in I^*$; if $\mathbf{i} \not\sqsubset \mathbf{j}$ and $\mathbf{j} \not\sqsubset \mathbf{i}$, $\mathbf{i}$ and $\mathbf{j}$ are *incomparable*.

For a multiindex $\mathbf{j} \in I^*$, we write $S_\mathbf{j} = S_{j_1 j_2 \ldots j_n} = S_{j_1} S_{j_2} \ldots S_{j_n}$, and for the set $A \subset X$, we denote $S_\mathbf{j}(A)$ by $A_\mathbf{j}$; we also denote by $G_\mathcal{S} = \{S_\mathbf{j}, \mathbf{j} \in I^*\}$ the semigroup, generated by $\mathcal{S}$;
$I^\infty = \{\alpha = \alpha_1 \alpha_2 \ldots, \ \ \alpha_i \in I\}$ denotes the index space; and $\pi : I^\infty \to K$ is the *index map* , which sends α to the point $\bigcap_{n=1}^\infty K_{\alpha_1 \ldots \alpha_n}$.
Along with a system $\mathcal{S}$, we will consider its nth refinement $\mathcal{S}^{(n)} = \{S_\mathbf{j}, \mathbf{j} \in I^n\}$, whose Hutchinson's operator is equal to T^n.

Definition 2 The system $\mathcal{S}$ satisfies the *open set condition* (OSC) if there exists a non-empty open set $O \subset X$ such that $S_i(O)$, $\{1 \le i \le m\}$ are pairwise disjoint and all contained in O.

Let $\mathcal{C}$ be the union of all $S_i(K) \cap S_j(K)$, $i, j \in I, i \ne j$. *The post-critical set* $\mathcal{P}$ of the system $\mathcal{S}$ is the set of all $\alpha \in I^\infty$ such that for some $\mathbf{j} \in I^*$, $S_\mathbf{j}(\alpha) \in \mathcal{C}$. In other words, $\mathcal{P} = \{\sigma^k(\alpha) : \alpha \in \mathcal{C}, k \in \mathbb{N}\}$, where the map $\sigma^k : I^\infty \to I^\infty$ is defined by $\sigma^k(\alpha_1 \alpha_2 \ldots) = \alpha_{k+1} \alpha_{k+2} \ldots$ A system $\mathcal{S}$ is called *post-critically finite* (PCF) [9] if its post-critical set $\mathcal{P}$ is finite. Thus, if the system $\mathcal{S}$ is post-critically finite, then there is a finite set $\mathcal{V} = \pi(\mathcal{P})$ such that for any non-comparable $\mathbf{i}, \mathbf{j} \in I^*$, $K_\mathbf{i} \cap K_\mathbf{j} = S_\mathbf{i}(\mathcal{V}) \cap S_\mathbf{j}(\mathcal{V})$.

1.2 Dendrites

A *dendrite* is a locally connected continuum containing no simple closed curve.

The order $Ord(p, X)$ of the point p with respect to a dendrite X is the number of components of the set $X \setminus \{p\}$. Points of order 1 in a dendrite X are called *end points* of X; a point $p \in X$ is called a *cut point* of X if $X \setminus \{p\}$ is disconnected; points of order at least 3 are called *ramification points* of X.

A continuum X is a dendrite iff X is locally connected and uniquely arcwise connected.

1.3 Contractible Polygonal Systems

Let $P \subset \mathbb{R}^2$ be a finite polygon homeomorphic to a disk, $\mathcal{V}_P = \{A_1, \ldots, A_{n_P}\}$ be the set of its vertices. Let also $\Omega(P, A)$ denote the angle with vertex A in the polygon P. We consider a system of similarities $\mathcal{S} = \{S_1, \ldots, S_m\}$ in $\mathbb{R}^2$ such that:
(D1) for any $i \in I$ set $P_i = S_i(P) \subset P$;
(D2) for any $i \neq j, i, j \in I$, $P_i \bigcap P_j = \mathcal{V}_{P_i} \bigcap \mathcal{V}_{P_j}$ and $\#(\mathcal{V}_{P_i} \bigcap \mathcal{V}_{P_j}) < 2$;
(D3) $\mathcal{V}_P \subset \bigcup\limits_{i \in I} S_i(\mathcal{V}_P)$;
(D4) the set $\widetilde{P} = \bigcup\limits_{i=1}^{m} P_i$ is contractible.

Definition 3 The system $\mathcal{S}$ satisfying the conditions **(D1–D4)** is called a *contractible P-polygonal system of similarities.*

This theorem was proved by the authors in ([16], Theorem 4,(g)) (or [18], Theorem 10,(g)):

Theorem 4 *Let $\mathcal{S}$ be a contractible P-polygonal system of similarities.*
(a) The system $\mathcal{S}$ satisfies (OSC).
(b) $P_{\mathbf{j}} \subset P_{\mathbf{i}}$ iff $\mathbf{j} \sqsupset \mathbf{i}$.
(c) If $\mathbf{i} \sqsubset \mathbf{j}$, then $S_{\mathbf{i}}(\mathcal{V}_P) \cap P_{\mathbf{j}} \subset S_{\mathbf{j}}(\mathcal{V}_P)$.
(d) For any incomparable $\mathbf{i}, \mathbf{j} \in I^*$, $\#(P_{\mathbf{i}} \cap P_{\mathbf{j}}) \leq 1$ and $P_{\mathbf{i}} \cap P_{\mathbf{j}} = S_{\mathbf{i}}(\mathcal{V}_P) \cap S_{\mathbf{j}}(\mathcal{V}_P)$.
(e) The set $G_{\mathcal{S}}(\mathcal{V}_P)$ of vertices of polyhedra $P_{\mathbf{j}}$ is contained in K.
(f) If $x \in K \setminus G_{\mathcal{S}}(\mathcal{V}_P)$, then $\#\pi^{-1}(x) = 1$.
(g) For any $x \in G_{\mathcal{S}}(\mathcal{V}_P)$, there is $\varepsilon > 0$ and a finite system $\{\Omega_1, ..., \Omega_n\}$, where $n = \#\pi^{-1}(x)$, of disjoint angles with vertex x, such that if $x \in P_{\mathbf{j}}$ and diam$P_{\mathbf{j}} < \varepsilon$, *then for some $k \leq n$, $\Omega(P_{\mathbf{j}}, x) = \Omega_k$. Conversely, for any Ω_k there is such $\mathbf{j} \in I^*$, that $\Omega(P_{\mathbf{j}}, x) = \Omega_k$.*

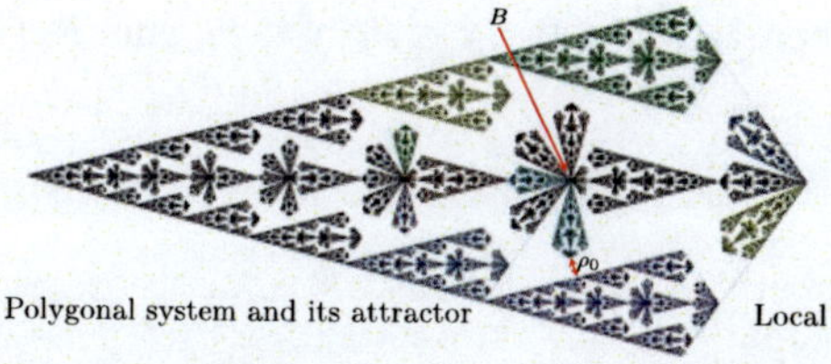

Polygonal system and its attractor

Local structure of K near the vertex B.(rotated)

Definition 5 Let $\mathcal{S}$ be a contractible P-polygonal system of similarities. The vertex $A \subset \mathcal{V}_P$ is called a cyclic vertex, if there is such multiindex $\mathbf{i} = i_1 i_2 \ldots i_k$, that $S_{\mathbf{i}}(A) = A$. The least number $k = |\mathbf{i}|$ among all $\mathbf{i}$ for which $S_{\mathbf{i}}(A) = A$ is called *the order* of the cyclic vertex A.

Definition 6 A point $B \in \cup_{i=1}^{m} \mathcal{V}_{P_i}$ is subordinate to a cyclic vertex A, if for certain multiindex $\mathbf{i}$, $S_{\mathbf{i}}(A) = B$.

Proposition 7 *Let $\mathcal{S}$ be a contractible P-polygonal system of similarities. Then:*
(1) Each vertex $B \in \mathcal{V}_P$ is subordinate to some cyclic vertex.
(2) There is such n, that in the system $\mathcal{S}^{(n)} = \{S_{\mathbf{i}}, \mathbf{i} \in I^n\}$ all the cyclic vertices have order 1.

Proof Notice that if $A \in \mathcal{V}_P$ is a cyclic vertex, then there is such $\mathbf{j} \in I^*$ that $S_{\mathbf{j}}(A) = A$. Therefore, if for some $\mathbf{j} \in I^*, A \in P_{\mathbf{j}}$, then for some n, $S_{\mathbf{j}}^n(P) \subset P_{\mathbf{j}} \subset P$, A being a vertex of each of these polygons. Since $\Omega(S_{\mathbf{j}}^n(P), A) = \Omega(P, A)$, for any $\mathbf{j} \in I^*$, for which $A \in P_{\mathbf{j}}$, $\Omega(P_{\mathbf{j}}, A) = \Omega(P, A)$. This implies that $\#\pi^{-1}(A) = 1$ and for any n there is unique $\mathbf{j} \in I^n$ such that $A \in P_{\mathbf{j}}$.

Conversely if for any $\mathbf{i} \in I^*$, for which $A \in P_{\mathbf{i}}$, $\Omega(P_{\mathbf{i}}, A) = \Omega(P, A)$, then $\#\pi^{-1}(A) = 1$ and A is a cyclic vertex of the system $\mathcal{S}$.

Then, by Theorem 4, for any vertex $B \in G_{\mathcal{S}}(\mathcal{V}_P)$, there is a finite set $\{\mathbf{i}_1, ..., \mathbf{i}_n\}$ of incomparable multiindices such that for any l, l', $P_{\mathbf{i}_l} \cap P_{\mathbf{i}_{l'}} = \{B\}$, the set $\bigcup_{l=1}^{k} K_{\mathbf{i}_l}$ is a neighborhood of the point B in K and for any $l = 1, ..., k$, the point $S_{\mathbf{i}_l}^{-1}(B) = A_l$ is a cyclic vertex. This proves (1).

Let now $A_1, ..., A_k$ be the full set of cyclic vertices in $\mathcal{V}_P$ and $p_1, ..., p_k$ be their respective orders. Let N be the least common multiple of $p_1, ..., p_k$. Then $\mathcal{S}^{(n)}$ is the desired P-polygonal system. ∎

1.4 Main Parameters of a Contractible Polygonal System

For any set $X \subset \mathbb{R}^2$ or point A by $V_\varepsilon(X)$ (resp.$V_\varepsilon(A)$), we denote ε-neighborhood of the set X (resp. of the point A) in the plane.

$\boldsymbol{\rho_0}$: Take such $\rho_0 > 0$ that:
(i) for any vertex $A \in \mathcal{V}_P$, $V_\rho(A) \bigcap P_k \neq \varnothing \Rightarrow A \in P_k$;

(ii) for any $x, y \in P$ such that there are $P_k, P_l : x \in P_k, y \in P_l$ and $P_k \bigcap P_l = \varnothing, d(x, y) \geq \rho_0$.

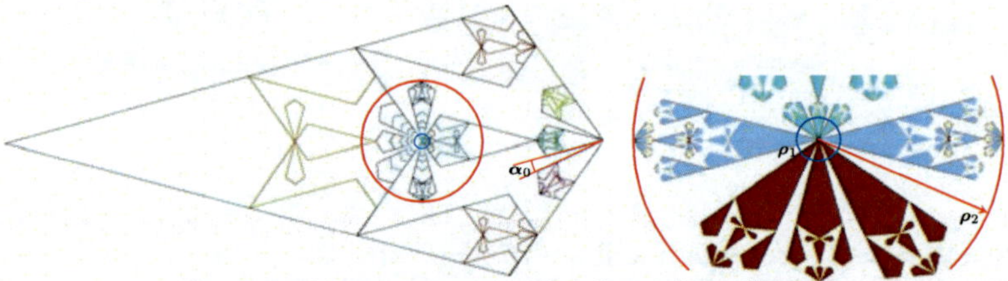

Choosing the parameters α_0, ρ_1 and ρ_2 for a polygonal system.

$\boldsymbol{\rho_1, \rho_2}$: As it follows from Theorem 4, for any vertex $B \in \mathcal{V}_{\widetilde{P}}$, there is a finite set of cyclic vertices $A_{i_1}, ..., A_{i_k} \in \mathcal{V}_P$, and multiindices $\mathbf{j}_1, ..., \mathbf{j}_k$ such that for any $l = 1, ..., k$, $S_{\mathbf{j}_l}(A_l) = B$ and $S_{i_l}(A_l) = A_l$ and the set $\bigcup\limits_{l=1}^{k} S_{\mathbf{j}_l} S_{i_l}^n(K)$ is a neighborhood of the point B in K for any $n \geq 0$.
Let ρ_1 and ρ_2 be such positive numbers that for for any vertex $B \in \mathcal{V}_{\widetilde{P}}$

$$(V_{\rho_1}(B) \cap K) \subset \bigcup_{l=1}^{k} S_{\mathbf{j}_l}(P_{i_l}) \quad \text{and} \bigcup_{l=1}^{k} P_{\mathbf{j}_l} \subset V_{\rho_2}(B). \tag{1}$$

$\boldsymbol{\alpha_0}$: Let α_0 denote the minimal angle between those sides of polygons $P_i, P_j, i, j \in I$, which have common vertex.
Arrangement of maps fixing cyclic vertices. Let $\mathcal{S}$ be a contractible P-polygonal system all of whose cyclic vertices have order 1. In this case, we can arrange the indices in I and enumerate the vertices in $\mathcal{V}_P$ in such a way that each cyclic vertex A_l will be the fixed point of $S_l \in \mathcal{S}$. Notice that S_l is a homothety $S_l(z) = q_l(z - A_l) + A_l$, and the polygon P lies inside the angle $\Omega(P, A_l)$ and $K \backslash \{A_l\} = \bigsqcup_{n=0}^{\infty} S_l^n(K \backslash K_l)$. The number of points in $\overline{K_l \backslash S_l(K_l)} \cap S_l(K_l)$ is finite and is equal to the ramification order of A_l in K.

2 Generalized Polygonal Systems

If we omit the condition **(D1)** in the definition of contractible P-polygonal system $\mathcal{S}$, we get the definition of a *generalized P-polygonal system*:

Definition 8 A system $\mathcal{S} = \{S_1, ..., S_m\}$, satisfying the conditions **D2-D4**, is called a generalized P-polygonal system of similarities.

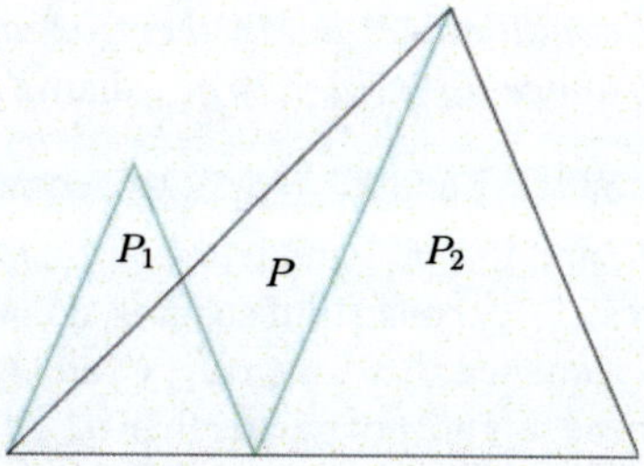

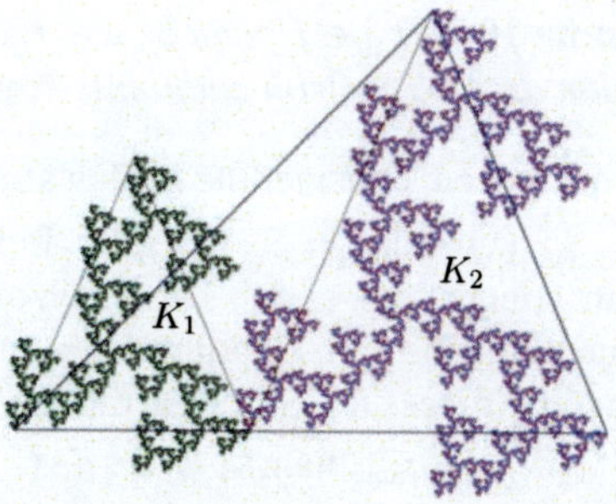

Theorem 9 *Let $\mathcal{S}$ be a generalized P-polygonal system. If for any $i, j \in I$*

$$S_i(K) \cap S_j(K) = P_i \cap P_j, \tag{2}$$

then the attractor K of the system $\mathcal{S}$ is a dendrite.

Proof Let $i, i' \in I$. By a (simple) chain of indices, connecting i and i', we mean a sequence $i = i_1, i_2, ..., i_l = i'$ of pairwise different indices such that $P_{i_k} \cap P_{i_{k'}} = \varnothing$ if $|k' - k| > 1$, and that for any $k = 1, ..., l-1$, $\quad P_{i_k} \cap P_{i_{k+1}} = \{x_k\}$, where x_k denotes a common vertex of the polygons P_{i_k} and $P_{i_{k+1}}$. The last condition also means, that $K_{i_k} \cap K_{i_{k+1}} \ni x_k$ for any $k \in I$. ■

Since in a generalized polygonal system for any two indices i, i', there is a chain of indices $i = i_1, i_2, ..., i_l = i'$ connecting them, then by [9, Theorem 1.6.2], the attractor K is connected, locally connected and arcwise connected. Thus, any two points of K can be connected by some Jordan arc in K.

Notice also that if the condition (2) holds, and the indices i, $i' \in I$ can be connected by a chain $i = i_1, i_2, ..., i_l = i'$, then for any points $x \in K_i, y \in K_{i'}$ there is some Jordan arc $\gamma_{xy} \in K$, consisting of subarcs

$$\gamma_{xx_1} \subset K_{i_1}, \ldots, \ \gamma_{x_{k-1}x_k} \subset K_{i_k}, \ldots, \ \gamma_{x_{l-1}y} \subset K_{i_l} \tag{3}$$

with disjoint interiors.

At the same time, if the condition (2) holds, and a Jordan arc $\gamma_{xy} \subset K$ with endpoints in x and y, meets sequentially the pieces K_{i_1}, ..., K_{i_l}, then it passes sequentially through the points x_k, where $\{x_k\} = K_{i_{k-1}} \cap K_{i_k}$ and consists of subarcs of the form (3) with disjoint interiors.

And vice versa, if the condition (2) holds, then for any Jordan arc γ_{xy} in K there is unique chain of indices $i_1, \ldots, i_l$, such that γ_{xy} consists of subarcs of the form (3).

We need a small Lemma to continue the proof:

Lemma 10 *Let* $\mathbf{j} \in I^*$ *and* $x, y \in K_{\mathbf{j}}$. *If the condition (2) holds, then for any two Jordan arcs* λ_1, λ_2 *with endpoints* x, y, *the distance* $d_H(\lambda_1, \lambda_2) \leq q_{max}\text{diam}K_{\mathbf{j}}$.

Proof Indeed, consider the Jordan arcs $\lambda_1' = S_{\mathbf{j}}^{-1}(\lambda_1)$ and $\lambda_2' = S_{\mathbf{j}}^{-1}(\lambda_2)$, connecting $x' = S_{\mathbf{j}}^{-1}(x)$ and $y' = S_{\mathbf{j}}^{-1}(y)$ in K. Let $x' \in K_i$ and $y' \in K_{i'}$, and let $i_1, i_2, ..., i_l$ be the chain, connecting i and i'. Then each of the arcs λ_1', λ_2' consists of subarcs, connecting sequentially the pairs of points x_k, x_{k+1} in the sequence $x', x_1, ..., x_{l-1}, y'$, and lying in respective pieces K_{i_k}. Since the diameters of these sets are not greater than $q_{max}\text{diam}K$, $d_H(\lambda_1', \lambda_2') \leq q_{\max}\text{diam}K$. Then $d_H(\lambda_1, \lambda_2) \leq q_{\max}\text{diam}K_{\mathbf{j}} \leq \text{diam}K q_{\max}^{|\mathbf{j}|+1}$. ■

Now we can finish the proof of the Theorem. Let λ and λ' be Jordan arcs in K with endpoints at x and y. Applying the Lemma 10 by induction to the subarcs of which the arcs λ and λ' consist, we get that for any $n > |\mathbf{j}|$, $d_H(\lambda_1, \lambda_2) \leq q_{\max}^n\text{diam}K$. Taking a limit with $n \to \infty$, we obtain that a Jordan arc, connecting the points x and y is unique. Therefore, K is a dendrite. ■

Remark 1 It is possible for a generalized P-polygonal system $\mathcal{S}$ not to satisfy the condition 2 and to have the attractor K which is a dendrite. The attractor K of a generalized polygonal system $\mathcal{S}$ on the picture below is a dendrite, but $P_7 \cap P_9 = \varnothing$, whereas $K_7 \cap K_9$ is a line segment.

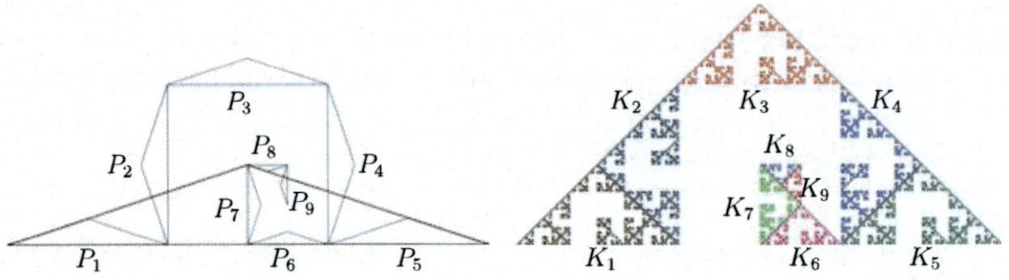

Corollary 11 *Let* $\mathcal{S}$ *be a generalized P-polygonal system, satisfying the condition(2). For any subarc* $\gamma_{xy} \subset K$ *and for any n, there is a unique chain of pairwise different multiindices* $\mathbf{i}_1, \mathbf{i}_2, ..., \mathbf{i}_l \in I^n$, *which divides* γ_{xy} *to sequential arcs* $\gamma_{xx_1} \subset K_{i_1}, \ldots, \gamma_{x_{k-1}x_k} \subset K_{i_k}, \ldots, \gamma_{x_{l-1}y} \subset K_{i_l}$. ■

3 δ-deformations of Contractible Polygonal Systems

Definition 12 Let $\delta > 0$. A generalized P'-polygonal system $\mathcal{S}' = \{S_1', ..., S_m'\}$ is called a δ-deformation of a P-polygonal system $\mathcal{S} = \{S_1, ..., S_m\}$, if there is a bijection $f : \bigcup_{k=1}^{m} \mathcal{V}_{P_k} \to \bigcup_{k=1}^m \mathcal{V}_{P_k'}$, such that

(a) $f|_{\mathcal{V}_P}$ extends to a homeomorphism $\tilde{f} : P \to P'$;

(b) $|f(x) - x| < \delta$ for any $x \in \bigcup_{k=1}^{m} \mathcal{V}_{P_k}$

(c) $f(S_k(x)) = S'_k(f(x))$ for any $k \in I$ and $x \in \mathcal{V}_P$.

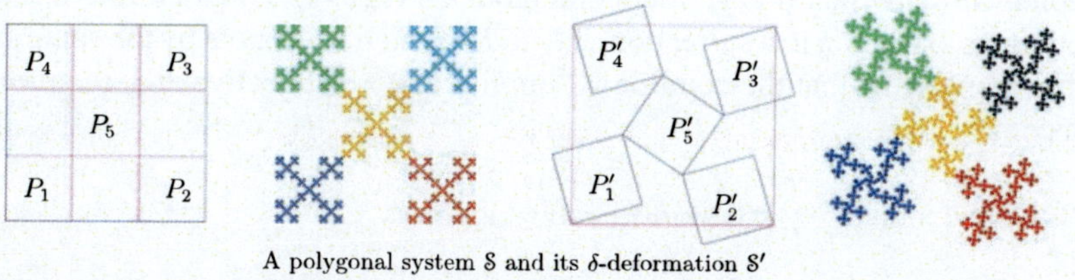

A polygonal system $\mathcal{S}$ and its δ-deformation $\mathcal{S}'$

Notice that by the Definition 12 if $z_1, z_2 \in \mathcal{V}_P$, $i, j \in I$ and $S_i(z_1) = S_j(z_2)$, then $S'_i(f(z_1)) = S'_j(f(z_2))$. Moreover, we have the following

Lemma 13 *If* $A_1, A_2 \in \mathcal{V}_P$, $\mathbf{i}, \mathbf{j} \in I^*$ *and* $S_{\mathbf{i}}(A_1) = S_{\mathbf{j}}(A_2)$, *then* $S'_{\mathbf{i}}(f(A_1)) = S'_{\mathbf{j}}(f(A_2))$.

Proof Suppose $S_{\mathbf{i}}(A) = B \in \mathcal{V}_{\tilde{P}}$ for some $A \in \mathcal{V}_P$ and let $\mathbf{i} = i_1 i_2 ... i_n$. Denote $S_{i_{k+1}...i_n}(A)$ by A_k.

Then we have a finite sequence of relations between $B \in \mathcal{V}_{\tilde{P}}$ and the vertices $A_k \in \mathcal{V}_P$:

$$B = S_{i_1}(A_1); \quad A_1 = S_{i_2}(A_2); \quad \ldots A_{n-1} = S_{i_n}(A) \tag{4}$$

Since, by (c), $f(S_k(A_k)) = S'_k(A'_k)$, $\quad A'_{k-1} = f(A_{k-1}) = f(S_k(A_k)) = S'_k(A'_k)$, therefore, the map f transforms the relations 4 to

$$B' = S'_{i_1}(A'_1); \quad A'_1 = S'_{i_2}(A'_2); \quad \ldots A'_{n-1} = S'_{i_n}(A') \tag{5}$$

which implies $S'_{\mathbf{i}}(A') = B'$

Therefore, if $S_{\mathbf{i}}(A_1) = S_{\mathbf{j}}(A_2) \in \mathcal{V}_{\tilde{P}}$, then $S'_{\mathbf{i}}(f(A_1)) = S'_{\mathbf{j}}(f(A_2))$.

Now suppose $S_{\mathbf{i}}(A_1) = S_{\mathbf{j}}(A_2)$ and $\mathbf{i} = \mathbf{l}\mathbf{i}'$, $\mathbf{j} = \mathbf{l}\mathbf{j}'$ and $S_{\mathbf{i}}(A_1) = S_{\mathbf{j}}(A_2) = S_{\mathbf{l}}(B)$ for some $B \in \mathcal{V}_{\tilde{P}}$. Then $S_{\mathbf{i}'}(A_1) = S_{\mathbf{j}'}(A_2) = B$, therefore $S'_{\mathbf{i}'}(f(A_1)) = S'_{\mathbf{j}'}(f(A_2)) = f(B)$ and $S'_{\mathbf{i}}(f(A_1)) = S'_{\mathbf{j}}(f(A_2)) = S'_{\mathbf{l}}(f(B))$. ■

Theorem 14 *Let K and K' be the attractors of a contractible P-polygonal system $\mathcal{S}$ and of its δ-deformation $\mathcal{S}'$, respectively, and $\pi : I^\infty \to K$, $\pi' : I^\infty \to K'$ be respective address maps.*

(i) There is unique continuous map $\hat{f} : K \to K'$ such that $\hat{f} \circ \pi = \pi'$.

(ii) If $\mathcal{S}'$ satisfies condition 2, then $\hat{f}$ is a homeomorphism.

Remark 2 Equivalent formulation of the statement (i) of the Theorem is:
There is unique continuous map $\hat{f} : K \to K'$ such that for any $z \in K$ and $\mathbf{i} \in I^*$,

$$\hat{f}(S_{\mathbf{i}}(z)) = S'_{\mathbf{i}}(\hat{f}(z)). \tag{6}$$

Proof The proof is similar to (cf. [1, Lemma 1.]). First, we define the function $\hat{f}$ which is a surjection of the dense subset $G_{\mathcal{S}}(\mathcal{V}_P) \subset K$ to the dense subset $G_{\mathcal{S}'}(\mathcal{V}_{P'}) \subset K'$. Second, we show that it is Hölder continuous on $G_{\mathcal{S}}(\mathcal{V}_P)$, and therefore has unique continuous extension to a surjection of K to K', which we denote by the same symbol $\hat{f}$. Third, we show that the condition 2 implies that $\hat{f}$ is injective and, therefore, is a homeomorphism.

1. Define a map $\hat{f}(z) : G_{\mathcal{S}}(\mathcal{V}_P) \to G_{\mathcal{S}'}(\mathcal{V}_{P'})$ by:

$$\hat{f}(z) = S'_{\mathbf{i}}(f(S_{\mathbf{i}}^{-1}(z)) \text{ if } z \in S_{\mathbf{i}}(\mathcal{V}_P) \tag{7}$$

As it follows from Lemma 13, if $S_{\mathbf{i}}(A_1) = S_{\mathbf{j}}(A_2) = z$ then $S'_{\mathbf{i}}(f(S_{\mathbf{i}}^{-1}(z))) = S'_{\mathbf{j}}(f(S_{\mathbf{j}}^{-1}(z)))$, so the map $\hat{f}$ is well-defined.
Obviously, $\hat{f}(G_{\mathcal{S}}(\mathcal{V}_P)) = G_{\mathcal{S}'}(\mathcal{V}_{P'})$ because if $A' \in \mathcal{V}_{P'}$ and $z' = S'_{\mathbf{i}}(A')$, then there is a vertex $A = f^{-1}(A') \in \mathcal{V}_P$, therefore $z' = \hat{f}(S_{\mathbf{i}}(A))$.

Moreover, for any $z \in G_{\mathcal{S}}(\mathcal{V}_P)$ and $\mathbf{i} \in I^*$, $\hat{f}(S_{\mathbf{i}}(z)) = S'_{\mathbf{i}}(\hat{f}(z))$ and if $z_1, z_2 \in G_{\mathcal{S}}(\mathcal{V}_P)$, $\mathbf{i}, \mathbf{j} \in I^*$ and $S_{\mathbf{i}}(z_1) = S_{\mathbf{j}}(z_2)$, then $S'_{\mathbf{i}}(\hat{f}(z_1)) = S'_{\mathbf{j}}(\hat{f}(z_2))$.

2. Let $q_k = \text{Lip} S_k$, $q'_k = \text{Lip} S'_k$, $\beta = \min_{k \in I} \dfrac{\log q'_k}{\log q_k}$.

Then, following the proof of [13, Theorem 27, step 4.], in which for our estimates we use K' instead of P', we see that for any $z_1, z_2 \in G_{\mathcal{S}}(\mathcal{V}_P)$,

$$|z'_1 - z'_2| \le \frac{2\,K'|}{(\rho_0 \cdot \sin(\alpha_0/2))^{\beta}} |z_1 - z_2|^{\beta}.$$

Therefore, the map $\hat{f}$ can be extended to a Hölder continuous surjective map of K to K'. Since for any $z \in K$ and any $k \in I$, $\hat{f}(S_k(z)) = S'_k(f(z))$, $\hat{f} \circ \pi = \pi'$.

3. Now, suppose the system $\mathcal{S}'$ satisfies the condition (2). Suppose for some $\boldsymbol{\sigma} = i_1 i_2 \ldots \in I^{\infty}$ and $\boldsymbol{\tau} = j_1 j_2 \ldots \in I^{\infty}$, $\hat{f} \circ \pi(\boldsymbol{\sigma}) = \hat{f} \circ \pi(\boldsymbol{\tau})$. Then, if $i_1 \neq j_1$, then, by condition 2, $P'_{i_1} \cap P'_{j_1} \neq \varnothing$, therefore $P_{i_1} \cap P_{j_1} = \{B\}$ for some $B \in \mathcal{V}_{\tilde{P}}$ and $\pi(\boldsymbol{\sigma}) = \pi(\boldsymbol{\tau}) = B$.
Suppose now $\boldsymbol{\sigma} = \mathbf{l}\boldsymbol{\sigma}'$ and $\boldsymbol{\tau} = \mathbf{l}\boldsymbol{\tau}'$ and $\hat{f} \circ \pi(\boldsymbol{\sigma}) = \hat{f} \circ \pi(\boldsymbol{\tau})$. Then, by formula 6, $\hat{f} \circ \pi(\boldsymbol{\sigma}') = \hat{f} \circ \pi(\boldsymbol{\tau}')$, so if first indices in $\boldsymbol{\sigma}'$ and $\boldsymbol{\tau}'$ are different, then $\pi(\boldsymbol{\sigma}) = \pi(\boldsymbol{\tau}) = S_{\mathbf{l}}(B)$ for some $B \in \mathcal{V}_{\tilde{P}}$.
This implies injectivity of the map $\hat{f}$. So $\hat{f}$ is a homeomorphism of compact sets K and K'. ■

4 Parameter Matching Theorem

The Definition 5 of cyclic vertices can be applied to generalized polygonal systems. In this case, if A is a cyclic vertex of a generalized P-polygonal system $\mathcal{S}$, the map $S_{\mathbf{i}}$ for which $S_{\mathbf{i}}(A) = A$, need not be a homothety and we have to define the rotation parameter for such map. Though the rotation angle α_i of the map $S_{\mathbf{i}}$ is defined up to $2n\pi$, the number n is uniquely defined by the set $\widetilde{P}$ and depends on its geometric configuration.

Lemma 15 *Let $\mathcal{S}$ be a generalized P-polygonal system, satisfying the condition(2). For any vertices $A, B \in \mathcal{V}_P$, there are $A', B' \in \mathcal{V}_P$ and a map $S_i \in \mathcal{S}$ such that $S_i(A') = A$ and $S_i(\gamma_{A'B'}) \subset \gamma_{AB}$.*

Proof Consider the unique arc γ_{AB}, connecting A and B.

For the arc γ_{AB}, we consider the chain $i_1, i_2, ..., i_l$, which partitions it to subarcs $\gamma_{Ax_1} \subset K_{i_1}, ..., \gamma_{x_{k-1}x_k} \subset K_{i_k}, ... \gamma_{x_{l-1}B} \subset K_{i_l}$ (possibly to the only arc γ_{AB} if $\gamma_{AB} \subset K_{i_1}$). Put $A' = S_{i_1}^{-1}(A)$, $B' = S_{i_1}^{-1}(x_1)$, and $\gamma(A'B') = S_{i_1}^{-1}(\gamma_{Ax_1})$. ■

Proposition 16 *Let $\mathcal{S}$ be a generalized P-polygonal system satisfying the condition (2) and let A be a cyclic vertex of the polygon P. Then there is such vertex $B \in V_P$ and a multiindex $\mathbf{i} \in I^*$, that $S_{\mathbf{i}}(A) = A$ and the Jordan arc $\gamma_{AB} \subset K$ satisfies the inclusion $S_{\mathbf{i}}(\gamma_{AB}) \subset \gamma_{AB}$.*

Proof Notice that if $\mathcal{S}$ is a contractible P-polygonal system then for any cyclic vertex A and for any n, there is *unique* multiindex $\mathbf{i} \in I^n$, and unique vertex $B \in V_P$, such that $S_{\mathbf{i}}(B) = A$. Therefore, if $S_{\mathbf{i}}(A) = A$, the piece $S_{\mathbf{i}}(K)$ separates the point A from the other part of the attractor K of the system $\mathcal{S}$, i.e., $A \notin \overline{K \setminus S_{\mathbf{i}}(K)}$ and each Jordan arc γ_{AB} where $B \in V_P \setminus \{A\}$, contains a point $B' \in S_{\mathbf{i}}(V_P \setminus \{A\})$.

In the case when $\mathcal{S}$ is a generalized polygonal system, the situation is more complicated. Since the attractor K is a dendrite in the plane which has one-point intersection property, it follows from [15] that the system $\mathcal{S}$ satisfies OSC and each vertex $A' \in V_P$ has finite ramification order. Let $U_1, ..., U_s$ be the components of $K \setminus \{A\}$. Since $S_{\mathbf{i}}$ fixes A, there is a permutation σ of the set $\{1, ..., s\}$, such that for any $l \in \{1, ..., s\}$, $S_{\mathbf{i}}(U_l) \subset U_{\sigma(l)}$. Therefore, there is such N that $\sigma^N = \mathrm{Id}$ and $S_{\mathbf{j}} = S_{\mathbf{i}}^N$ sends each U_l to a subset of U_l. Each of those components U_l which have non-empty intersection with $V_P \setminus \{A\}$ has also non-empty intersection with $S_{\mathbf{j}}(V_P \setminus \{A\})$, therefore each arc γ_{AB}, $B \in V_P$ contains a point $B' \in S_{\mathbf{j}}(V_P)$.

Let us enumerate the vertices of P so that $A = A_1$ and other vertices are $A_2, ..., A_{n_P}$. For each vertex A_k, $k \geq 2$, there is unique vertex $A_{k'}$ such that $\gamma_{A_1A_k} \cap S_{\mathbf{j}}(V_P) = S_{\mathbf{j}}(A_{k'})$. The formula $\phi(k) = k'$ defines a map ϕ of $\{2, 3..., n_P\}$ to itself. There is some N' such that $\phi^{N'}$ has a fixed point k. Therefore, $S_{\mathbf{j}}^{N'}(\gamma_{A_1A_k}) \subset \gamma_{A_1A_k}$. ■

Definition 17 The arc γ_{AB} is called an *invariant arc* of the cyclic vertex A.

Let A be a cyclic vertex and γ_{AB} be its invariant arc and $S_{\mathbf{i}}(A) = A$. Let $B' = S_{\mathbf{i}}(B)$. We denote by α the total change of argument of $z - A$ when z travels along γ_{AB} from B to B'. This gives unique representation $S_{\mathbf{i}}(z) = q_{\mathbf{i}}e^{i\alpha}(z - A) + A$.

Remark 3 The following picture shows how the angle α depends on the geometric configuration of the system $\mathcal{S}$, though the similarity which fixes A and sends B to B' is the same.

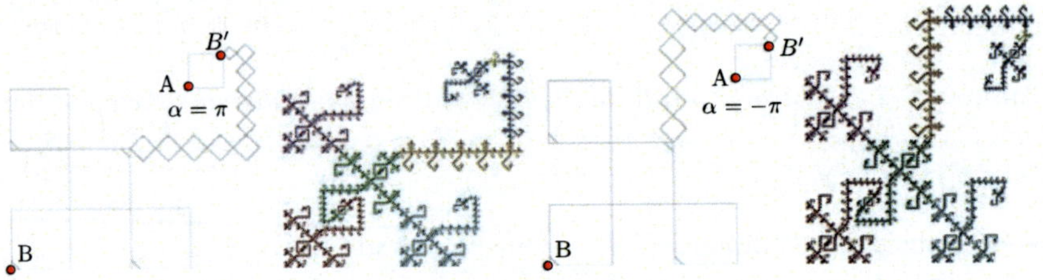

Definition 18 The number $\lambda_A = \dfrac{\alpha}{\ln r}$ is called the parameter of the cyclic vertex A.

Definition 19 Generalized P-polygonal system $\mathcal{S}$ of similarities satisfies the *parameter matching condition*, if for any $B \in \cup_{i=1}^{m} \mathcal{V}_{P_i}$ and for any cyclic vertices A, A' such that for some $\mathbf{i}, \mathbf{j} \in I^*$, $S_{\mathbf{i}}(A) = S_{\mathbf{j}}(A') = B$, the equality $\lambda_A = \lambda_{A'}$ holds.

From Propositions 7 and 16 and V.V.Aseev's Lemma about disjoint periodic arcs [1, Lemma 3.1], we come to the following parameter matching theorem:

Theorem 20 *Let the generalized P'-polygonal system $\mathcal{S}'$ be a δ-deformation of a contractible P-polygonal system $\mathcal{S}$ and the attractor K' of the system $\mathcal{S}'$ be a dendrite. Then the system $\mathcal{S}'$ satisfies parameter matching condition.*

Proof Let $\mathcal{S}$ be a generalized polygonal system whose attractor K is a dendrite. Let $C \in \cup_{i=1}^{m} \mathcal{V}_{P_i}$ and $A, A' \in \mathcal{V}_P$ be such cyclic vertices that for some $i, j \in I$, $S_i(A) = S_j(A') = C$. Denote the images $S_i(K)$ and $S_j(K)$ by K_i, K_j, respectively. Without loss of generality we can suppose that the point C has coordinate 0 in $\mathbb{C}$. Since for some $\mathbf{i}, \mathbf{j} \in I^*$, $S_{\mathbf{i}}(A) = A$ and $S_{\mathbf{j}}(A') = A'$, the maps $S_{b1} = S_i S_{\mathbf{i}} S_i^{-1}$ and $S_{b2} = S_j S_{\mathbf{j}} S_j^{-1}$ have C as their fixed point and $S_{b1}(K_i) \subset K_i$ and $S_{b2}(K_j) \subset K_j$. Let $S_{b1}(z) = q_{\mathbf{i}} e^{i\alpha_{\mathbf{i}}} z$ and $S_{b2}(z) = q_{\mathbf{j}} e^{i\alpha_{\mathbf{j}}} z$. So the parameters of the vertices A and A' will be $\lambda_1 = \dfrac{\alpha_{\mathbf{i}}}{\log q_{\mathbf{i}}}$ and $\lambda_2 = \dfrac{\alpha_{\mathbf{j}}}{\log q_{\mathbf{j}}}$. Let $\gamma_{AB} \subset K$ and $\gamma_{A'B'} \subset K$ be invariant arcs for the vertices A and A'. Let also $\gamma_1 = S_i(\gamma_{AB})$ and $\gamma_2 = S_j(\gamma_{A'B'})$. Then $S_{b1}(\gamma_1) \subset \gamma_1$ and $S_{b2}(\gamma_2) \subset \gamma_2$. By V.V.Aseev's Lemma on disjoint periodic arcs [1, Lemma 3.1] it follows that if $\gamma_1 \cap \gamma_2 = \{C\}$, then $\lambda_1 = \lambda_2$. ■

5 Main Theorem

5.1 Some Assumptions

From now on, we will use the following convention: $\mathcal{S} = \{S_1, ..., S_m\}$ will denote a contractible P-polygonal system and $\mathcal{S}' = \{S'_1, ..., S'_m\}$ will be a P'-polygonal system which is a δ-deformation of $\mathcal{S}$ defined by a map f.
For any $k \in I$, $S_k(z) = q_k e^{i\alpha_k}(z - z_k) + z_k$ and $S'_k(z) = q'_k e^{i\alpha'_k}(z - z'_k) + z'_k$, where $z_k = \text{fix}(S_k)$. We also suppose by default that $\text{diam}P = 1$. We suppose that

$$\delta < q_{min}/8 \quad \text{and} \quad \delta < (1 - q_{max})/8 \tag{8}$$

Lemma 21 *Let $\mathcal{S}' = \{S'_1, ..., S'_m\}$ be a δ-deformation of a contractible P-polygonal system $\mathcal{S}$. For sufficiently small δ, and for any $k \in I$,*

$$\frac{q_k - 2\delta}{1 + 2\delta} \leq q'_k \leq \frac{q_k + 2\delta}{1 - 2\delta} \quad \text{and} \quad |\alpha'_k - \alpha_k| \leq \arcsin 2\delta + \arcsin \frac{2\delta}{q_k}. \tag{9}$$

Proof Let A, B be such vertices of P that $|B - A| = 1$. Let us write $S_k(A) = A_k$ and $f(A) = A'$ and use the similar notation for all vertices so by definition, $S'_k(A') = A'_k = f(A_k)$. Notice that $\dfrac{B_k - A_k}{B - A} = q_k e^{i\alpha_k}$ and $\dfrac{B'_k - A'_k}{B' - A'} = q'_k e^{i\alpha'_k}$.

Since the map f moves A, B, A_k, B_k to a distance $\leq \delta$, so $|(B - A) - (B' - A')| \leq 2\delta$ and $|(B_k - A_k) - (B'_k - A'_k)| \leq 2\delta$. Therefore $|(B_k - A_k)| - 2\delta \leq |(B'_k - A'_k)| \leq |(B_k - A_k)| + 2\delta$ and

$$\alpha'_i - \alpha_i = \arg \frac{B'_k - A'_k}{B' - A'} \frac{B - A}{B_k - A_k} = \arg \frac{B'_k - A'_k}{B_k - A_k} - \arg \frac{B' - A'}{B - A} \tag{10}$$

This implies the inequalities (9). ∎

Under the Assumptions (8), $3q_{min}/5 < \dfrac{q_{min} - 2\delta}{1 + 2\delta} < q'_k < \dfrac{q_{max} + 2\delta}{1 - 2\delta} < \dfrac{1 + 3q_{max}}{3 + q_{max}}$; taking into account that $q_k < 1$ and $1 - 2\delta > 3/4$, and that if $0 < x < .5$, then $\arcsin x < 1.05x$, we have

$$\Delta q_k = |q'_k - q_k| < \frac{2\delta(1 + q_k)}{1 - 2\delta} < 6\delta \quad \text{and} \quad \Delta\alpha_k = |\alpha'_k - \alpha_k| < C_\alpha \delta \tag{11}$$

where $C_\alpha = 2.1(1 + 1/q_{min})$.

Let $V_\delta(P)$ denote δ-neighborhood of the polygon P.

Lemma 22 *Let $\mathcal{S}' = \{S'_1, ..., S'_m\}$ be a δ-deformation of a contractible P-polygonal system $\mathcal{S}$. The set $U = V_{\delta_1}(P)$, where $\delta_1 = \dfrac{8\delta}{1 + 3q_{max}}$, satisfies the condition*

$$\text{for any} k \in I, \quad S_k(U) \subset U \quad \text{and} \quad S'_k(U) \subset U \tag{12}$$

Proof By Definition 12, $V_\delta(P_k) \supset P'_k$, $V_\delta(P'_k) \supset P_k$ and since vertices of P are also moved at distance less than δ, $V_\delta(P) \supset P'$ and $V_\delta(P') \supset P$.

So we can write $S'_k(P') \subset V_\delta(P_k) \subset V_\delta(P)$ from which it follows that $S'_k(P) \subset V_{2\delta}(P_k) \subset V_{2\delta}(P)$.

For any positive ρ we have the inclusion $S'_k(V_\rho(P)) \subset V_{2\delta + q'_k \rho}(P)$. In the case when $\rho = 2\delta + q'_k \rho$ this implies $S'_k(V_\rho(P)) \subset V_\rho(P)$ where $\rho = \dfrac{2\delta}{1 - q'_k}$. Since $q'_k \le q_k + 2\delta$, $q'_{max} \le q_{max} + 2\delta < \dfrac{3q_{max} + 1}{4}$, we come to inclusions (12). ■

Lemma 23 *For any $z \in V_{\delta_1}(P)$, $|S'_k(z) - S_k(z)| < C_\Delta \delta$, where $C_\Delta = 14 + 2C_\alpha$.*

Proof Take $z \in V_{\delta_1}(P)$ and consider the difference $S'_k(z) - S_k(z)$. It can be represented in the form $S'_k(A) - S_k(A) + (q'_k e^{i\alpha'_k} - q_k e^{i\alpha_k})(z - A)$. So

$$|S'_k(z) - S_k(z)| < |S'_k(A) - S_k(A)| + (|q'_k - q_k| + q_k |e^{i\alpha'_k} - e^{i\alpha_k}|)|z - A|. \tag{13}$$

Since $|z - A| < 1 + \delta_1 < 2$ and $|S'_k(A) - S_k(A)| < 2\delta$, the right hand side of (13) is no greater than $2\delta + 2(6\delta + C_\alpha \delta)$. ■

Proposition 24 *Let $\pi : I^\infty \to K$ and $\pi' : I^\infty \to K'$ be the address maps for the systems $\mathcal{S}$ and $\mathcal{S}'$, respectively.*
1.Under the assumptions (8), for any $\sigma \in I^\infty$,

$$|\pi'(\sigma) - \pi(\sigma)| < C_K \delta \text{ where } C_K = \frac{2C_\Delta}{1 - q_{max}} \tag{14}$$

2. For any n, if the system $\mathcal{S}'^{(n)}$ is a generalized polygonal system, then it is $C_K\delta$-deformation of the system $\mathcal{S}^{(n)}$. ■

Remark 4 Let $\mathcal{S}' = \{S'_1, ..., S'_m\}$ be a δ-deformation of a contractible P-polygonal system $\mathcal{S}$. Let $A \in S_j(\mathcal{V}_P)$ for some $j \in I$. Let $g(z) = z - A + A'$ and $\hat{S}''_k = g \circ S'_k \circ g^{-1}$. Then $\mathcal{S}'' = \{S''_1, ..., S''_m\}$ is a 2δ-deformation of the system $\mathcal{S}$, for which $A'' = A$, $K'' = g(K')$, $P''_j = g(P_j)$. Since g is a translation, the estimates (9) and (11) for $\mathcal{S}''$ remain the same with the same δ, while $|\pi''(\sigma) - \pi(\sigma)| < (C_K + 1)\delta$.Thus, we will denote $\delta_2 = (C_K + 1)\delta$.

Taking into account the propositions 7 and 24, it is sufficient to prove the theorem for the case when all cyclic vertices of the system $\mathcal{S}$ have order 1.

Proposition 25 *Let P'-polygonal system $\mathcal{S}'$ be a δ-deformation of a contractible P-polygonal system $\mathcal{S}$. Let $A \in \mathcal{V}_P$ be a cyclic vertex (of order 1) and $S_k(z) = q_k e^{i\alpha_k}(z - A) + A$. Then the rotation angle α_k of the map S'_k does not exceed $\arcsin 2\delta + \arcsin \dfrac{2\delta}{q_k}$ and the parameter λ_k of the map S'_k satisfies the inequality*

$$|\lambda_k| \le \frac{\arcsin 2\delta + \arcsin \dfrac{2\delta}{q_k}}{|\log(q_k + 2\delta) - \log(1 - 2\delta)|} \tag{15}$$

Proof The formula (15) follows directly from Lemma 21. ■

Under the assumptions (8),

$$|\lambda_k| < C_\lambda \delta, \text{ where } C_\lambda = \frac{2.1(1 + 1/q_{max})}{\log(3 + q_{max}) - \log(3q_{max} + 1)}. \tag{16}$$

Lemma 26 *Let $\mathcal{S}$ be a contractible P-polygonal system whose cyclic vertices have order 1 and $\mathcal{S}'$ be its δ-deformation. Then if*

$$2.1\frac{\delta_2}{\rho_1} + \lambda \log \frac{\rho_2 + \delta_2}{\rho_1 - \delta_2} < \alpha_0 \text{ and } 2\delta_2 < \rho_0, \tag{17}$$

then the system $\mathcal{S}'$ satisfies the Condition (2)

Proof Take a vertex $B \in V_{\tilde{P}}$. We may suppose for convenience that $B = 0$ and, following Remark 4, we can suppose that the mapping f fixes the vertex $B = 0$, so $B' = B = 0$. Let $W_l = S_{\mathbf{j}_l}(K \backslash K_{i_l})$. The maps $\bar{S}_l = S_{\mathbf{j}_l} S_{i_l} S_{\mathbf{j}_l}^{-1}$ are homotheties with a fixed point B such that

$$K_{\mathbf{j}_l} \backslash \{B\} = \bigsqcup_{n=0}^{\infty} \bar{S}_l^n(W_l) \tag{18}$$

Similarly, let $W'_l = \hat{f}(W_l)$ and $\bar{S}'_l = S'_{\mathbf{j}_l} S'_{i_l} S'^{-1}_{\mathbf{j}_l}$. Then

$$K'_{\mathbf{j}_l} \backslash \{B\} = \bigsqcup_{n=0}^{\infty} \bar{S}'^n_l(W'_l) \tag{19}$$

Notice that for any l, $\bar{S}_l(z) = q_{i_l} z$ and $\bar{S}'_l(z) = q'_{i_l} e^{i\alpha_{i_l}} z$, and due to parameter matching condition, there is such λ, that for any l, $\alpha_{i_l} = \lambda \log q'_{i_l}$.

Consider the map $z = exp(w)$ of the plane $(w = \varrho + i\varphi)$ as universal cover of the punctured plane $\mathbb{C} \backslash \{0\}$.

Consider polygons $P_{\mathbf{j}_l}$ and choose their liftings in the plane $(w = \varrho + i\varphi)$. We may suppose these liftings lie in respective horizontal strips $\theta_l^- \le \varphi \le \theta_l^+$, where $0 < \theta_l^- < \theta_l^+ < 2\pi$ and $\theta_l^+ + \alpha_0 < \theta_{l+1}^-$ for any $l < k$ and $\theta_k^+ + \alpha_0 < \theta_1^- + 2\pi$. We also consider liftings of $K_{\mathbf{j}_l}$, W_l, $K'_{\mathbf{j}_l}$ and W'_l. We denote these liftings by $\mathcal{K}_{\mathbf{j}_l}$, $\mathcal{W}_l$, $\mathcal{K}'_{\mathbf{j}_l}$ and $\mathcal{W}'_l$. It follows from the Eqs. 18 and 19, that

$$\mathcal{K}_{\mathbf{j}_l} = \bigsqcup_{n=0}^{\infty} \bar{T}_l^n(\mathcal{W}_l) \quad \text{and} \quad \mathcal{K}'_{\mathbf{j}_l} = \bigsqcup_{n=0}^{\infty} \bar{T}'^n_l(\mathcal{W}'_l) \tag{20}$$

where $T_l(w) = w + \log q_l$ and $T_l'(w) = w + (1 + i\lambda) \log q_l'$ are parallel translations for which $T_l(\mathcal{K}_l) \subset \mathcal{K}_l$ and $T_l'(\mathcal{K}_l') \subset \mathcal{K}_l'$.

The sets $\mathcal{K}_l$ lie in the half-strips $\varrho \leq \log \rho_2, \theta_l^- \leq \varphi \leq \theta_l^+$, while the sets $\mathcal{W}_l$ are contained in rectangles $R_l = \{\log \rho_1 \leq \varrho \leq \log \rho_2, \theta_l^- \leq \varphi \leq \theta_l^+\}$.

Then the sets $\mathcal{W}_l'$ lie in a rectangle

$$R_l' = \left\{ \log(\rho_1 - \delta_2) \leq \varrho \leq \log(\rho_2 + \delta_2), \theta_l^- - 1.05\frac{\delta_2}{\rho_1} \leq \varphi \leq \theta_l^+ + 1.05\frac{\delta_2}{\rho_1} \right\} \tag{21}$$

Each union $\bigcup\limits_{n=0}^{\infty} T_l'^{n}(R_l')$ lies in a half-strip

$$\begin{cases} \varrho \leq \log(\rho_2 + \delta_2) \\ \theta_l^- - 1.05\dfrac{\delta_2}{\rho_1} - \lambda \log(\rho_2 + \delta_2) \leq \varphi - \lambda\varrho \leq \theta_l^+ + 1.05\dfrac{\delta_2}{\rho_1} - \lambda \log(\rho_1 - \delta_2) \end{cases} \tag{22}$$

Therefore, the set $\mathcal{K}_{\mathbf{j}_l}'$ also lies in this half-strip. So, if

$$\theta_{l-1}^+ + 1.05\frac{\delta_2}{\rho_1} - \lambda \log(\rho_1 - \delta_2) < \theta_l^- - 1.05\frac{\delta_2}{\rho_1} - \lambda \log(\rho_2 + \delta_2) \tag{23}$$

then $\mathcal{K}_{\mathbf{j}_{l-1}}' \cap \mathcal{K}_{\mathbf{j}_l}' = \varnothing$.

We can guarantee that such inequality holds for any l if $2.1\dfrac{\delta_2}{\rho_1} + \lambda \log \dfrac{\rho_2 + \delta_2}{\rho_1 - \delta_2} < \alpha_0$.

If, moreover, $2\delta_2 < \rho_0$, then for any $i_1, i_2 \in I$ such that $P_{i_1} \cap P_{i_2} = \varnothing$, $P_{i_1}' \cap P_{i_2}' = \varnothing$ and $K_{i_1}' \cap K_{i_2}' = \varnothing$ which implies the condition (2). ■

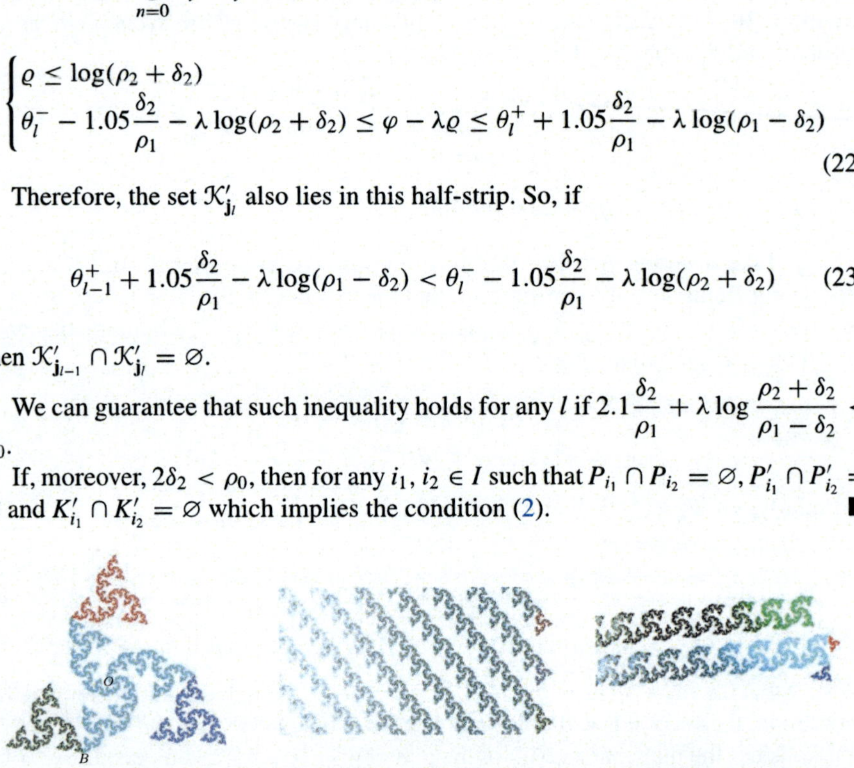

The images of the set K' under the map $w = \log(z - O)$ and the map $w = \log(z - B)$.

Theorem 27 *Let $\mathcal{S}$ be a contractible P-polygonal system. There is such $\delta > 0$ that for any δ-deformation $\mathcal{S}'$ of the system $\mathcal{S}$, satisfying parameter matching condition, the attractor $K(\mathcal{S}')$ is a dendrite, homeomorphic to $K(\mathcal{S})$.*

Proof Let all the cyclic vertices of the P-polygonal system $\mathcal{S}$ have order 1. If we suppose that $\delta_2 < \rho_1/4$, and $\delta_2 < (1 - \rho_2)/4$ and combine the inequalities 11, 14, 16, 17, we see that if the following inequalities hold:

1.$\delta < \dfrac{q_{min}}{8}$; 2. $\delta < \dfrac{1 - q_{max}}{8}$; 3. $\delta < \dfrac{\rho_0}{2(C_K + 1)}$; 4. $\delta < \dfrac{\rho_1}{4(C_K + 1)}$;

5. $\delta < \dfrac{1-\rho_2}{4(C_K+1)}$; and 6. $\delta < \dfrac{\alpha_0}{\dfrac{2.1(C_K+1)}{\rho_1} + C_\lambda \log \dfrac{1+3\rho_2}{3\rho_1}}$,

then the attractor K' of δ-deformation $\mathcal{S}'$ of the system $\mathcal{S}$ satisfies the condition (2). Therefore K' is a dendrite. By Theorem 14, the map $\hat{f} : K \to K'$ is a bijection and therefore it is a homeomorphism.

Suppose now that $\mathcal{S}$ has cyclic vertices of order greater than 1 and let $M = 12 + 4.2\left(1 + \dfrac{1}{q_{min}}\right)$. There is such n, that the system $\mathcal{S}^{(n)}$ has cyclic vertices of order 1. Suppose any δ-deformation of the system $\mathcal{S}^{(n)}$ generates a dendrite. Then for any δ/M-deformation deformation $\mathcal{S}'$ of the system $\mathcal{S}$, the system $\mathcal{S}'^{(n)}$ is a δ-deformation of the system $\mathcal{S}^{(n)}$. ■

References

1. V.V. Aseev, A.V. Tetenov, A.S. Kravchenko, Self-similar Jordan curves on the plane. Siberian Math. J. **44**(3), 379–38, MR1984698 (2003)
2. C. Bandt, J. Stahnke, Interior distance on deterministic fractals, in *Self-similar sets 6* (preprint, Greifswald, 1990)
3. J. Charatonik, W. Charatonik, Dendrites. Aportaciones Mat. Comun. **22**, 227–253 (1998)
4. L.L. Cristea , B. Steinsky, Curves of infinite length in labyrinth fractals. Proc. Edinb. Math. Soc., II. Ser., 54(2011)(2), 329–344
5. L.L. Cristea, B. Steinsky, Curves of infinite length in 4 × 4-labyrinth fractals. Geom. Dedicata **141**, 1–17 (2009)
6. L.L. Cristea, B. Steinsky, Mixed labyrinth fractals. Topology Appl. **229**, 112–125 (2017)
7. D. Croydon, *Random fractal dendrites* (St. Cross College, University of Oxford, Trinity, 2006). Ph.D. thesis
8. M. Hata, On the structure of self-similar sets. Jpn. J. Appl. Math **3**, 381–414 (1985)
9. J. Kigami, *Analysis on fractals*. Cambridge Tracts in Mathematics 143 (Cambridge University Press, 2001), 233 p
10. J. Kigami, Harmonic calculus on limits of networks and its application to dendrites. J. Funct. Anal. **128**(1), 48–86 (1995)
11. K. Kuratowski, *Topology* (Academic Press and PWN, New York, 1966)
12. S.B. Nadler, Jr., *Continuum theory: an introduction* (Dekker, 1992)
13. M. Samuel, A. Tetenov, D. Vaulin, Self-similar dendrites generated by polygonal systems in the plane. Sib. El. Math. Rep. 14, 737–751. MR3693741 (2017)
14. R.S. Strichartz, Isoperimetric estimates on Sierpinski gasket type fractals. Trans. Amer. Math. Soc. **351**, 1705–1752 (1999)
15. A.V. Tetenov , D. Mekhontsev, D. Vauilin, On weak separation property for plane dendrites (to appear)
16. A.V. Tetenov , M. Samuel, D. Mekhontsev, On dendrites generated by symmetric polygonal systems: the case of regular polygons. In: *Trends in Mathematics, Advances in Algebra and Analysis: International Conference on Advances in Mathematical Sciences, Vellore, India, Dec 2017*, vol. I (Springer, 2019), pp. 17–25

17. A.V. Tetenov , M. Samuel, D. Vauilin, On dendrites generated by polyhedral systems and their ramification points. In: *Proceedings of Krasovskii Institute of Mathematics and Mechanics of the UB RAS*, vol. 23(4), 281–291 (2017).https://doi.org/10.21538/0134-4889-2017-23-4-281-291
18. A.V. Tetenov , M. Samuel, D. Vauilin, On dendrites generated by polyhedral systems and their ramification points. www.arXiv.org/math.MG/1707.02875

General Position Theorem and Its Applications

Vladislav Aseev, Kirill Kamalutdinov, and Andrei Tetenov

Abstract We introduce some general and special formulations of general position theorem for parametrized families of fractals and explain the techniques of its application to prove the existence of self-similar sets with prescribed special properties.

Keywords Self-similar dendrite · Generalized polygonal system · Attractor · Postcritically finite set

1 Introduction

Consider the following problem:

Let K be the attractor of a system $\mathcal{S} = \{S_1, \ldots, S_m\}$ of contraction maps in $\mathbb{R}^n$ and let $\dim_H K < n/2$. Suppose that the intersection $S_i(K) \cap S_j(K)$ is nonempty for some i, j. Is it possible to change the maps $S_k \in \mathcal{S}$ slightly to maps S'_k to get a system $\mathcal{S}' = \{S'_1, \ldots, S'_m\}$ with the attractor K', such that the set $S'_i(K') \cap S'_j(K')$ is empty? To find the answer to this question, we consider the system $\mathcal{S} = \mathcal{S}_0$ as an element of a parametrized family $\mathcal{S}_t = \{S_{1,t}, \ldots, S_{m,t}\}$, where the parameter t assumes the values from some subset D in $\mathbb{R}^n$. We denote the attractor of the system $\mathcal{S}_t$ by K_t. We search

Supported by Russian Foundation of Basic Research projects 18-01-00420 and 18-501-51021.

V. Aseev · A. Tetenov (✉)
Sobolev Institute of Mathematics, Siberian Branch, Russian Academy of Sciences, Novosibirsk 630090, Russia
e-mail: a.tetenov@g.nsu.ru

V. Aseev
e-mail: aseevvv@yandex.ru

K. Kamalutdinov · A. Tetenov
Novosibirsk State University, Novosibirsk 630090, Russia
e-mail: k.kamalutdinov@g.nsu.ru

A. Tetenov
Gorno-Altaisk State University, Gorno-Altaisk 649000, Russia

R. L. Devaney et al. (eds.), *Topological Dynamics and Topological Data Analysis*, Springer Proceedings in Mathematics & Statistics 350,
https://doi.org/10.1007/978-981-16-0174-3_12

for the conditions under which $S_{i,t}(K_t) \cap S_{j,t}(K_t)$ is empty for almost all $t \in D$. In this case, we say that $S_{i,t}(K_t)$ and $S_{j,t}(K_t)$ are disjoint in general position.

Particularly, this occurs when Hausdorff dimension of the set $\Delta = \{t \in D : S_{i,t}(K_t) \cap S_{j,t}(K_t) \neq \infty\}$ is less than $\dim_H(D)$.

It is possible to make an estimate of $\dim_H(\Delta)$ in terms of upper bound for similarity dimensions of the systems $\{\mathcal{S}_t : t \in D\}$. The method for finding such estimates is based on General Position Theorem [6], which was initially introduced in [10].

2 Definitions and Notations

Let (X, d) be a complete metric space. A mapping $S : X \to X$ is a contraction if $\operatorname{Lip} S < 1$, and it is called a similarity if $d(S(x), S(y)) = rd(x, y)$ for all $x, y \in X$ and some fixed r.

Let $\mathcal{S} = \{S_1, \ldots, S_m\}$ be a system of contractions in a complete metric space (X, d). A nonempty compact set $K \subset X$ is called the attractor of the system $\mathcal{S}$, if $K = \bigcup\limits_{i=1}^{m} S_i(K)$. By Hutchinson's Theorem [5], the attractor K is uniquely defined by the system $\mathcal{S}$. We also call the set K *self-similar* with respect to $\mathcal{S}$, when all S_i are similarities.

Multiindices. Given a system $\mathcal{S} = \{S_1, \ldots, S_m\}, I = \{1, \ldots, m\}$ is the set of indices, $I^* = \bigcup\limits_{n=1}^{\infty} I^n$ is the set of all finite I-tuples, or multiindices $\mathbf{j} = j_1 j_2 \ldots j_n$. By $\mathbf{ij}$ we denote the concatenation of the corresponding multiindices; we write $\mathbf{i} \sqsubset \mathbf{j}$, if $\mathbf{j} = \mathbf{ik}$ for some $\mathbf{k} \in I^*$; we say that $\mathbf{i}$ and $\mathbf{j}$ are *incomparable*, if neither $\mathbf{i} \sqsubset \mathbf{j}$ nor $\mathbf{j} \sqsubset \mathbf{j}$; by $\mathbf{i} \wedge \mathbf{j}$ we mean the maximal $\mathbf{k}$ for which $\mathbf{k} \sqsubset \mathbf{i}$ and $\mathbf{k} \sqsubset \mathbf{j}$; by $|\mathbf{i}|$ we denote the length of $\mathbf{i}$.

We write $S_{\mathbf{j}} = S_{j_1 j_2 \ldots j_n} = S_{j_1} S_{j_2} \ldots S_{j_n}$, and for the set $A \subset X$, we denote $S_{\mathbf{j}}(A)$ by $A_{\mathbf{j}}$; given a set of m ratios $\{r_k, k \in I\}$ we write $r_{\mathbf{j}} = r_{j_1} r_{j_2} \ldots r_{j_n}$.

The Index Space. $I^\infty = \{\mathbf{i} = i_1 i_2 \ldots : \ i_k \in I\}$ is the index space; $\pi : I^\infty \to K$ is the *index map*, which sends $\mathbf{i} \in I^\infty$ to the point $\bigcap\limits_{n=1}^{\infty} K_{i_1 \ldots i_n}$. For a given vector $\mathbf{r} = (r_1, \ldots, r_m) \in (0, 1)^m$, we define a metrics $\rho_{\mathbf{r}}$ on I^∞ by $\rho_{\mathbf{r}}(\boldsymbol{\alpha}, \boldsymbol{\beta}) = r_{\boldsymbol{\alpha} \wedge \boldsymbol{\beta}}$. The set I^∞ supplied with this metrics will be denoted by $I^\infty_{\rho_{\mathbf{r}}}$. Let $s_{\mathbf{r}}$ denote the unique solution of the Moran equation $r_1^s + \cdots + r_m^s = 1$. Then, by [3, Theorem 6.4.3], $\dim_H I^\infty_{\rho_{\mathbf{r}}} = s_{\mathbf{r}}$.

Separation conditions. Denote $\mathcal{F} = \{S_{\mathbf{i}}^{-1} S_{\mathbf{j}} : \mathbf{i}, \mathbf{j} \in I^*\}$. Then the system $\mathcal{S} = \{S_1, \ldots, S_m\}$ of contraction similarities has the weak separation property (WSP) iff $\mathrm{Id} \notin \overline{\mathcal{F} \setminus \mathrm{Id}}$ [11]. The system $\mathcal{S}$ satisfies open set condition (OSC) if there is an open set V such that for any $i \in I$, $S_i(V) \in V$ and for any non-equal $i, j \in I$, $S_i \cap S_j(V) = \varnothing$. The system satisfies strong separation condition (SSC), if for any non-equal $i, j \in I$, $K_i \cap K_j = \varnothing$. There are well-known implications (SSC)$\to$(OSC) and (OSC)$\to$(WSP) [1, 8, 11]

3 General Position Theorem

We begin with a simple example. Let A, B be compact subsets in $\mathbb{R}^n$, and the set B is being translated by a vector $t \in D$, where $D \subset \mathbb{R}^n$. We wish to understand, how large can be the set of parameters $\Delta = \{t \in D : A \cap (B + t) \neq \varnothing\}$, which we will call the set of exceptional parameters.

It is easy to see that $A \cap (B + t) \neq \varnothing$ is equivalent to: " there are such $a \in A$, $b \in B$ that $a = b + t$". Finding t from this equation, we see that $\Delta = \{a - b : a \in A, \ b \in B\}$. How to evaluate the Hausdorff dimension of the set Δ in terms of A and B?

For that reason, we introduce the map $f : A \times B \to \Delta, f(a, b) = a - b$. Since f is Lipschitz, $\dim_H \Delta \leq \dim_H(A \times B)$, and if the product $A \times B$ has the dimension less than $\dim_H D$, then A and $B + t$ are disjoint for almost all $t \in D$.

We will extend this approach to a very general situation, taking a normed linear space $\mathcal{M}$ instead of $\mathbb{R}^n$, replacing A and B by metric spaces (L_1, σ_1), (L_2, σ_2) and finding the set Δ for parametrized families $A_t = \varphi_1(t, L_1)$ and $B_t = \varphi_2(t, L_2)$ instead of A and $B + t$ [6]:

Theorem 1 *Let the Cartesian products of metric spaces (D, ρ), (L_1, σ_1), (L_2, σ_2) be supplied with the canonical metrisation (see [7], §21.VI, (1)). Let continuous maps $\varphi_1 : D \times L_1 \to \mathcal{M}$ and $\varphi_2 : D \times L_2 \to \mathcal{M}$ to the normed linear space $(\mathcal{M}, \|.\|)$ be such that:*
(a) there are $C_0 > 0$ and $\alpha > 0$ such that for any $i = 1, 2$ and for all (ξ, x), (ξ, y) in $D \times L_i$ the estimate holds

$$\|\varphi_i(\xi, x) - \varphi_i(\xi, y)\| \leq C_0[\sigma_i(x, y)]^\alpha$$

(uniform α-Hölder continuity condition);
(b) there are such $M_0 > 0$ and $\beta > 0$ that for any $(x_1, x_2) \in L_1 \times L_2$ and $\xi, \xi' \in D$ the function

$$\Phi(\xi, x_1, x_2) := \varphi_1(\xi, x_1) - \varphi_2(\xi, x_2)$$

on the set $D \times L_1 \times L_2$ satisfies the condition

$$\|\Phi(\xi', x_1, x_2) - \Phi(\xi, x_1, x_2)\| \geq M_0[\rho(\xi', \xi)]^\beta \ . \tag{1}$$

Then Hausdorff dimension of the set $\Delta := \{\xi \in D : \varphi_1(\xi, L_1) \cap \varphi_2(\xi, L_2) \neq \varnothing\}$ satisfies

$$\dim_H \Delta \leq \min\{(\beta/\alpha) \dim_H(L_1 \times L_2), \dim_H D\} \ . \tag{2}$$

Moreover, if the spaces (L_1, σ_1), (L_2, σ_2) are compact, Δ is closed in D.

Proof Put $\tilde{\Delta} := \{(\xi, x_1, x_2) \in D \times L_1 \times L_2 : \varphi_1(\xi, x_1) = \varphi_2(\xi, x_2)\} = \{(\xi, x_1, x_2) \in D \times L_1 \times L_2 : \Phi(\xi, x_1, x_2) = 0\}$ and notice that $\Delta = \mathrm{pr}_1 \tilde{\Delta}$, where $\mathrm{pr}_1 : D \times L_1 \times L_2 \to D$ is the canonical projection.

Applying canonical projection $\mathrm{pr}_2 : D \times (L_1 \times L_2) \to L_1 \times L_2$ we obtain a set $\Delta_L := \mathrm{pr}_2(\tilde{\Delta})$, that is,

$$\Delta_L = \{(x_1, x_2) \in L_1 \times L_2 |\ \exists \xi \in D : \varphi_1(\xi, x_1) = \varphi_2(\xi, x_2)\}.$$

The maps $\pi_D = \mathrm{pr}_1|_{\tilde{\Delta}} : \tilde{\Delta} \to \Delta$ and $\pi_L = \mathrm{pr}_2|_{\tilde{\Delta}} : \tilde{\Delta} \to \Delta_L$ are continuous open maps (by properties of canonical projections). Let us show that π_L is a bijection. Indeed, if for $(\xi', x_1', x_2') \in \tilde{\Delta}$ and $(\xi'', x_1'', x_2'') \in \tilde{\Delta}$ the equality $\pi_L(\xi', x_1', x_2') = \pi_L(\xi'', x_1'', x_2'')$ holds, then $(x_1', x_2') = (x_1'', x_2'') = (x_1, x_2)$, whereas $\Phi(\xi', x_1, x_2) = 0 = \Phi(\xi'', x_1, x_2)$. Then from (1) it follows that $0 = \|\Phi(\xi', x_1, x_2) - \Phi(\xi'', x_1, x_2)\| \geq M_0[\rho(\xi', \xi'')]^\beta$, that is, $\rho(\xi', \xi'') = 0$. This means that $\xi' = \xi''$.

Since every open bijective continuous map is a homeomorphism (see [7, §13.XIII]), the maps π_L and π_L^{-1} are homeomorphisms.

Now we find Hölder continuity estimate for a map $g = \pi_D \circ \pi_L^{-1} : \Delta_L \to \Delta$. Let $\xi' = g(x_1', x_2')$ and $\xi = g(x_1, x_2)$. Then $\Phi(\xi', x_1', x_2') = 0 = \Phi(\xi, x_1, x_2)$ and, particularly, $\varphi_1(\xi', x_1') = \varphi_2(\xi', x_2')$. The inequality (1) gives an estimate

$$M_0[\rho(\xi', \xi)]^\beta \leq \|\Phi(\xi', x_1, x_2) - \Phi(\xi, x_1, x_2)\| = \|\Phi(\xi', x_1, x_2) - 0\|$$

$$= \|\varphi_1(\xi', x_1) - \varphi_2(\xi', x_2)\| \leq \|\varphi_1(\xi', x_1) - \varphi_1(\xi', x_1')\| + \|\varphi_1(\xi', x_1') - \varphi_2(\xi', x_2)\|$$

$$= \|\varphi_1(\xi', x_1) - \varphi_1(\xi', x_1')\| + \|\varphi_2(\xi', x_2') - \varphi_2(\xi', x_2)\|.$$

Applying the condition (a), we get the inequality

$$M_0[\rho(\xi', \xi)]^\beta \leq C_0[\sigma_1(x_1, x_1')]^\alpha + C_0[\sigma_2(x_2, x_2')]^\alpha \leq 2C_0\left[\sqrt{\sigma_1(x_1, x_1')^2 + \sigma_2(x_2, x_2')^2}\right]^\alpha.$$

Denoting by $\tilde{\sigma}$ the metrics of Cartesian product of the spaces (L_1, σ_1) and (L_2, σ_2), we get Hölder continuity estimate of the map g:

$$\rho(g(x_1', x_2'), g(x_1, x_2)) \leq (2C_0/M_0)^{1/\beta}[\tilde{\sigma}((x_1', x_2'), (x_1, x_2))]^{\alpha/\beta}.$$

Applying [4, Proposition 2.3] and the inequality $\dim_H \Delta_L \leq \dim_H(L_1 \times L_2)$, we get the desired relation (2):

$$\dim_H \Delta = \dim_H g(\Delta_L) \leq (\beta/\alpha)\dim_H(L_1 \times L_2) \text{ and } \dim_H \Delta \leq \dim_H D.$$

Since the maps φ_i are continuous, Φ is continuous too. The set $\tilde{\Delta}$ is closed in $D \times L_1 \times L_2$ as a set of zeros of Φ, so the set $\Delta = \pi_D\tilde{\Delta}$ is closed in D (by properties of canonical projections). ■

Remark 1. We see from the inequality (2) that if the product $L_1 \times L_2$ has sufficiently small dimension, then the sets $\varphi(t, L_1)$ and $\psi(t, L_2)$ do not intersect for almost all $t \in D$. The proof of the inequality (2) in the Theorem does not use the condition that

the functions φ_1 and φ_2 are continuous with respect to the metrization of product spaces, so this condition may be omitted. It is needed only to show that Δ is closed in D.
2. The condition (b) in the Theorem may be considered as a form of transversality condition [9], where $D \subset \mathbb{R}^n$ is an open set, $\beta = 1$ and φ_i $(i = 1, 2)$ are the address maps to different copies of a self-similar set, depending of a parameter $\xi \in D$.
3. Notice that the only information required of the parameter space D is its Hausdorff dimension. Moreover, if $\dim_H D = s$ but the measure $H^s(D)$ is zero, we take some s' satisfying $\dim_H \Delta < s' < s$ to see that Δ is negligible in D in a sense that $H^{s'}(D) = \infty$ and $H^{s'}(\Delta) = 0$.

For more easy understanding of the main idea ot the Theorem 1, we apply it to much more simplified settings. Nevertheless, even the following simplified form will be useful for many applications:

Corollary 2 *Let A, B, D be some subsets of $\mathbb{R}^n$. Let the map $\varphi : D \times B \to \mathbb{R}^n$ be such that:*
(a) there is $C_0 > 0$ such that for any $x, y \in B$ and $t \in D$, $\|\varphi(t, x) - \varphi(t, y)\| \le C_0\|x - y\|$
(b) there is such $M_0 > 0$ that for any $x \in B$ and $t, t' \in D$

$$\|\varphi(t', x) - \varphi(t, x)\| \ge M_0\|t' - t\| \,. \tag{3}$$

Then Hausdorff dimension of the set $\Delta := \{t \in D : \varphi(t, B) \cap A \neq \varnothing\}$ satisfies

$$\dim_H \Delta \le \min\{\dim_H(A \times B), \dim_H D\} \tag{4}$$

Moreover, if A and B are compact and the map φ is continuous, then Δ is closed in D. ■

One can consider several specific applications which may be derived from the Corollary 2:

Example 1 If $A, B \subset \mathbb{C}$ and $0 \notin \bar{A}$ and $\dim_H A \times B < 2$ then for Lebesgue almost all $z \in \mathbb{C}$: $B \cap zA = \varnothing$.

Indeed, let $M_0 = \inf\{|z| : z \in A\}$ and for some $C_0 > 0$, let $D = \{z : |z| < C_0\}$. Then the conditions (a) and (b) of the Corollary 2 are fulfilled. Therefore, if $\dim_H(A \times B) < 2$ then for Lebesgue almost all $z \in D$ the sets A and B are disjoint. Letting C_0 tend to infinity, we get that the statement is true for Lebesgue almost all $z \in \mathbb{C}$.

Example 2 If $A, B \subset \mathbb{R}^n$, $M_2 > M_1 > 0$, a map $f : B \times \mathbb{R}^n \to \mathbb{R}^n$ is M_1-Lipschitz, and $\dim_H(A \times B) < n$, then the set $\Delta = \{t \in \mathbb{R}^n : M_2 t + f(B, t) \cap A \neq \varnothing\}$ has zero measure in $\mathbb{R}^n$.

In this case, the conditions (a), (b) are fulfilled with $C_0 = M_1$ and $M_0 = M_2 - M_1$. Since the set Δ can be represented also as $\{t \in \mathbb{R}^n : f(B, t) \cap M_2 t + A \neq \varnothing\}$ this means that if A moves faster that the set B is deformed, for almost all t the set A escapes the intersection with the set $f(B, t)$.

Example 3 Suppose $A, B \subset \mathbb{R}^n$, a map $F : \mathbb{R}^n \to \mathbb{R}^n$ is bi-Lipschitz, and $f : B \times \mathbb{R}^m$ is defined by $f(x, t) = F(x + t)$. $\dim_H(A \times B) < n$, then the set $\Delta = \{t \in \mathbb{R}^n : f(B, t) \cap A \neq \varnothing\}$ has zero measure in $\mathbb{R}^n$.

In this case, we can interpret $f(B, t)$ as a bi-Lipschitz distortion of a translation of the set B by a vector t.

4 Application of General Position Theorem to Self-similar Sets

The General Position Theorem is a tool for treating more complicated cases than those in which one of the sets undergoes simple rigid motions or similarities or translations in some curvilinear coordinates. It works with the attractors K_t of parametrized systems $\mathcal{S}_t$ of contraction maps. These attractors need not be even homeomorphic to each other for different values of the parameter t.

To analyze transformations of the attractors of such systems, we define the following settings for parametrized families:
(S1). Let $\mathcal{S}_t = \{S_{1,t}, \ldots, S_{m,t}\}$ be a system of contraction maps in $\mathbb{R}^n$, depending on the parameter $t \in D \subset \mathbb{R}^n$ and let K_t be its attractor.
(S2) Suppose there is a compact set V such that for any $k \in I$ and any $t \in D$, $S_{k,t}(V) \subset V$.
(S3) There is a vector $\mathbf{r} = (r_1, \ldots, r_m)$ such that for any $t \in D$ and for any $k \in I$, $\operatorname{Lip} S_{k_t} \leq r_k < 1$. Let $\bar{r} = \max\{r_1, \ldots, r_m\}$.
(S4) There is such $C > 0$ that for any $x \in V$, $k \in I$ and for any $t, t' \in D$, $\|S_{k,t'}(x) - S_{k,t}(x)\| \leq C\|t' - t\|$.

4.1 Moving Subpieces Apart from Each Other.

First notice that it follows from the settings **(S1), (S3)** that all the address maps are Lipschitz with a constant equal to $\operatorname{diam}(K)$:

Lemma 3 *If the settings* **(S1), (S3)** *are fulfilled then the map* $\pi : I^\infty_{\rho_\mathbf{r}} \to K$ *is* $\operatorname{diam}(K)$*–Lipschitz.*

Proof (cf. [3], Ex. 4.2.4). Suppose $\boldsymbol{\alpha} \wedge \boldsymbol{\beta} = \mathbf{j}$, so $\boldsymbol{\alpha} = \mathbf{j}\boldsymbol{\alpha}'$ and $\boldsymbol{\beta} = \mathbf{j}\boldsymbol{\beta}'$. From $\rho_\mathbf{r}(\boldsymbol{\alpha}', \boldsymbol{\beta}') = 1$ we get $\|\pi(\boldsymbol{\alpha}) - \pi(\boldsymbol{\beta})\| = \|S_\mathbf{j}(\pi(\boldsymbol{\alpha}')) - S_\mathbf{j}(\pi(\boldsymbol{\beta}'))| \leq r_\mathbf{j} \operatorname{diam}(K) = \operatorname{diam}(K)\rho_\mathbf{r}(\boldsymbol{\alpha}, \boldsymbol{\beta})$. ■

To evaluate the distance between the points in K_t and $K_{t'}$ having the same addresses, we use the Displacement Theorem for parametrized families (cf. [6, Theorem 17]):

Theorem 4 *Suppose the settings* **(S1)—(S4)** *hold. Then for any* $\boldsymbol{\alpha} \in I^\infty$ *and any* $t, t' \in D$ *we have*

$$\|\pi_{t'}(\boldsymbol{\alpha}) - \pi_t(\boldsymbol{\alpha})\| \le \frac{C\|t'-t\|}{1-\bar{r}}. \tag{5}$$

Proof Take $\boldsymbol{\alpha} = i_1 i_2 \ldots$ and denote $\boldsymbol{\alpha}_k = i_k i_{k+1} \ldots$.
Since $\pi_t(\boldsymbol{\alpha}_k) = S^t_{i_k}\pi_t(\boldsymbol{\alpha}_{k+1})$, $\|\pi_t(\boldsymbol{\alpha}_k) - \pi_{t'}(\boldsymbol{\alpha}_k)\| \le \|S^t_{i_k}\pi_t(\boldsymbol{\alpha}_{k+1}) - S^t_{i_k}\pi_{t'}(\boldsymbol{\alpha}_{k+1})\| + \|S^t_{i_k}\pi_{t'}(\boldsymbol{\alpha}_{k+1}) - S^{t'}_{i_k}\pi_{t'}(\boldsymbol{\alpha}_{k+1})\|$, so $\|\pi_t(\boldsymbol{\alpha}_k) - \pi_{t'}(\boldsymbol{\alpha}_k)\| \le r_{i_k}\|\pi_t(\boldsymbol{\alpha}_{k+1}) - \pi_{t'}(\boldsymbol{\alpha}_{k+1})\| + C\|t'-t\|$ for any $k \in \mathbb{N}$.

Therefore, $\|\pi_t(\boldsymbol{\alpha}) - \pi_{t'}(\boldsymbol{\alpha})\| \le \bar{r}^{n+1}\|\pi_t(\boldsymbol{\alpha}_{n+1}) - \pi_{t'}(\boldsymbol{\alpha}_{n+1})\| + C\|t'-t\| \sum_{k=0}^{n} \bar{r}^k$,
which becomes (5) as k tends to ∞. ■

The following theorem gives the conditions under which the pieces $K_{\mathbf{j},t}$ and $K_{\mathbf{k},t}$ are disjoint for almost all $t \in D$:

Theorem 5 *Suppose the settings* **(S1)—(S4)** *hold. Let* $\mathbf{j}, \mathbf{k} \in I^*$ *be incomparable multiindices.*
Suppose there are such $c_\mathbf{j} > 0$, $C_\mathbf{k} > 0$ *that for any* $x \in V$ *and for any* $t, t' \in D$,

$$\|S^{t'}_\mathbf{k}(x) - S^t_\mathbf{k}(x)\| \le C_\mathbf{k}\|t'-t\| \text{ and } \|S_{\mathbf{j},t'}(x) - S_{\mathbf{j},t}(x)\| \ge c_\mathbf{j}\|t'-t\| \tag{6}$$

If

$$c_\mathbf{j} - C_\mathbf{k} - \frac{(r_\mathbf{j} + r_\mathbf{k})C}{1-\bar{r}} > 0 \tag{7}$$

and $s_\mathbf{r} < \dim_H(D)/2$, *then* $K_\mathbf{j} \cap K_\mathbf{k} = \varnothing$ *for almost all* $t \in D$.

Proof Let $\varphi(t,x) = S_{\mathbf{k},t}(\pi_t(x))$, $\psi(t,x) = S_{\mathbf{j},t}(\pi_t(x))$, $\Phi(t,x,y) = \varphi(t,x) - \psi(t,y)$,
$\Delta = \{t \in D : K_\mathbf{j} \cap K_\mathbf{k} \ne \varnothing\}$. Note that

$$\|\Phi(t',x,y) - \Phi(t,x,y)\| \ge \|\psi(t',y) - \psi(t,y)\| - \|\varphi(t',x) - \varphi(t,x)\|;$$
$$\|\varphi(t',x) - \varphi(t,x)\| \le \|S_{\mathbf{k},t'}(\pi_{t'}(x)) - S_{\mathbf{k},t}(\pi_{t'}(x))\| + \|S_{\mathbf{k},t}(\pi_{t'}(x)) - S_{\mathbf{k},t}(\pi_t(x))\|;$$
$$\|\psi(t',x) - \psi(t,x)\| \ge \|S_{\mathbf{j},t'}(\pi_{t'}(x)) - S_{\mathbf{j},t}(\pi_{t'}(x))\| - \|S_{\mathbf{j},t}(\pi_{t'}(x)) - S_{\mathbf{j},t}(\pi_t(x))\|.$$

From Theorem 4, we have upper estimates

$$\|S_{\mathbf{k},t}(\pi_{t'}(x)) - S_{\mathbf{k},t}(\pi_t(x))\| \le \frac{r_\mathbf{k}C\|t'-t\|}{1-\bar{r}} \quad \text{and} \quad \|S_{\mathbf{j},t}(\pi_{t'}(x)) - S_{\mathbf{j},t}(\pi_t(x))\| \le \frac{r_\mathbf{j}C\|t'-t\|}{1-\bar{r}}$$

Combining them with inequalities (6), we obtain

$$\|\Phi(t',x,y) - \Phi(t,x,y)\| \ge \left(c_\mathbf{j} - C_\mathbf{k} - \frac{C(r_\mathbf{k} + r_\mathbf{j})}{1-\bar{r}}\right)\|t'-t\| \tag{8}$$

Applying the Theorem 1 with $\alpha = \beta = 1$ we get $\dim_H \Delta < 2\dim_H(I^\infty_{\rho_\mathbf{r}}) = 2s_\mathbf{r}$.
Since $s_\mathbf{r} < \dim_H(D)/2$, we get $H^{2s_\mathbf{r}}(\Delta) = 0$ and at the same time $H^{2s_\mathbf{r}}(D) = \infty$. ■

4.1.1 The Case When the Parameters Are Translation Vectors.

We consider the case is when the initial system $\mathcal{S} = \{S_1, ..., S_m\}$ consists of the contraction maps S_k in $\mathbb{R}^n$, and we consider a parametrized system $\mathcal{S}_t = \{S_{1,t}, ..., S_{m,t}\}$ where each $S_{k,t}$ is defined by the formula $S_{k,t}(x) = S_k(x) + t_k$, where $\boldsymbol{t} = (t_1, ..., t_m) \in (\mathbb{R}^n)^m$. Translations have no effect upon the contraction ratios, therefore Lip $S_{k,t} = r_k$ for any t.

First we allow only one map, say $S_{m,t}$, to depend on the parameter t, leaving all others unchanged.

Corollary 6 *Let $\mathcal{S}_t = \{S_1, \ldots, S_{m-1}, S_{m,t}(x) = S_m(x) + t\}$ be a system of contraction maps in $\mathbb{R}^n$, depending on the parameter $t \in \mathbb{R}^n$ and let K_t be its attractor. Let $1 \le k < m$. If $r_k + r_m + \bar{r} < 1$ and $s_{\mathbf{r}} < n/2$, then $K_{k,t} \cap K_{m,t} = \varnothing$ for almost all $t \in \mathbb{R}^n$.*

Proof For any open bounded $D \subset \mathbb{R}^n$, there is such $V \subset \mathbb{R}^n$ that the system $\mathcal{S}^t$ satisfies the settings **(S1)—(S4)**; since $C = 1$ the condition 7 of the Theorem 5 becomes equivalent to $r_k + r_m + \bar{r} < 1$. Therefore $K_{k,t} \cap K_{m,t} = \varnothing$ for almost all $t \in D \subset \mathbb{R}^n$. The result does not depend on the choice of $D \subset \mathbb{R}^n$, so it holds for the whole $\mathbb{R}^n$. ■

Now, if we apply a translation by some vector $t_k \in \mathbb{R}^n$ to each map $S_k \in \mathcal{S}$, we obtain the following:

Corollary 7 *Let $\mathcal{S} = \{S_1, \ldots, S_{m-1}, S_m\}$ be a system of contraction maps in $\mathbb{R}^n$. Let $\boldsymbol{t} = \{t_1, ..., t_m\}$, where $t_k \in \mathbb{R}^n$. Let $S_{k,t}(x) = S_k(x) + t_k$. Let K_t be the attractor of the system $\mathcal{S}_t = \{S_{1,t}, ..., S_{m,t}\}$. If for any non-equal $j, k \in I$, $r_j + r_k + \bar{r} < 1$ and $s_{\mathbf{r}} < n/2$, then for almost all $\boldsymbol{t} \in \mathbb{R}^{mn}$, the system $\mathcal{S}$ satisfies Strong Separation Condition.*

Proof Notice that by Theorem 4, the maps $\pi_{j,t} : I^\infty \times \mathbb{R}^{nm} \to \mathbb{R}^n$ are continuous with respect to $\boldsymbol{t}$. Therefore the function $\rho_{jk}(\boldsymbol{t}) = \min\{\|\pi_{j,t}(\boldsymbol{\alpha}) - \pi_{k,t}(\boldsymbol{\beta})\|, \boldsymbol{\alpha}, \boldsymbol{\beta} \in I^\infty\}$ is continuous with respect to $\boldsymbol{t}$. Therefore, the set $\Delta_{jk} = \rho^{-1}(\{0\})$ is closed in $\mathbb{R}^{nm}$. Since all of its k-slices $\{(t_1, .., t_{k-1}, t, t_{k+1}, ..., t_m) \in \Delta_{jk}; t \in \mathbb{R}^n\}$ have zero Lebesgue n-dimensional measure, the set Δ_{jk} has zero measure in $\mathbb{R}^{mn}$. Thus, the set $\Delta = \bigcup_{j,k \in I} \Delta_{jk}$ also has zero measure in $\mathbb{R}^{mn}$. Therefore, for almost all $\boldsymbol{t} \in \mathbb{R}^{mn}$, the system $\mathcal{S}_t$ satisfies Strong Separation Condition. ■

4.2 *Non-empty Overlaps of Prescribed Type*

If we we get rid of all overlaps in a self-similar set, we obtain a system $\mathcal{S}$, which satisfy strong separation condition and whose attractor K is just a Cantor set. There is a mush more interesting case, when we use our techniques to obtain a system $\mathcal{S}$ of contraction maps which has the attractor K such that the intersections of its pieces K_j strictly follow some predefined pattern. The attractors of such systems possess a set of interesting properties, and often, they do not satisfy WSP. In this subsection, we will see

(a) how to find systems $\mathcal{S}$ for which two maps S_1 and S_2 commute and for which $S_1(K) \cap S_2(K)$ is exactly equal to $S_{12}(K)$ and
(b) how to find systems $\mathcal{S}$ which do not satisfy OSC though all the pieces $S_i(K)$ are disjoint except $S_1(K) \cap S_2(K)$ which is a single point.

4.2.1 Exact Overlaps: An Example

First we consider the systems $\mathcal{S}$ in which two maps S_1, S_2 have a common fixed point and commute (cf. [2]). Let the system $\mathcal{S}_t$ in $[0, 1]$ consist of 3 maps: $S_1(x) = tx$, $S_2(x) = bx$, $S_3(x) = \dfrac{x+8}{9}$ in $\mathbb{R}$, where $b, t \in (0, 1/9)$. It depends on the parameter t, while b is a fixed value.

Since the maps $S_{1,t}$ and S_2 commute, we have the following inclusion:

$$S_{1,t}S_2(K_t) \subseteq S_{1,t}(K_t) \cap S_2(K_t) \tag{9}$$

We want to study for which $t \in (0, 1/9)$ the inclusion (9) becomes equality. In this case, we say the system $\mathcal{S}_t$ has exact overlap $S_1(K) \cap S_2(K) = S_{12}(K)$.

Notice that the same way as in ([6, Proposition 2(v)]),

$$K_t \setminus \{0\} = \bigcup_{m,n=0}^{\infty} S_1^m S_2^n(K_{3,t}) \tag{10}$$

Since $t, b < 1/9$ and $K_3 \subset [8/9, 1]$, for any $m \neq n$, $S_i^m(K_3) \cap S_i^n(K_3) = \varnothing$ for $i = 1, 2$.

Following the argument of [6, Proposition 3], we obtain

Proposition 8 *For the system $\mathcal{S}_t$, the following statements are equivalent:*
(i) For any $m, n \in \mathbb{N}$, $S_1^m(K_3) \cap S_2^n(K_3) = \varnothing$;
(ii) $K = \{0\} \cup \bigsqcup_{m,n=0}^{\infty} S_1^m S_2^n(K_3)$;
(iii) For any $m, n \in \mathbb{N}$, $S_1^m(K) \cap S_2^n(K) = S_1^m S_2^n(K)$. ■

Proposition 9 *The system $\mathcal{S}_t$ has exact overlap $S_1(K) \cap S_2(K) = S_{12}(K)$ for Lebesgue almost all $t \in (0, 1/9)$.*

Proof By proposition 8, it suffices to find the set of those t, for which $S_1^m(K_3) \cap S_2^n(K_3) = \varnothing$ for any $m \neq n$.

Take non-equal $m, n \in \mathbb{N}$ and let $D_{mn} = \{t \in (0, 1/9) : S_{1,t}^m([8/9, 1]) \cap S_2^n([8/9, 1]) \neq \varnothing\}$.

If $t \in D_{mn}$ then $\dfrac{8b^n}{9} \leq t^m \leq \min\left\{\dfrac{9b^n}{8}, \dfrac{1}{9^m}\right\}$. Put $\bar{t} = \left(\min\left\{\dfrac{9b^n}{8}, \dfrac{1}{9^m}\right\}\right)^{1/m}$.

To apply the Theorem 5, we interpret the case under consideration in terms of its settings:
The system $\mathcal{S}_t$ depends on the parameter $t \in D_{mn}$.
The set $V = [0, 1]$, the constant $C = 1$. Since the vector $\mathbf{r} = (\bar{t}, b, 1/9)$, we have $s_{\mathbf{r}} < 1/2$.
Further, $S_{\mathbf{j}} = S^m_{1,t}$, $S_{\mathbf{k}} = S^n_2$, therefore $r_{\mathbf{j}} = \bar{t}^m < \dfrac{9b^n}{8}$, $r_{\mathbf{k}} = b^n$.
By definition, $c_{\mathbf{j}} = \inf\limits_{t,t' \in D_{mn}} \dfrac{t'^m - t^m}{t' - t} = \inf\limits_{t \in D_{mn}} mt^{m-1} \geq \inf\limits_{t \in D_{mn}} \dfrac{t^m}{t}$.
Replacing t^m by $\dfrac{8b^n}{9}$ and t in denominator by 1/9, we get $c_{\mathbf{j}} > 8b^n$.
Since $C_{\mathbf{k}} = 0$, we have $c_{\mathbf{j}} - C_{\mathbf{k}} - \dfrac{r_{\mathbf{j}} + r_{\mathbf{k}}}{1 - \bar{r}} > \left(8 - \dfrac{9/8 + 1}{8/9}\right) b^n$.
Therefore by Theorem 5, the set $\Delta_{mn} = \{t \in D : S^m_{1,t}(K_{3,t}) \cap S^n_2(K_{3,t}) \neq \varnothing\}$ is a closed subset of D_{mn} and $\dim_H(\Delta_{mn}) < 1$.

Let Δ be the union of all Δ_{mn}, where $m, n \in \mathbb{N}$ and $m \neq n$.
Then $\dim_H(\Delta) \leq 2s_{\mathbf{r}} < 1$ which implies the statement of the proposition. ■

For almost all t, the systems $\mathcal{S}_t$ possess several remarkable properties:

1. Violation of WSP. Consider the set D^* of those values of the parameter $t \in D \setminus \Delta$ for which $\dfrac{\log t}{\log b}$ is irrational. The set D^* has full measure in D. For each $t \in D^*$, there are sequences of positive integers l_k,n_k such that the sequence $t^{l_k} b^{-n_k}$ converges to 1. Therefore, the system $\mathcal{S}_t$ does not satisfy weak separation property.

2. Measure and dimension. The Hausdorff dimension s of the attractor K_t, $t \in D^*$ is equal to the solution of the equation $t^x + b^x - t^x b^x + 9^{-x} = 1$. Since the weak separation property is violated, the Hausdorff measure $H^s(K_{t_0}) = 0$.

3. All K_t are isomorphic. For any two sets $K_{t_1}, K_{t_2}, t_i \in D^*$, there is a homeomorphism $\varphi : K_{t_1} \to K_{t_2}$, which agrees with the systems $\mathcal{S}_1$ and $\mathcal{S}_2$, i.e. for any $k = 1, ..., 4$ and for any $x \in K_t$, $\varphi(S_{k,t}(x) = S_{k,t'}(\varphi(x))$.

We refer the reader to [6] for detailed proofs of the properties of such type of self-similar sets.

4.2.2 One-Point Intersections: An Example

Take p, q, r in $(0, 1/36)$ and put $h = \dfrac{1}{2}$, $a = \dfrac{1}{3}$. Consider a system $\mathcal{S} = \{S_1, S_2, ..., S_6\}$ of contractions in [0, 1] whose equations are

$$S_1(x) = px, \quad S_2(x) = a + rx, \quad S_3(x) = h - qx, \quad S_4(x) = h - r + rx,$$

$$S_5(x) = 1 - a - rx, \quad S_6(x) = 1 - r + rx$$

The similarity dimension for any such system is strictly less than 1/2.

Let K be the attractor of the system $\mathcal{S}$ and $K_i = S_i(K)$ be its pieces. By the construction, $\{0, 1\} \subset K \subset [0, 1]$ and the pieces K_i, $i \in \{1, 2, 3, 5, 6\}$ are contained in disjoint segments of length $1/36$, while $K_3 \cup K_4 \subset [h - 1/36, h]$ and $K_3 \cap K_4 \ni \{h\}$ which is the only possible non-empty intersection of the pieces.

We wish to know the set of those p, q, r for which $K_3 \cap K_4 = \{h\}$. In this case, we say that the system $\mathcal{S}$ has *unique one-point intersection.*

If $\dfrac{\log p}{\log r} \notin \mathbb{Q}$, then the system $\mathcal{S}$ does not have WSP for any q. Indeed, consider the maps $H_m(x) = S_3 S_1^m S_5(x)$ and $G_n(x) = S_4 S_6^n S_2(x)$. Notice that for any $q > 0$, there is a sequence $(m_k, n_k) \in \mathbb{N}^2$, such that $p^{-m_k} r^{n_k+1}$ converges to q as $k \to \infty$. Easy computation shows that if we choose such a sequence (m_k, n_k), then the sequence

$$G_{n_k}^{-1} H_{m_k}(x) = \frac{(r^{n_k+1} - p^{m_k} q)(1 - a)}{r^{n_k+2}} + \frac{p^{m_k} q}{r^{n_k+1}} x$$

converges to identity, which means violation of WSP.

Therefore, we fix some $p, r \in (0, 1/36)$ such that $\log_r p$ is irrational and consider a 1-parameter family of systems $\mathcal{S}_q$, $q \in (0, 1/36)$, for which we show that for Lebesgue almost all $q \in (0, 1/36)$ the system $\mathcal{S}_q$ has unique one-point intersection and does not have weak separation property.

For the simplicity of notation, we denote the system under consideration by $\mathcal{S}$, keeping in mind that it depends on the parameter q whenever it does not cause any ambiguity.

From the representation of the pieces K_3 and K_4 as unions of infinite sequences

$$K_3 = \{h\} \cup \bigcup_{m=0}^{\infty} S_3 S_1^m (K \setminus K_1), \quad K_4 = \{h\} \cup \bigcup_{n=0}^{\infty} S_4 S_6^n (K \setminus K_6),$$

we see that $K_3 \cap K_4 = \{h\}$ iff

$$\text{for any } m, n \in \mathbb{N} \cup \{0\} \text{ and any } i \in I \setminus \{6\},\ j \in I \setminus \{1\}, \quad S_3 S_1^m (K_j) \cap S_4 S_6^n (K_i) = \varnothing \tag{11}$$

Note that if $p^m[aq, q] \cap r^{n+1}[a, 1] = \varnothing$ then for any $i \in I \backslash \{6\}, j \in I \backslash \{1\}$ the intersections $S_3 S_1^m S_j(K) \cap S_4 S_6^n S_i(K)$ are empty. Therefore, in search of those q for which $S_3 S_1^m S_j(K)$ and $S_4 S_6^n S_i(K)$ may intersect, we can restrict the values of q to the intervals

$$D_{mn}(p, r) := \left(\frac{ar^{n+1}}{p^m}, \min \left(\frac{r^{n+1}}{ap^m}, 1/36 \right) \right)$$

We apply the Theorem 5 to the family $\mathcal{S}_q$ with the parameter set $D_{mn}(p, r)$ and to $S_{\mathbf{j}} = S_3 S_1^m$ and $S_{\mathbf{k}} = S_4 S_6^n$. We take $\mathbf{r} = (p, r, 1/36, r, r, r)$, therefore $s_{\mathbf{r}} < 1/2$ and $\bar{r} = 1/36$. We have $C = 1$, $C_{\mathbf{k}} = 0$ and $r_{\mathbf{k}} = r^{n+1}$. Now since the set K_j lies in the interval $[a, 1]$, for $x \in K_j$ and $q', q \in D_{mn}(p, r)$ we have $|S_{\mathbf{j},q'}(x) - S_{\mathbf{j},q}(x)| = |q' - q|p^m x \geq |q' - q|p^m a$, so $c_{\mathbf{j}} = p^m/3$. Notice also that $r^{n+1} < 3p^m q$. Therefore,

$$c_{\mathbf{j}} - C_{\mathbf{k}} - \frac{r_{\mathbf{j}} + r_{\mathbf{k}}}{1 - \bar{r}} > p^m \left(\frac{1}{3} - \frac{1}{35} - \frac{3}{35} \right) > \frac{p^m}{4}$$

Therefore, the set $\Delta_{mn}(p, r) = \{q : S_3 S_1^m(K \backslash K_1) \cap S_4 S_6^n(K \backslash K_6)$ has the dimension less than $2s_{\mathbf{r}}$. The same is true for the set $\Delta(p, r)$ which is a countable union of the sets $\Delta_{mn}(p, r)$.

This shows that

if $p, r \in (0, 1/36)$ and $\dfrac{\log p}{\log r}$ is irrational then for Lebesgue almost all $q \in (0, 1/36)$ the system $\mathcal{S}$ has totally disconnected attractor with unique one-point intersection, and at the same time, it does not satisfy weak separation property.

The reader may see that the properties similar to The properties **1. 2. 3.** in the previous subsection are also valid for the systems, described above.

References

1. C. Bandt, S. Graf, Self-similar sets 7. A characterization of self-similar fractals with positive Hausdorff measure. Proc. Am. Math. Soc. **114**(4), 995–1001. MR1100644 (1992)
2. B. Barany, Iterated function systems with non-distinct fixed points. J. Appl. Math. Anal. Appl. **383**(1), 244–258 (2011)
3. G. Edgar, *Measure, Topology, and Fractal Geometry*, 2nd edn. (Springer, New York, 2008), p. 272
4. K.J. Falconer, *Fractal Geometry: Mathematical Foundations and Applications* (Wiley, New York, 1990)
5. J. Hutchinson, Fractals and self-similarity. Indiana Univ. Math. J. **30**(5), 713–747 (1981)
6. K.G. Kamalutdinov, A.V. Tetenov, Twofold cantor sets in $\mathbb{R}$. Siberian Electron. Math. Rep. **15**, 801–814 (2018)
7. K. Kuratowski, *Topology*, vol. 1 (PWN and Acad. Press, 1966). MR0217751
8. K.S. Lau, S.M. Ngai, Multifractal measures and a weak separation condition. Adv. Math. **141**, 45–96. MR1667146 (1999)
9. K. Simon, B. Solomyak, M. Urbański, Hausdorff dimension of limit sets for parabolic IFS with overlaps. Pacific J. Math. **201**(2), 441–478 (2001)
10. A. Tetenov, K. Kamalutdinov, D. Vaulin, Self-similar Jordan arcs which do not satisfy OSC (2015). http://arxiv.org/abs/1512.00290 arXiv:1512.00290
11. M.P.W. Zerner, Weak separation properties for self-similar sets. Proc. Am. Math. Soc. **124**(11), 3529–3539 (1996)

The nth Itarate of a Map with Dense Orbit

P. Amalraj and P. B. Vinod Kumar

Abstract Suppose that X is a Hausdorff space space having no isolated points and $f : X \to X$ is continuous. We show that the orbit of a point $x \in X$ under f is dense in X while the orbit of x under $f^n = f \circ f \circ \circ \cdots \circ f$, n times is not for some $n \geq 2$, then the set $D = \{x, \text{orb}(f, x)\text{is dense in}X\}$ is disconnected. As a consequence of this, we show that the set $D = \{x, \text{orb}(f, x)\text{is dense in}X\}$ is connected, then $orb(f^n, x)$ is dense for all $x \in X$.

Keywords Chaotic functions · Dense orbit · Decomposition

2000 Mathematics Subject Classification. Chaos Theory

1 Introduction

Suppose X is a Hausdorff space having no isolated points and $f : X \to X$ is continuous. In [1], it is proved that the orbit of a point $x \in X$ under f is dense in X while the orbit of x under $f \circ f$ is not, then the space X is decomposes in to three sets realtive to which the dynamics of f are easy to describe. And also he proves that f acts chaotically on X and that the closure of the set of periodic points of X having odd period under f has nonempty interior, then $f \circ f$ is chaotic on X. They conclude their paper with the question "For $n \geq 2$, what kind of decomposition of X may be obtained if one assumes that f is toplogically transitive on X while f^n is not ?"

P. Amalraj (✉)
Department of Mathematics, Sanatana Dharma Collge, Alappuzha, Kerala, India

Rajagiri School of Engineering and Technology, A P J Abdul Kalam Technological University, Thiruvananthapuram, Kerala, India

P. B. Vinod Kumar
Department of Mathematics Centre for Topology and Applications, Rajagiri School of Engineering and Technology, A P J Abdul Kalam Technological University, Thiruvananthapuram, Kerala, India
e-mail: vinod_kumar@rajagiritech.edu.in

R. L. Devaney et al. (eds.), *Topological Dynamics and Topological Data Analysis*, Springer Proceedings in Mathematics & Statistics 350,
https://doi.org/10.1007/978-981-16-0174-3_13

This motivated us to find a solution to this problem. We are solving a part of this problem. We show that if orb(f^n, x) is not dense for some $n \geq 2$, then the set $D = \{x, \text{orb}(x, f) \text{is dense in} X\}$ is disconnected. Also we show that the set $D = \{x, \text{orb}(x, f) \text{ is dense in} X\}$ is connected, then orb(f^n, x) is dense for all $x \in X$. Also a consequence of this we proved the fact that if T is a linear function on a complex Banach space B and that the orbit of $b \in B$ under T is dense in B;then for each positive integer n,the orbit of B under T^n is also dense. (This is S.I Ansari's remarkable theorem. It is just a corollary of our result) ([2],Theorem1)

1.1 The General Separation Theorem

In this section, X denotes a Hausdorff topological space having no isolated points and $f : X \to X$ is continuous.

<u>Notation</u>

$f^n = f \circ f \circ f \circ f \cdots \circ f$, n times with $f^0 = i_d$

$f^{-(n)}(A) = \{x \in X, f^n(x) \in A\}$

$D = \{x \in X, \overline{\text{orb}(f, x)} = X\}$

$D^n = \{x \in X, \overline{\text{orb}(f^n, x)} = X\}$

Theorem 1.1 *Suppose D^n is nonempty and f is onto. Then D^n is dense subset of X, for all n and D^n is invariant under f^n, for all n.*
i.e., $f^n(D^n) \subset D^n$ and $f^{-(n)}(D^n) \subset D^n$.

Proof Assume that D^n is nonempty.
Let $x \in D^n \Longrightarrow \overline{\text{orb}(f, x)} = X$

$$\begin{aligned}\text{orb}(f^n, x) &= \{x, f^n(x), f^{2n}, \ldots, \}\\ \text{orb}(f^n, f(x)) &= \{f(x), f^{n+1}, f^{2n+1} \ldots\}\\ &= f(\text{orb}(f^n, x))\end{aligned}$$

so, $\overline{\text{orb}(f^n, f(x))} = \overline{f(\text{orb}(f^n, x)} \supseteq f(\overline{\text{orb}(f^n, x)}$
i.e., $f(X) \subseteq \overline{\text{orb}(f^n, f(x)}$
Since f is onto, $X \subseteq \overline{\text{orb}(f^n, f(x)}$
i.e., $f(x) \in D^n$
i.e., D^n is invariant under f and so is under f^n.
Next we show that D^n is a dense subset of X.
given x in D^n, orb$(f^n, x) = \{x, f(x), f^2(x), \ldots\} \subseteq D^n$
,however, orb(f^n, x) is dense in X, and thus, D^n is dense in X as well.
Now we show that $f^{-(n)} \subset D^n$
Let $y \in f^{-(n)}(D^n)$
i.e., $f^n(y) \in D^n$
So, $f^2 n(y), f^3 n(y), \cdots \in D^n$ since D^n is invariant underf^n

$\Longrightarrow$ orb(f^n, y) contains orb($f^n, f^n(y)$). But orb($f^n, f^n(y)$) is dense in X, since $f^n(y) \in D^n$
ie,orb(f^n, y) is dense in X.
$\Longrightarrow y \in D^n$
i.e., $f^{-(n)}(D^n) \subset D^n$. Hence, the theorem.

Theorem 1.2 *Suppose that $x \in X$, $h : X \to X$ is continuous and G is the complement of the closure of orb(h, x). Then for every non negative integer k,$h^{-(k)} \subseteq G$.*

Proof See [1].

Theorem 1.3 (Generalized Separation Theorem) *Suppose $x \in X$ such that orb(f, x) is dense in X.*

(1) orb(f^n, x) is not dense in X for some $n \geq 2$

(2) $D = \{x \in X, \overline{orb(f, x)} = X\}$ is disconnected.

we have, (1)$\Longrightarrow$ (2).

Proof Assume (1) holds
$G = \left(\overline{\text{orb}(f^n, x)}\right)^c$, therefore, G is not empty and is open since (1) holds.
therefore for each non-negative integer$k$,$f^{-(nk)}(G) \subseteq G$.
We claim that $f^{-(1)}(G) \cap f^{-(2)}(G) \cap f^{-(3)}(G) \cap f^{-(4)}(G) \cdots \cap f^{-(n-1)}(G)$ must be contained in the closure of G.
Suppose $f^{-(1)}(G) \cap f^{-(2)}(G) \cap f^{-(3)}(G) \cap f^{-(4)}(G) \cdots \cap f^{-(n-1)}(G)$ intersects G. $f^{-(1)}(G) \cap f^{-(2)}(G) \cap f^{-(3)}(G) \cap f^{-(4)}(G) \cdots \cap f^{-(n-1)}(G) \cap G$ is open and orb(f, x) is dense, there is a non-negative integer j such that $f^j(x) \in . f^{-(1)}(G) \cap f^{-(2)}(G) \cap f^{-(3)}(G) \cap f^{-(4)}(G) \cdots \cap f^{-(n-1)}(G) \cap G$.

$$f^j(x) \in G \Longrightarrow j \neq \text{multiple of } n$$
$$f^j(x) \in f^{-(1)}(G) \Longrightarrow j + 1 \neq \text{multiple of } n$$
$$f^j(x) \in f^{-(2)}(G) \Longrightarrow j + 2 \neq \text{multiple of } n$$
$$\ldots\ldots$$
$$f^j(x) \in f^{-(n-1)}(G) \Longrightarrow j + n - 1 \neq \text{multiple of } n$$

is a contradiction.
Therefore,$f^{-(1)}(G) \cap f^{-(2)}(G) \cap f^{-(3)}(G) \cap f^{-(4)}(G) \cdots \cap f^{-(n-1)}(G)$ is contained in the complement of G.
Let $S_1 = G$ and $S_2 = f^{-(1)}(G) \cap f^{-(2)}(G) \cap f^{-(3)}(G) \cap f^{-(4)}(G) \cdots \cap f^{-(n-1)}(G)$
Then S_1 and S_2 are open and disjoint.
Let w be in D. G is open and w has a dense orbit under$f$, and there is a non-negative integer$m$ such that $f^m(w) \in G$.
Thus,$w \in f^{-m}(G)$ and is either in S_1 (by using theorem 1.2 if m is a multiple of n) or is S_2 (if $m \equiv r(mod n)$, for $1 \leq r \leq n - 1$).

, therefore, $D \subset S_1 \cup S_2$.
Because D is dense,$S_1 \cap D$ and $S_2 \cap D$ are non empty.
Thus the pairs $S_1 \cap D = D_1$ and $S_2 \cap D = D_2$ is a separation of D. ie, D is disconnected.

Theorem 1.4 *Let D_1 and D_2 are sets mentioned in the Generalized Separation Theorem, then $f^{n-1}(D_2) \subseteq D_1$*

Proof Suppose $t \in D_2$.
In particular $t \in f^{-(n-1)}(G)$
ie, $f^{n-1}(t) \in G = S_1$
$\Longrightarrow f^{n-1}(t) \in S_1 \cap D = D_1$.
Hence the result.

Theorem 1.5 *Let $f : X \to X$ be chaotic and $D = \{x \in X, \overline{orb(f, x)} = X\}$ is connected.Then orb(f^n, x) is dense in X for all x.*

Proof Clear from Generalized Separation Theorem.

Theorem 1.6 *Suppose B is a complex Banach space and $T : B \to B$ is bounded and linear. If for some $b \in B$, orb(T, b) is dense in B, then orb(T^n, b) is also dense in B,for all n.*

Proof Suppose orb(T, b) is dense in B.
Then the set $E = \{P(T)b : p \text{ is a polynomial}\} \backslash \{0\}$ is a dense set of vectors in B, each element of which has dense orbit [3]. Because E is connected and dense , the set D of vectors in B having dense orbit under T cannot be separated. Thus by the Generalized Separation Theorem, orb(T^n, b) is also dense in B, for all n.

Acknowledgements The authors would like to express their sincere gratitude to APJ Abdul Kalam Technological University, Kerala, India; the management of Sanatana Dharma College, Alappuzha, Kerala, India and the management of Rajagiri School of Engineering and Technology, Kerala, India for their support.

References

1. P.S. Bourdon, The second iterate of a map with dense orbit. Proc. Am. Math. Soc. **124**, 1577–1581 (1996)
2. S.I. Ansari, Hypercyclic and cyclic vectors. J. Funct. Anal. **128**, 374–383 (1995)
3. P.S. Bourdon, Invariant manifolds of hypercyclic vectors. Proc. Am. Math. Soc. **118**, 845–847 (1993)

Periodic Points of N-Dimensional Toral Automorphisms

K. Ali Akbar and T. Mubeena

Abstract In this article, subsets of $\mathbb{T}^n$ that can arise as sets of all periodic points of a continuous n-dimensional toral automorphism are characterized. Here, the torus $\mathbb{T}^n$ is viewed as $[0, 1) \times \cdots \times [0, 1)$ (n-times) as a group under coordinate-wise addition modulo 1.

Keywords Periodic points · Toral automorphism · Triangular matrix

2010 Mathematics Subject Classification. Primary: 54H20 · secondary: 20K30, 15A04

1 Introduction

There have been some papers discussed about the sets of periodic points for continuous self-maps of intervals on $\mathbb{R}$ (see [3–5]). It is natural to ask: Which subsets will arise as the set of all periodic points of these self maps? In the case of n-dimensional toral automorphism, we have a neat answer.

A dynamical system is simply a pair (X, f), where X is a metric space, and $f : X \to X$ is a continuous function. For $x \in X$, the orbit of x under f is the sequence $x, f(x), f^2(x), \cdots$, where $f^n = f \circ f \circ \cdots \circ f$ is the composition of f with itself n times. A point $x \in X$ is said to be periodic with period n if $f^n(x) = x$ for some $n \in \mathbb{N}$, and $f^m(x) \neq x$ for $1 \leq m < n$. We denote the set of all periodic points of f by $P(f)$. We refer ([5, 6]) for preliminaries from topological dynamics.

Let $\mathbb{Q}_1$ be the set of all rational points in $[0, 1)$ and $\mathbb{Q}_1^n$ be $\mathbb{Q}_1 \times \cdots \times \mathbb{Q}_1$ (n-times). Our main results prove that set of all periodic points of a continuous n-dimensional

K. Ali Akbar (✉)
Department of Mathematics, Central University of Kerala, Kasaragod 671320, Kerala, India

T. Mubeena
Department of Mathematics, Government College, Kasaragod 671123, Kerala, India

R. L. Devaney et al. (eds.), *Topological Dynamics and Topological Data Analysis*, Springer Proceedings in Mathematics & Statistics 350,
https://doi.org/10.1007/978-981-16-0174-3_14

toral automorphism has to be either $\mathbb{Q}_1^n$ or $\mathbb{T}^n$ or a finite intersection of (atmost n) sets of the form $S_{r_1,\ldots,r_n}$ for some $r_i \in \mathbb{Q}$; where $S_{r_1,\ldots,r_n} = \{(x_1, \ldots, x_n) \in \mathbb{T}^n : r_1x_1 + \cdots + r_nx_n$ is rational$\}$. In this article, we generalize our results in [7] to a more general setting and provide a more general proof.

2 Basic Results

Let $GL(n, \mathbb{Z})$ be the set of all $n \times n$ matrices A with integer entries and $\text{Det}(A) = \pm 1$, where $\text{Det}(A)$ denotes the determinant of A. Each such matrix A gives an invertible linear map on $\mathbb{R}^n$ by $X \to AX$. We define an automorphism on the torus $T_A : \mathbb{T}^n \to \mathbb{T}^n$ by $T_AX \equiv AX(mod 1)$, coordinate-wise addition modulo 1.

Let $Aut(\mathbb{T}^n)$ denotes the set of all continuous automorphisms on $\mathbb{T}^n$. The following proposition says that every automorphism T_A on the torus is continuous, and every continuous automorphism is induced by a matrix from $GL(n, \mathbb{Z})$.

Proposition 1 (see [4, 7]) *The above map $A \to T_A$ from $GL(n, \mathbb{Z})$ to $Aut(\mathbb{T}^n)$ is a group isomorphism.*

Note that, for a toral automorphism T_A, the periodic points with period n are solutions of the congruent equation $A^nX \equiv X(mod 1)$. Now, we state the following well-known lemma.

Lemma 1 (see [2]) *If $T : \mathbb{R}^n \to \mathbb{R}^n$ is an invertible linear transformation, then for every Riemann measurable set, $S \subset \mathbb{R}^n$, $T(S)$ is Riemann measurable, and the Riemann measure of $T(S)$ is equal to $|Det(T)|$ times the Riemann measure of S.*

The following propositions may be known. But we provide a proof here. See [5] for $n = 2$.

Proposition 2 *Let $A \in GL(n, \mathbb{Z})$.*

(1) The number of solutions of $A^nX \equiv X(mod 1)$ in $\mathbb{T}^n$ is $|Det(A^n - I)|$, provided $|Det(A^n - I)| \neq 0$.
(2) If $|Det(A^n - I)| = 0$, then $A^nX \equiv X(mod 1)$ has infinitely many solutions in $\mathbb{T}^n$.

Proof (1) Suppose $\text{Det}(A^n - I) \neq 0$. The number of solutions of the equation, $A^nX \equiv X(mod 1)$ in $\mathbb{T}^n$, is equal to the number of integer points in the image of $\mathbb{T}^n$ under $A^n - I$, treated as a linear map from $\mathbb{R}^n$ to $\mathbb{R}^n$. Note that the number of integer points in the image is equal to its measure, which is equal to $|\text{Det}(A^n - I)|$ by Lemma 1.
(2) If $\text{Det}(A^n - I) = 0$, then the system $(A^n - I)X = 0$ has infinitely many solutions in $\mathbb{T}^n$.

□

Proposition 3 *For each $A = (a_{ij})_{n\times n} \in GL(n, \mathbb{Z})$, the set $P(T_A)$ is dense in $\mathbb{T}^n$.*

Proof We prove that $P(T_A)$ contains $\mathbb{Q}_1^n$, and so it is dense. A general element in $\mathbb{Q}_1^n$ is of the form $X = (\frac{p_1}{q}, \ldots, \frac{p_n}{q})$, $p_1, p_2, \ldots, p_n, q \in \mathbb{Z}$ with $0 \le p_i < q$. Now, $T_A(X) =$ (fractional part of the sum $a_{11}(\frac{p_1}{q}) + \cdots + a_{1n}(\frac{p_n}{q}), \ldots,$ fractional part of the sum $a_{n1}(\frac{p_1}{q}) + \cdots + a_{nn}(\frac{p_n}{q})$) = an element of the form $(\frac{m_1}{q}, \ldots, \frac{m_n}{q})$, $0 \le m_i < q$. Observe that, for a fixed $q \in \mathbb{N}$, the set $\{(\frac{m_1}{q}, \ldots, \frac{m_n}{q}) : 0 \le m_1, \ldots, m_n < q, m_i \in \mathbb{N}\}$ is T_A-invariant and finite. Then, the orbit of X is finite and therefore eventually periodic. Hence, the result follows from the fact that for invertible maps, the eventually periodic points are periodic. □

Remark 1 A continuous toral automorphism, T_A, $A \in GL(n, \mathbb{Z})$, is said to be hyperbolic if A has no eigenvalue with absolute value 1. In this case, $\mathrm{Det}(A^n - I) \neq 0$ for all $n \in \mathbb{N}$. Hence, $P(T_A) = \mathbb{Q}_1^n$ (see [5]).

Observe that, for any continuous toral automorphisms T_A, the set $P(T_A)$ is a subgroup of the torus. We now ask: Which subgroups of $\mathbb{T}^n$ arise in this way?

3 Main Results

For $n \in \mathbb{N}$, define a sub-class $\mathcal{A}_{1,n}$ of $GL(n, \mathbb{Z})$ such that each member of $\mathcal{A}_{1,n}$ is of the form $\begin{pmatrix} 1 & \bar{k} \\ \bar{0} & I_{n-1} \end{pmatrix}$ for some vector $\bar{k} = (k_1, \ldots, k_{n-1})$ with integer coordinates, and I_{n-1} denotes the identity matrix of size $n-1$, $\bar{0}$ is the zero vector in $\mathbb{R}^{n-1}$. Also, we define $S_{r_1,\ldots,r_n} := \{(x_1, \ldots, x_n) \in \mathbb{T}^n : r_1x_1 + \cdots + r_nx_n \text{ is rational}\}$ for $r_i \in \mathbb{Q}$. If $A \in \mathcal{A}_{1,n}$ then A and its powers $A^2, A^3, \ldots$ share the same set of periodic points. Note that, for any $j \in \mathbb{N}$, the periodic points of T_A with period j are contained in $P(T_{A^j})$. Hence, $P(T_A)$ is a finite intersection of sets of the form $S_{r_1,\ldots,r_n}$ for some $r_i \in \mathbb{Q}$.

Now, we consider our main theorem.

Theorem 1 (*Main theorem*) *For any continuous toral automorphism $T_A : \mathbb{T}^n \to \mathbb{T}^n$, the set $P(T_A)$ of periodic points of T_A is one of the following:*

(1) $\mathbb{Q}_1^n$.
(2) A finite intersection of atmost n sets of the form $S_{r_1,\ldots,r_n}$ for some $r_i \in \mathbb{Q}$.
(3) $\mathbb{T}^n$.

Proof Let $A = (a_{ij}) \in GL(n, \mathbb{Z})$. Then, $(A - I)X \equiv 0 (mod 1)$ if and only if $a_{i1}x_1 + \cdots + (a_{ii} - 1)x_i + \cdots + a_{in}x_n \in \mathbb{Z}$ for all $1 \le i \le n$, where $X = [x_1, \ldots, x_n]^T$. This fact will be used often in the proof.

Case 1: $\mathrm{Det}(A^m - I) \neq 0$ for all $m \in \mathbb{N}$.

By Cramer's rule, $P(T_A) = \mathbb{Q}_1^n$.

Case 2: $\mathrm{Det}(A^m - I) = 0$ for some $m \in \mathbb{N}$.

Let $S = \{s \in \mathbb{N} : \mathrm{Det}(A^s - I) = 0\}$ and consider a $k \in S$.

If $A^k \in \mathcal{A}_{1,n}$, then A^k and its powers $A^{2k}, A^{3k}, \ldots$ share the same set of periodic points. Note that, for any $j \in \mathbb{N}$, the periodic points of T_A with period j are contained in $P(T_{A^j})$. Hence, $P(T_A)$ is a finite intersection of sets of the form $S_{r_1,\ldots,r_n}$ for some $r_i \in \mathbb{Q}$. In particular, if $A = I$ then $P(T_A) = \mathbb{T}^n$.

Now, we have to prove that no other subset of $\mathbb{T}^n$ can come as the set of periodic points. In general, A^k need not be in $\mathcal{A}_{1,n}$ for $k \in S$. This general situation can be handled as follows.

First, we prove that if $P(T_A) = \bigcap_{m\in\mathbb{N}} S_{r_1^{(m)},\ldots,r_n^{(m)}}$, then it is a finite intersection of sets of the form $S_{r_1,\ldots,r_n}$. For this, consider $\bigcap_{m\in\mathbb{N}} S_{r_1^{(m)},\ldots,r_n^{(m)}}$ for $r_i^{(m)} \in \mathbb{Q}$. Without loss of generality assume that $(r_1^{(m)}, \ldots, r_l^{(m)})$ is a rational multiple of $(r_1^{(1)}, \ldots, r_l^{(1)})$ but $(r_{l+1}^{(m)}, \ldots, r_n^{(m)})$ is not a rational multiple of $(r_{l+1}^{(1)}, \ldots, r_n^{(1)})$ and l is maximum with respect to this property. Otherwise, there is a permutation σ on $\{1, 2, \ldots, n\}$ such that $(r_{\sigma(1)}^{(m)}, \ldots, r_{\sigma(l)}^{(m)})$ is a rational multiple of $(r_{\sigma(1)}^{(1)}, \ldots, r_{\sigma(l)}^{(1)})$ but $(r_{\sigma(l+1)}^{(m)}, \ldots, r_{\sigma(n)}^{(m)})$ is not a rational multiple of $(r_{\sigma(l+1)}^{(1)}, \ldots, r_{\sigma(n)}^{(1)})$, and l is maximum with respect to this property. It is possible to find such a permutation to arrange the n-tuples $(r_1^{(m)}, \ldots, r_n^{(m)})$ simultaneously as we required. Therefore, if $X = [x_1, \ldots, x_n]^T \in \bigcap_{m\in\mathbb{N}} S_{r_1^{(m)},\ldots,r_n^{(m)}}$ for $r_i^{(m)} \in \mathbb{Q}$, then $X \in S_{r_1^{(1)},\ldots,r_l^{(1)}} \times \mathbb{Q}_1^{n-l}$. From this, it follows that if $P(T_A) = \bigcap_{m\in\mathbb{N}} S_{r_1^{(m)},\ldots,r_n^{(m)}}$, then it is a finite intersection of sets of the form $S_{r_1,\ldots,r_n}$ because $S_{0,\ldots,0,r_i,0,\ldots,0} = [0, 1) \times \ldots \times [0, 1) \times \mathbb{Q}_1 \times [0, 1) \times \ldots \times [0, 1)$ ($\mathbb{Q}_1$ is in the ith position).

Next suppose that $P(T_A)$ is a set which is not of the form $\mathbb{T}^n$ or $\mathbb{Q}_1 \times \ldots \times \mathbb{Q}_1$ or finite intersection of sets of the form $S_{r_1,\ldots,r_n}$. Then, there exists $X = [x_1, \ldots, x_n]^T \in P(T_A)$ such that $r_1x_1 + \cdots + r_nx_n \notin \mathbb{Q}$ for all $r_i \in \mathbb{Q} \setminus \{0\}$. This is because, if $X = [x_1, \ldots, x_n]^T \in S_{r_1,\ldots,r_n} \cap P(T_A)$, then $S_{r_1,\ldots,r_n} \subset P(T_A)$., and hence, otherwise, $P(T_A)$ becomes countable intersection of sets of the form $S_{r_1,\ldots,r_n}$. It is not possible. Also $A^m X \equiv X (mod 1)$ for some m, by our assumption. Hence, there exists $s_i \in \mathbb{Q}_1 \setminus \{0\}$ such that $s_1x_1 + \cdots + s_nx_n \in \mathbb{Q}$. Which is a contradiction. Hence, the proof follows.

Remark 2 If $r_i = 0$ for some $1 \le i \le n$, then $S_{r_1,r_2,\ldots,r_n} = \{(x_1, x_2, \ldots, x_n) \in \mathbb{T}^n : r_1x_1 + \cdots + r_{i-1}x_{i-1} + r_{i+1}x_{i+1} + \cdots + r_nx_n \in \mathbb{Q}\}$. Hence, $\{x_i : (x_1, x_2, \ldots, x_i, \ldots, x_n) \in S_{r_1,\ldots,r_n}\} = [0, 1)$.

The following result is an immediate corollary of Theorem 1. In [7], a different proof is given.

Corollary 1 *If $A \in GL(2, \mathbb{Z})$, then for any continuous toral automorphism T_A, the set $P(T_A)$ of periodic points of T_A is one of the following:*

1. $\mathbb{Q}_1^2$.
2. $\mathbb{Q}_1 \times [0, 1)$ *or* S_r *for some* $r \in \mathbb{Q}$; *where* $S_r = \{(x, y) \in \mathbb{T}^2 \colon rx + y$ *is rational*$\}$.
3. $\mathbb{T}^2$.

Remark 3 For $A, B \in GL(n, \mathbb{Z})$, we say that $A \sim B$ if there exists $P \in GL(n, \mathbb{Z})$ such that $A = P^{-1}BP$. If $A \sim B$, then $P(T_A) = P(T_B)$. Hence, if we know a nice representative from each equivalence class of $GL(n, \mathbb{Z})$ with respect to the equivalence relation $\sim$, then the proof will be so easy. It seems to be too difficult to find the nice representatives for $n > 2$. But for $GL(2, \mathbb{Z})$, we have nice representatives (See [1]). Define $A_{m,n} = \begin{pmatrix} m & n \\ \frac{-(m-1)^2}{n} & 2-m \end{pmatrix}$ for $n \neq 0$ and n divides $m-1$, and $B_{m,n} = \begin{pmatrix} m & n \\ \frac{-(m+1)^2}{n} & -2-m \end{pmatrix}$ for $n \neq 0$ and n divides $m+1$. Then, the set $\{A_{1,j} : j \in \mathbb{Z} \setminus \{0\}\}$ contains exactly one representative from each conjugacy class of $A_{m,n}$ for $(m, n) \in \mathbb{Z} \times (\mathbb{Z} \setminus \{0\})$. Also the set $\{B_{-1,j} : j \in \mathbb{Z} \setminus \{0\}\}$ contains exactly one representative from each conjugacy class of $B_{m,n}$ for $(m, n) \in \mathbb{Z} \times \mathbb{Z} \setminus \{0\}$. From this representation, we can give an independent proof for Corollary 1. When $n = 1$, $GL(n, \mathbb{Z})$ is isomorphic to $\mathbb{Z}_2$ and which is equal to $\mathrm{Aut}(S^1)$, the automorphism group of S^1. So the only subset of S^1 that can arise as set of all periodic points of an automorphism of S^1 is S^1 itself.

4 Summary

For each self-map f on a set X, we associate a subset of X as follows: $P(f) = \{x \in X : f^n(x) = x \text{ for some } n \in \mathbb{N}\}$. If f belongs to a certain nice class of functions, then, not all subsets of X may arise as the set of all periodic points of f. It is natural to ask: Which subsets of X arise as $P(f)$, for some f in that class? We answer this question, for all continuous n-dimensional toral automorphisms. For $n \geq 2$, even though there are apparently $nC_1 + nC_2 + \cdots + nC_n$ kinds of subsets which can appear as the set of periodic points for some continuous toral automorphism, there are only $n+1$ up to homeomorphism.

References

1. K. Ali Akbar, Some results in linear, symbolic and general topological dynamics. Ph.D Thesis, University of Hyderabad (2010)
2. D.C. Aliprantis, (Principles of Real Analysis, Academic Press, Owen Burkinshaw, 1998)
3. I.N. Baker, Fixpoints of polynomials and rational functions, J. London Math. Soc. 39(1964), 615–622
4. Bodil Branner, Poul Hjorth, *Real and Complex Dynamical Systems, NATO Advanced Science Institutes Series C: Mathematical and Physical Sciences*, vol. 464 (Kluwer Academic Publishers, Dordrecht, 1995)
5. M. Brin, G. Stuck, *Introduction to Dynamical Systems* (Cambridge University Press, 2002)
6. R.L. Devaney, *An Introduction to Chaotic Dynamical Systems*, 2nd edn. (Addison-wesley Publishing Company Advanced Book Program, Redwood City, CA, 1989)
7. I. Subramania Pillai, K. Ali Akbar, V. Kannan, B. Sankararao, Sets of all periodic points of a Toral automorphism. J. Math. Anal. Appl. **366**, 367–371 (2010)

Julia Sets in Topological Spaces

Sanil Jose and P. B. Vinod Kumar

Abstract In this paper, a study of Julia sets as generalization of classical Julia sets on the complex plane is attempted. Interpreting Julia sets in various forms, we generalize them to topological spaces.

Keywords Julia sets · T_2 space

1 Introduction

The theory of iterated functions on the complex plane is well studied from the times of Fatou and Julia onwards. The interest in this area got another flavour by the introduction of Fractals in 1980s by Benoit Mandelbrot. See [1, 2].
The filled in Julia set was defined in the extended complex plane $\mathbf{C} \cup \{\infty\}$ for the function $f(z)$ as $K(f) = \{z \in \mathbf{C}./f^k(z) \to \infty\}$, and the corresponding Julia set is defined as $J(f) = \partial K(f)$, i.e. Julia set is the boundary of the set $K(f)$.

Example 1 Consider the function $f(z) = z^2$ in the complex plane.
For all points z inside the unit circle $|z| = 1$, we can easily see that $f^n(z)$ tends to 0 as n tends to ∞. Also all points with $|z| \geq 1$, $f^n(z)$ tends to ∞ as n tends to ∞. For all points on the boundary of the circle $|z| = 1$, we can see that $f^n(z)$ remains bounded as n tends to ∞. Hence, the Julia set of the function $f(z) = z^2$ is clearly the the boundary f the circle $i.e.$ $J(f) = \{z \in \mathbf{C} \backslash |z| = 1\}$.

S. Jose (✉)
Department of Mathematics, Sacred Heart College, Kochi, Kerala, India

Rajagiri School of Engineering and Technology, A P J Abdul Kalam Technological University, Thiruvananthapuram, Kerala, India
e-mail: saniljose@shcollege.ac.in

P. B. Vinod Kumar
Department of Mathematics, Centre for Topology and Applications, Rajagiri School of Engineering and Technology, Ernakulam, Kerala, India
e-mail: vinod_kumar@rajagiritech.edu.in

R. L. Devaney et al. (eds.), *Topological Dynamics and Topological Data Analysis*, Springer Proceedings in Mathematics & Statistics 350,
https://doi.org/10.1007/978-981-16-0174-3_15

Note that $z = 0$ and $z = \infty$ are the only fixed points of $f(z) = z^2$.
Julia set of a Complex rational function is non empty, perfect, compact and closed [2].

Not much study of Julia sets was done in general topological space. We consider the extended complex plane as a one point compactication as $\mathbf{C} \cup \{\infty\}$. Naturally, the question arries as can we define Julia sets in general topological space and weather the corresponding function which is chaotic in the Julia set.

2 Basin of Attractors to a Point

Classical Julia set.
In the classical Julia sets, the basin of attractors of a fixed pointz_0 is defined as the $\{z \in \mathbf{C} \backslash f^n(z)$ converges to $z_0\}$ In a general topological space X, we will take any point $x \in X$ and a function $f(x)$ which we define the basin of attractor of f to x as $B_f(x) = \{y \in X \backslash f^n(y) \longrightarrow x\}$.

Results
1. $B_f(x) \neq \phi$ only for fixed points.

Proof If possible, there exist a point x which is not a fixed point such that $B_f(x) \neq \phi$
i.e. $y \in B_f(x) \Rightarrow f^n(y) \longrightarrow x$ as $n \Rightarrow \infty$
i.e. $f(f^n(y)) \Rightarrow f(x)$ as $n \Rightarrow \infty$, i.e. $f^{n+1}(y) \Rightarrow f(x)$, Since $\lim_{x\to\infty} f^n(y) = \lim_{x\to\infty} f^{n+1}(y)$, we get $f(x) = x$,
i.e. x is a fixed point.

2. $B_f(x) \cap B_f(y) \neq \phi \Rightarrow x = y$, where x and y are fixed points.

Proof Given that $B_f(x) \cap B_f(y) \neq \phi$, i.e. $\exists a \in B_f(x) \cap B_f(y)$
$\Rightarrow a \in B_f(x)$ and $a \in B_f(y) \Rightarrow f^n(a) \longrightarrow x$ and $f^n(a) \longrightarrow y$
Uniquence of limit gives $x = y$.

3 Fatou and Julia Sets in General Topological Space

Let (X, τ) be any topological space, and let $x \in X$ be any point. We define $K_f(x) = (B_f(x))^c$
i.e. $K_f(x) = \{y \in X / f^n(y) \not\longrightarrow x\}$.

Example 2 Consider the topological space $[0, 1]$ and the function $f(x) = x^2$. We know that the fixed points of the function are 0 and 1.

Now, $B_f(0) = [0, 1)$, and hence, $K_f(0) = \{1\}$.
Also $B_f(1) = \{1\}$, and hence, $K_f(1) = [0, 1)$. For all other points, the set $B_f(x) = \phi$, and hence, $K_f(x) = X = [0, 1]$

Note 1 : The example clearly shows that we must concentrate only on fixed points of the function, and also if the space is compact, then the set $K_f(x)$ is of not much exciting for us.

Note 2 : The point ∞ is the one point compactification of the complex plane **C**. We will think about spaces which T_2.

T_2 Space or Hausdroff Space
A topological space Let (X, τ) is said to be T_2 space or *Hausdroff* if for every pair of distinct points x and y, and in X, there exists disjoint open sets U and V such that $x \in U$ and $y \in V$.

Theorem 1 *Let X be a T_2 space, and $x \in X$ is any point. Let $K_f(x) = \{y \in X / f^n(y) \not\longrightarrow x\}$. If x is not a fixed point, then $K_f(x) = X$.*

Proof We need to show that $K_f(x) = X$, if x is not a fixed point.
i.e. we need to show that $\{y \in X / f^n(y) \not\longrightarrow x\} = X$
i.e. we need to show that $\{y \in X / f^n(y) \longrightarrow x\} = \phi$
If possible assume that there exist $y \in X$ such that $f^n(y) \longrightarrow x$, i.e. the sequence $(y, f(y), f^2(y), \dots f^n(y) \dots)$ converges to x.
i.e. the sequence $(f(y), f^2(y), f^3(y) \dots f^{n+1}(y) \dots)$ converges to $f(x)$, and the two sequences differ only in the first term.
i.e. $f(x) = x$, i.e. x is a fixed point of f, which is a contradiction.
Hence the result.

Remark 1 The condition (X, τ) is T_2 is important
For
Let $X = \{a, b, c\}$ and $\tau = \{X, \phi, \{a\}, \{a, b\}\}$. Clearly (X, τ) is not Hausdroff since for b and c, we cannot find two distinct open sets.
Define $f : X \longrightarrow X$ as $f(a) = a,\ f(b) = c,\ f(c) = b$
Clearly, b and c are not fixed points. Also $B_f(a) = \{a\}$ and so $K_f(a) = \{b, c\} \neq X$
$B_f(b) = \{a\}$ and so $K_f(b) = \{b, c\} \neq X$
$B_f(c) = \{a, b, c\}$ and so $K_f(a) = \phi \neq X$.

4 Locally Compact and T_2 Space

Theorem 2 *Let X be locally compact and T_2. Let $\hat{X} = X \cup \{\infty\}$ be the one point compactification of X. Then, $K_f(\infty) = \{x \in X / O_f(x) \subset K$, where K is any compact set contained in X$\}$*

Proof Let $x \in \{x \in X/O_f(x) \subset K\}$
$\Rightarrow O_f(x) \subset K$ where K is a compact subset of X
$\Rightarrow (x, f(x), f^2(x), f^3(x), \ldots f^n(x)\ldots) \subset K \subset X$
$\Rightarrow (x, f(x), f^2(x) \ldots f^n(x) \ldots)$ does not converge to ∞
$\Rightarrow x \in K_f(\infty)$ Hence $\{x \in X/O_f(x) \subset K\} \subset K_f(\infty)$
Conversly let $x \in K_f(\infty)$
$\Rightarrow$ the sequence $(x, f(x), \ldots f^n(x) \ldots)$ does not converge to ∞
$\Rightarrow$ Either $f^n(x)$ converges to $y \in X$ or $(x, f(x), f^2(x), \ldots, f^n(x) \ldots)$ is bounded in some compact set K subset of X.
If $f^n(x)$ converges to $y \in X$ then $\{x, f(x), f^2(x) \ldots f^n(x) \ldots y\}$ is compact and is contained in X.
i.e. in both cases $O_f(x) \subset K$
Hence, $K_f(\infty) \subset \{x \in X/O_f(x) \subset K\}$
Thus, $K_f(\infty) = \{x \in X/O_f(x) \subset K$, where K is a compact set contained in $X\}$.

Example 3 Let $X = (0, 1]$, then $\hat{X} = X \cup \{\infty\}$is a one-point compactification
Let $f(x) = x^2$ and $K_f(0) = \{x \in X/f^n(x) \not\longrightarrow 0\}$
clearly for all $x \in (0, 1)$, $f^n(x) \longrightarrow 0$
$B_f(0) = (0, 1)$ and $K_f(0) = \{1\}$, which is closed and bounded and hence is a compact subset of X

5 Julia Sets

Theorem 3 *Let X is not compact but locally compact and T_2. Define $f : \hat{X} \longrightarrow \hat{X}$ such that $f(\infty) = \infty$ Define $J_f(\infty) = \{x \in \hat{X}/f^n(x) \not\longrightarrow \infty\}$. Then*
1. $J_f(\infty)$ is perfect.
2. $J_f(\infty)$ is closed.
3. $J_f(\infty)$ is not always compact
4. $J_f(\infty)$ is non-empty.

Proof We have $J_f(\infty) = \{x \in \hat{X}/f^n(x) \not\longrightarrow \infty\}$
1. First, we will prove that $J_f(\infty)$ is perfect. For that we need to prove that $\overline{J_f(\infty)} \subset J_f(\infty)$.
Assume that x is a limit point of a sequence $\{x_1, x_2, \ldots x_n \ldots\}$ of elements in $J_f(\infty)$.
Since each $x_i \in J_f(\infty)$, by definition of $J_f(\infty)$, $f^n(x_i) \not\longrightarrow \infty \forall i$ as $n \longrightarrow \infty$.
Now, $x_i \longrightarrow x \Rightarrow f(x_i) \longrightarrow f(x) \Rightarrow f^2(x_i) \longrightarrow f^2(x) \Rightarrow \ldots \Rightarrow f^n(x_i) \longrightarrow f^n(x)$
Since each $f^n(x_i) \not\longrightarrow \infty$, $\exists$ an open ball $B(\infty)$ containing ∞ which does not contain $f^n(x_i) \forall n$
Hence, $f^n(x) \not\longrightarrow \infty$.
$x \in J_f(\infty)$ i.e. $\overline{J_f(\infty)} \subset J_f(\infty)$
i.e. $J_f(\infty)$ is perfect.

2. Since $J_f(\infty)$ is perfect $J_f(\infty)$ is closed.

3. Consider $\mathbb{N}$ under discrete topology. Consider $f(n) = \begin{cases} n/2 & \text{if } n \text{ is even} \\ (n+1)/2 & \text{if } n \text{ is odd} \end{cases}$
Clearly, $J_f(\infty) = \mathbb{N}$ which is not compact
$\therefore$ $J_f(\infty)$ is not compact always.

4. Here, we need to prove that $J_f(\infty)$ is non-empty. We use contradiction method to do it.
If possible assume that $J_f(\infty) = \phi, \forall x \in X, f^n(x) \longrightarrow \infty$
, i.e $O_f(x) \not\subset K, \forall$ compact set K
Given any compact set $K, \exists n_k \in \mathbb{Z}_+$ such that $f^{n_k}(x) \notin K$.
Let U be any neighbourhood of x, and for every compact set $F \supset U$, there exist K such that $F \supset K \supset U$ (Since X is locally compact)
But $f^{n_k}(x) \notin K$ Hence, K is not compact, which is a contradiction.
Hence $J_f(\infty)$ is non empty

Result
Let $K(f) = \{x \in \hat{X}/O_f(x) \subset K,$ where K is compact set $\}$. If X is compact $i.e.\hat{X} = X$, does $\exists x \in X$ such that $K(f) = (B_f(x))^c$

Proof $z \in (K(f))^c \Rightarrow O_f(z) \not\subset K$ for every compact subset of X.
$\Rightarrow$ for every $K \subset X, \exists m$ such that $f^m(z) \in K^c$
Let $x \in K^c$, which is open. Also let $B(x)$ be any open ball containing x. $(B(x))^c$is closed and since X is compact, and every closed subset of X is also compact; $(B(x))^c$ is compact.
But $z \in (K(f))^c \Rightarrow \exists$ some $f^m(z) \notin (B(x))^c$
$\Rightarrow f^m(z) \in B(x)$
$f^n(z) \longrightarrow x \Rightarrow z \in B_f(x)$
$\therefore (K(f))^c \subset B_f(x) \Rightarrow (B_f(x))^c \subset K(f)$ Conversly let $z \in K(f) \Rightarrow O_f(z) \subset K$
$\Rightarrow \{f^n(z)\} \subset K$, has a limit point say $y \in K$
$f^n(z) \longrightarrow y$
$z \in B_f(y)$ for some y
$\Rightarrow z \notin B_f(x)$ for $x \notin K$
$\Rightarrow z \in (B_f(x))^c$
$K(f) \subset (B_f(x))^c$

6 Conclusion

In this paper, we tried to generalize the classical Julia sets which was defined in the extended complex plain to a general topological space. But we restricted the defintion to locally compact and Hausdroff space so that the Julia set has some properties of the classical Julia sets.

References

1. M.P. Blanchard,*Complex analytic dynamics on the Reiman sphere*. Bull. Am. Math. Soc. (N.S), 11 (1984)
2. L. Carleson, T. Gamelin, *Complex Dynamics* (Springer, 1993)
3. C. McMillen, Complex dynamics and renormalization. Ann. Math. Stud. **135** (1996)

Julia Set of Some Graphs Using Independence Polynomials

K. U. Sreeja, P. B. Vinod Kumar, and P. B. Ramkumar

Abstract Graph polynomial is a graph invariant whose values are polynomials and found many applications in different fields of science. The goal of this paper is to connect the theory of fractal geometry to the theory of the much broader class graph theory using independence polynomial as basis of our fractals. We are particularly interested in Julia sets and Mandelbrot sets. The various relations between independence polynomial, energy, Julia set and Hausdorff dimension of different classes of graphs are closely examined. The paper concludes with a discussion on Petersen graph and its connectivity.

Keywords Graph · Independence polynomial · Fractal · Julia set · Hausdorff dimension · Mandelbrot · Petersen graph

1 Introduction

The independence polynomial is introduced by Gutman and Harary in 1983 [1]. Let s_k denote the number of independent sets of size k, which are induced subgraphs of G, then $I(G, x) = \sum_{k=0}^{\alpha(G)} s_k x^k$ where $\alpha(G)$ is the independence number of G. The independence polynomials are almost everywhere, but it is an NP complete problem to determine the independence polynomial of a graph [1].

K. U. Sreeja (✉)
Department of Mathematics, K.K.T.M. Government College, Thrissur, Kerala, India

Research Scholar, Rajagiri School of Engineering and Technology, APJ Abdul Kalam Technological University, Thiruvananthapuram, Kerala, India

P. B. Vinod Kumar · P. B. Ramkumar
Department of Mathematics, Centre for Topology and Applications, Rajagiri School of Engineering and Technology, APJ Abdul Kalam Technological University, Thiruvananthapuram, Kerala, India
e-mail: vinod_kumar@rajagiritech.edu.in

P. B. Ramkumar
e-mail: ramkumar_pb@rajagiritech.edu.in

R. L. Devaney et al. (eds.), *Topological Dynamics and Topological Data Analysis*, Springer Proceedings in Mathematics & Statistics 350,
https://doi.org/10.1007/978-981-16-0174-3_16

2 Preliminaries

Definition 2.1 A graph G consists of a set $V(G)$ of vertices along with an edge set $E(G)$, where each edge consists of a pair of vertices. A pair of vertices (x, y) is in $E(G)$, and then, x is adjacent to y.

Definition 2.2 An independent set in a graph G is a vertex subset $S \subseteq V(G)$ that contains no edge of G. The independence number of a graph is the maximum size of an independent set of vertices.

Lemma 2.3 [2]: *The independence polynomial of an empty graph G of order n is given by* $\boldsymbol{I}(G; x) = (1 + x)^n$.

Theorem 2.4 [2]: *Let G be a simple graph. Let* $v \in V(G)$ *and* $N[v]$ *be the closed neighborhood of* v. *Then,* $\boldsymbol{I}(G; x) = \boldsymbol{I}(G - v; x) + x\boldsymbol{I}(G - N[v]; x)$.

Definition 2.5 [3]: The ***reduced independence polynomial*** of G is the function $R(G, z) = \mathbf{I}(G, z) - 1$, since every independence polynomial has constant term 1.

Definition 2.6 [3]: ***Julia set*** is defined on extended complex plane. The filled-in Julia set of the polynomial f is defined as $K(f) = \{z \in C : f^n(z) \nrightarrow \infty\}$. The Julia set is defined as the boundary of the filled-in Julia set, i.e., $J(f) = \partial K(f)$. The **Fatou set** $F(f)$ is the complement of $J(f)$ in C. The Julia set of a polynomial typically has a complicated, self-similar structure. The dimension of a Julia set is ***Hausdorff dimension*** that gives a reasonable way of assigning appropriate non-integer dimension to such sets.

3 Computing the Independence Polynomial

By applying the above theorem, we have the following results (Table 1):

Definition 3.1 [4]: The energy $E(G)$ of G is defined as the sum of the absolute values of the eigen values of an adjacency matrix of a graph. $E(G) = \sum_{i=1}^{n} |\lambda_i|$.

Energy of standard graphs is listed in Table 2.

4 Complete Graph

A complete graph is a simple undirected graph in which every pair of distinct vertices is connected by a unique edge.

Table 1 Recurrence relations and independence polynomial of standard graphs [2]

No.	Graph type	Recurrence relations	Independence polynomial $\mathbf{I}(G, z)$
1	Complete graph K_n	$\mathbf{I}(K_n; z) = \mathbf{I}(K_{n-1}; z) + z$	$(1 + nz)$
2	Star graph S_n	$\mathbf{I}(S_n; z) = (1 + z)\mathbf{I}(S_{n-1}; z) - z^2$	$(1 + z)^n + z$
3	Path graph P_n	$\mathbf{I}(P_n; z) = \mathbf{I}(P_{n-1}; z) + z\mathbf{I}(P_{n-2}; z)$	$\frac{1}{2^{n+1}}[(1 + 2z + s)(1 + s)^n + (s - 1 - 2z)(1 - s)^n]$ where $s = \sqrt{1 + 4z}$
4	Cycle graph C_n	$\mathbf{I}(P_{n-1}; z) + z\mathbf{I}(P_{n-3}; z)$	$\frac{1}{2^{n+1}}[(1 + 2z + s)(1 + s)^{n-2} + (1 + 2z - s)(1 - s)^{n-2}]$ where $s = \sqrt{1 + 4z}$.

Table 2 Energy of standard graphs [4]

No.	Graph type	Energy
1	Complete graph K_n	$2(n - 1)$
2	Star graph S_n	$2\sqrt{n - 1}$
3	Path graph P_n	$2\sum_{j=1}^{n} \lvert\cos(\frac{\pi j}{n+1})\rvert$
4	Cycle graph C_n	$2\sum_{j=0}^{n-1} \lvert\cos(\frac{2\pi j}{n})\rvert$

4.1 Relation of Hausdorff Dimension and Energy of $J(I(G, z))$ of Complete Graph

Independence polynomial of a complete graph K_n is $(1 + nz)$, and energy of a complete graph is $2(n - 1)$. So when $z = 2$, $E(K_n) = \mathbf{I}\ (K_n, 2)$. For complete graph, we have $R(G, z) = nz$. Since any nonzero point has an unbounded forward orbit, its Julia set is $\{0\}$. Therefore, $J(R(G, z)) = \{0\}$ if $G = K_n$. Also, $dim_H(J(R(K_n)) = 0$ gives $E(K_n) \geq dim_H(J(R(K_n)))$.

4.2 Results on Complete Graph

- Zeros of $I(G, z)$ lie outside of $J(I(G, z))$.
- Zeros of $I(G, z)$ are stable for all values of z since all the roots are negative and lie in the negative half plane.
- Periodic points of $I(G, z)$ are not chaotic on C because periodic points are not dense.
- The kth power of a graph G is another graph that has the same set of vertices, but in which two vertices are adjacent when their distance in G is atmost k. But when the powers of complete graph are complete, G^k satisfies all the above results.

5 Mandelbrot Graph

A graph G is called a Mandelbrot graph if $J(R(G; z))$ is connected [5].

Mandelbrot graph is useful for the connectivity of a Julia set of independence polynomial. We denote $M = \{G/G$ is a Mandelbrot graph$\}$.

Theorem 5.1 [5]: *If G is a non-empty graph with independence number 2 having n vertices and m non-edges, then (i)* $-\frac{n}{m} \leq Re(z) \leq 0$ *and (ii)* $Im(z)=0$ *unless n=3, in which case* $-\frac{\sqrt{3}}{2m} \leq Im(z) \leq \frac{\sqrt{3}}{2m}$.

Theorem 5.2 [5]: *If G is a graph with independence number 2 having n=4 vertices and m non-edges, then* $J(R(G,z)) \subseteq [\frac{-4}{m}, 0]$

Corollary 5.3 [5]: *If G is a non-empty graph with independence number 2 having* $n \geq 5$ *vertices and m non-edges, then it lies outside the Mandelbrot set.*

5.1 Classification of Mandelbrot Graphs

- Clearly $K_n \in M$.
- If G is a non-empty graph with independence number m having n vertices denoted by $G_{m,n}$, then we have the following results. (i) $G_{2,2} \in M$ (ii) $G_{2,3} \in M$ (iii) $G_{2,4} \in M$ (iv) $G_{2,n} \notin M$, where $n \geq 5$.

5.2 Julia Set of Reduced Independence Polynomial of Some Graphs

- For complete graph K_n, $J(R(K_n, z)) = J(nz) = \{0\}$.
- For path graph on three vertices P_3, $J(R(P_3, z))=J(z^2 + 3z) \subseteq [-3, 0] \times [\frac{-\sqrt{3}}{2}, \frac{\sqrt{3}}{2}]$.
- For path graph on four vertices P_4, $J(R(P_4, z)) = J(3z^2 + 4z) \subseteq [\frac{-4}{3}, 0]$ since P_4 has three non-edges.
- For cycle graph on four vertices C_4, $J(R(C_4, z)) = J(2z^2 + 4z) \subseteq [-2, 0]$ since two non-edges.

6 Independence Polynomial of Second Degree of Graphs

We will study some graphs whose independence polynomial of second degree is as follows:

Table 3 Relation of Hausdorff dimension and energy of Barbell graph

Values of n	$J(\mathbf{I}(Bar_n, z))$	Hausdorff dimension of $J(I(Bar_n, z)$	Energy of $I(Bar_n, z)$
1	z^2	1	1
2	z^2+1	0.6791	4
3	z^2+2	0.3514	8.2926

Table 4 Relation of Hausdorff dimension and energy of Cocktail party graph

Values of n	$J(\mathbf{I}(CP_n, z))$	Hausdorff dimension of $J(I(CP_n, z)$	Energy of $I(CP_n, z)$
1	z^2+1	0.6791	2
2	z^2	1	4
3	z^2-3	0.4187	6

1. Barbell Graph
   Barbell graph of order n is a graph on $2n$ vertices which is formed by joining two copies of K_n by a single edge, known as a bridge. We denote this graph by Bar_n [2].
   Independence polynomial of Barbell graph of order n is given by a second-degree polynomial in z. $I(Bar_n, z) = z^2(n^2-1) + 2nz + 1$. If $f(z) = z^2(n^2-1) + 2nz + 1$, then it is conjugate to another polynomial of the form $g(z) = z^2 + (n-1)$.
   The relations between Julia set, Hausdorff dimension and energy of Barbell graph are listed in Table 3.
   From the table, it follows that if n$\geq$ 2, $J(z^2 + (n-1))$ is not connected, therefore only $J(I(Bar_1, z)) \in M$. As order increases, energy increases. Therefore, comparing Hausdorff dimension and energy, we have Hausdorff dimension of independence polynomial of Barbell graph that is less than or equal to energy of Barbell graph.
2. Cocktail Party Graph
   The Cocktail party graph n is a graph on $2n$ vertices. The graph is formed by taking n pairs of vertices such that the vertices in any one pair are adjacent to both vertices in any other pair. There is no edge between the two vertices within any given pair. We denote this graph by CP_n [2] (Table 4).
   Independence polynomial of Cocktail party graph of order n is given by a second-degree polynomial in z.
   $I(CP_n, z) = nz^2a = 2nz + 1$. If $f(z) = nz^2 + 2nz + 1$, then it is conjugate to another polynomial $g(z) = z^2 + 2n - n^2$.
   If $n = 2$, $J(z^2 + 2n - n^2$ is connected, therefore $J(I(CP_2) \in M$.

Table 5 Relation of Hausdorff dimension and energy of complete bipartite graph

Values of n	$J(\mathbf{I}(K_{2,2},, z))$	Hausdorff dimension of $J(I(K_{2,2}, z)$	Energy of $I(K_{2,2}, z)$
1	z^2	1	4

Table 6 Relation of Hausdorff dimension and energy of cycle graph

Values of n	$J(\mathbf{I}(C_5, z))$	Hausdorff dimension of $J(I(C_5, z)$	Energy of $I(C_5, z)$
1	$z^2 + \frac{5}{4}$	0.4346	5

Hausdorff dimension of independence polynomial of Cocktail graph of order n is less than energy of Cocktail graph.

3. Complete Bipartite Graph
   A complete bipartite graph is a bipartite graph (i.e., a set of graph vertices decomposed into two disjoint sets such that no two graph vertices within the same set are adjacent) such that every pair of graph vertices in the two sets are adjacent. If there are p and q graph vertices in the two sets, the complete bipartite graph is denoted $K_{p,q}$ [2] (Table 5).
   Independence polynomial of complete bipartite graph of order 2, $K_{2,2}$ is given by a second-degree polynomial in z.$I(K_{2,2}, z) = 2z^2 + 4z + 1$. It is same as independence polynomial of square graph C_4. If $f(z) = 2z^2 + 4z + 1$, then it is conjugate to another polynomial $g(z) = z^2$.
   $J(I(K_{2,2}, z) = J(z^2)$ is a unit circle and is connected. Therefore, $J(I(K_{2,2}, z) \in M$. Comparing Hausdorff dimension and energy, we have the following result: Hausdorff dimension of independence polynomial of complete graph $K_{2,2}$ or square graph is less than energy of complete graph $K_{2,2}$.
4. Cycle Graph
   A simple graph with n vertices (n $\geq$ 3) and n edges is called a cycle graph if all its edges form a cycle of length n. If the degree of each vertex in the graph is two, then it is called a cycle graph. We denote cycle graph by C_n [2].
   Independence polynomial of cycle graph of order 5 is given by a second-degree polynomial in z.
   $I(C_5, z) = 5z^2 + 5z + 1$. If $f(z) = 5z^2 + 5z + 1$, then it is conjugate to another polynomial $g(z) = z^2 + \frac{5}{4}$.
   $J(I(C_5, z) = J(I(z^2 + \frac{5}{4})$ is not connected. Therefore, $J(I(C_5, z) \notin M$.
   Comparing Hausdorff dimension and energy, we have the following result (Table 6).
   Hausdorff dimension of independence polynomial of cycle graph C_5 is less than energy of cycle graph C_5.

Table 7 Relation of Hausdorff dimension and energy of path graph of order 3

$J(\mathbf{I}(P_3, z))$	Hausdorff dimension of $J(I(P_3, z)$	Energy of $I(P_3, z)$
$z^2 + \frac{1}{4}$	1.0812	2.8285

Table 8 Relation of Hausdorff dimension and energy of path graph of order 4

$J(\mathbf{I}(P_4, z))$	Hausdorff dimension of $J(I(P_4, z)$	Energy of $I(P_4, z)$
$z^2 + 1$	0.6791	4.47206

5. Path Graph

The path graph is a tree with two nodes of vertex degree 1 and the other nodes of vertex degree 2. A path graph is therefore a graph that can be drawn so that all of its vertices and edges lie on a single straight line [2].

5.1 Path Graph of Order 3

It is denoted by P_3. Independence polynomial of path graph of order 3 is given by a second-degree polynomial in z (Table 7).

$I(P_3, z) = z^2 + 3z + 1$. It is same as that of independence polynomial of star graph of order 3, $S_3 . f(z) = z^2 + 3z + 1$, then it is conjugate to another polynomial $g(z) = z^2 + \frac{1}{4}$. $J(I(P_3, z) = J(z^2 + \frac{1}{4})$. $J(z^2 + \frac{1}{4})$ is connected, and therefore, $J(I(P_3, z)) \in M$.

Comparing Hausdorff dimension and energy, we have the following result. Hausdorff dimension of independence polynomial of path graph of order 3 is less than energy of path graph.

5.2 Path Graph of Order 4

It is denoted by P_4. Independence polynomial of path graph of order 4 is given by a second-degree polynomial in z.

$I(P_4, z) = 3z^2 + 4z + 1$. $f(z) = 3z^2 + 4z + 1$, then it is conjugate to another polynomial $g(z) = z^2 + 1$. $J(I(P_4, z) = J(z^2 + 1)$. $J(z^2 + 1)$ is totally disconnected, and therefore, $J(I(P_4, z)) \notin M$.

Comparing Hausdorff dimension and energy, we have the following result (Table 8):

Hausdorff dimension of independence polynomial of Path graph of order 4 is less than energy of path graph.

6. Wheel Graph

The wheel graph of order n is a graph on n+1 vertices. This graph is formed by taking a copy of C_n and adding a central vertex which is adjacent to every vertex in C_n. We denote the wheel graph of order n by W_n [2].

Table 9 Relation of Hausdorff dimension and energy of wheel graph of order 5

$J(\mathbf{I}(W_5, z))$	Hausdorff dimension of $J(I(W_5, z)$	Energy of $I(W_5, z)$
$z^2 - \frac{7}{4}$	1.1632	9.37

Table 10 Relation of Hausdorff dimension and energy of wheel graph of order 6

$J(\mathbf{I}(W_6, z))$	Hausdorff dimension of $J(I(W_6, z)$	Energy of $I(W_6, z)$
$z^2 - 1$	1.26835	11.92

6.1 Wheel Graph of Order 5
It is denoted by W_5. Independence polynomial of wheel graph of order 5 is given by a second-degree polynomial in z.
$I(W_5, z) = 2z^2 + 5z + 1$. If $f(z) = 2z^2 + 5z + 1$, then it is conjugate to another polynomial $g(z) = z^2 - \frac{7}{4}$. $J(I(W_5, z)) = J(z^2 - \frac{7}{4})$ is connected, and therefore, $J(I(W_5, z)) \in M$ (Table 9).
Comparing Hausdorff dimension and energy, we have the following result. Hausdorff dimension of independence polynomial of path graph W_5 is less than energy of path graph W_5.

6.2 Wheel Graph of Order 6
It is denoted by W_6. Independence polynomial of wheel graph of order 6 is given by a second-degree polynomial in z.
$I(W_6, z) = 5z^2 + 6z + 1$. If $f(z) = 5z^2 + 6z + 1$, then it is conjugate to another polynomial $g(z) = z^2 - 1$.
$J(I(W_6, z) = J(z^2 - 1)$ is connected, and therefore, $J(I(W_6, z) \in M$ (Table 10).
Comparing Hausdorff dimension and energy, we have the following result: Hausdorff dimension of independence polynomial of wheel graph W_6 is less than energy of wheel graph W_6.

7. Petersen Graph
The Petersen graph is an undirected graph with 10 vertices and 15 edges. It is a small graph that serves as a useful example and counterexample for many problems in graph theory. It is denoted by $P_{5,2}$ and is 3 regular [6].

The independence polynomial of Petersen graph is a fourth-degree polynomial and is given by $\mathbf{I}\,(\mathbf{P},\ \mathbf{z}) = 1 + 10z + 30z^2 + 30z^3 + 5z^4$. Its characteristic polynomial is given by $(t-1)^5(t+2)^4(t-3)$, making it an integral graph whose spectrum consists entirely of integers, and its spectrum is $-2, -2, -2, -2, 1, 1, 1, 1, 1, 3$. So, the energy of Petersen graph is 16. It is conjugate to another polynomial of the form $z^4 + d$ where $d = 29.0696$. It meets the real axis at $(-1, .5)$. So, its $J(z^4 + 29.0696)$ is disconnected, and its Hausdorff dimension lies between 0 and 2 (Fig. 1).

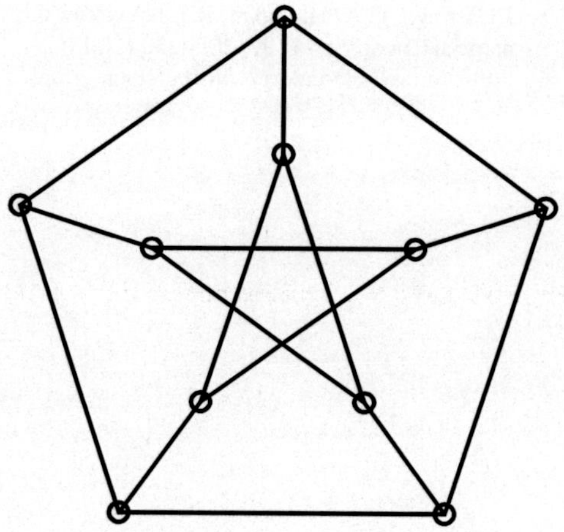

Fig. 1 Petersen graph $P_{5,2}$ [6]

7 Conclusion

Summarizing the research pertaining to the graphs, the salient observations are listed as follows:

- The relationship between a graph and its independence fractal still remains a question.
- Connectivity of a fractal does not depend on the connectivity of the graph.
- The Julia set of graphs with independence number 2 is studied, and for graphs with independence number 3 and higher, the same methods can be used with modifications.
- Hausdorff dimension of a Julia set of independence polynomial of second degree of graphs is less than the energy of corresponding graph.
- Juia set of independence polynomial of Bar_1, CP_2, $K_{2,2}$, P_3, W_5 and W_6 are all connected and therefore element of Mandelbrot set.
- As a special graph, Petersen graph connectivity examined and found that its Julia set is disconnected.

References

1. I. Gutman, F. Harary, Generalizations of the Matching polynomial. Utilitas Mathematica 24, 97–106 (1983)
2. G. Ferrin, *Independence polynomials*, Master Dissertation, University of South Carolina - Columbia (2014)
3. L. Kaskowitz, *The independence fractal of a graph*, Master Dissertation (2003)
4. O. Jones, *Spectra of Simple Graphs*, Whitman College (2013)

5. J.I. Brown, C.A. Hickman, R.J. Nowakowski, The independence fractal of a graph. J. Combinatorial Theor. Ser. B **87**, 209–230 (2003)
6. https://en.wikipedia.org/wiki/Petersen_graph

An Introduction to the Notion of Natural Pseudo-distance in Topological Data Analysis

Patrizio Frosini

Abstract The natural pseudo-distance d_G associated with a group G of self-homeomorphisms of a topological space X is a pseudo-metric developed to compare real-valued functions defined on X, when the equivalence between functions is expressed by the group G. In this paper, we illustrate d_G, its role in topological data analysis, its main properties and its link with persistent homology.

2000 Mathematics Subject Classification. (Primary) 55N31, (Secondary) 57R19 · 65D18 · 68U05 · 62R40

1 Introduction

In topological data analysis, data are frequently expressed by continuous real-valued (or vector-valued) functions defined on a topological space X, and two such functions are considered equivalent if they can be obtained from each other by composition with a suitable self-homeomorphism of X. This happens, e.g., when we are interested in comparing images with respect to the group of plane isometries, or ECG traces with respect to the group of translations in time, or temperature distributions on the earth with respect to rotations around the north pole-south pole axis. Such functions are called *filtering functions*. In order to compare this kind of data, a pseudo-distance is available, quantifying the infimum of the cost of matching two functions φ_1, φ_2 by composition with a homeomorphism in the considered group G, where the cost is defined by the L^∞ norm. According to this pseudo-metric, the measurements $\varphi, \varphi \circ g \in C^0(X, \mathbb{R})$ are considered equivalent to each other for every $g \in G$. In many applications, this property is important and useful, since it allows to choose the data equivalence the user is interested in. For the sake of simplicity, in this survey, we will only consider the case of data represented by real-valued functions. This paper is devoted to illustrate this pseudo-metric, called the *natural pseudo-*

P. Frosini (✉)
Dipartimento di Matematica, Università di Bologna, Bologna, Italy
e-mail: patrizio.frosini@unibo.it

R. L. Devaney et al. (eds.), *Topological Dynamics and Topological Data Analysis*, Springer Proceedings in Mathematics & Statistics 350,
https://doi.org/10.1007/978-981-16-0174-3_17

distance d_G associated with the group G. After recalling the definition of d_G (Sect. 2), we present some theoretical results concerning the values that $d_G(\varphi_1, \varphi_2)$ can take, showing that they are strictly related with the critical values of φ_1 and φ_2, provided that these functions are regular enough (Sect. 3). Secondly, we observe that while d_G represents a clear ground truth in our setting, it is usually quite difficult to compute, due to the size of the group G to be examined. Therefore, efficient methods to get information about d_G are needed. The most relevant method to study the natural pseudo-distance is based on its link with persistent homology and the theory of group equivariant non-expansive operator. Section 4 is devoted to describe this link and its main consequences. In Sect. 5, we conclude the paper by illustrating an open problem concerning d_G.

1.1 Related Literature and Historical Notes

This survey presents the main results obtained about the natural pseudo-distance in the last three decades. These results appeared in several papers and are reported here without proof. For every statement, the paper where the interested reader can find a precise proof is referred. The concept of natural pseudo-distance appeared for the first time in the paper [1], where the distance $\|A - B\|$ between pairs (A, B) of points in a submanifold $\mathcal{M}$ of a Euclidean space was considered as a filtering function and the group G was chosen to be the group of isometries of $\mathcal{M}$. A different but strictly related distance between real-valued functions defined on a manifold had already been presented in [2], referring to the group of similarities of $\mathbb{E}^n$.

The description given in this survey is mainly based on the paper [3]. The reader can find there definitions and proofs concerning the natural pseudo-distance d_G associated with a group G, together with its link with persistent homology and the theory of group equivariant non-expansive operators. The problem of obtaining lower bounds for $d_{\mathrm{Homeo}(X)}$ by means of persistent homology in degree 0 (size functions) has been investigated in [4–6]. Lower bounds for d_G obtained by means of persistent homotopy in the case $G = \mathrm{Homeo}(X)$ and via G-invariant persistent homology in the general case have been presented in [7] and [8], respectively. A study of d_G as a quotient pseudo-metric has been done in the paper [9]. The proofs of the results concerning the link between the values that d_G can take and the critical values of the filtering functions can be found in [10–12]. The proof of the result concerning the possible values of the natural pseudo-distance in the case $X = G = S^1$ can be found in [13]. The results concerning optimal homeomorphisms are illustrated in the papers [6, 10, 13, 14]. A survey about the natural pseudo-distance in the case $G = \mathrm{Homeo}(X)$ has appeared in [15].

2 The Definition of d_G

Let (X, d) and G be a finitely triangulable metric space and a subgroup of the group Homeo(X) of all homeomorphisms from X to X, respectively. If φ_1, φ_2 are two continuous and bounded functions from X to $\mathbb{R}$, we can consider the value $\inf_{g \in G} \|\varphi_1 - \varphi_2 \circ g\|_\infty$. This value is called *the natural pseudo-distance* $d_G(\varphi_1, \varphi_2)$ between φ_1 and φ_2 with respect to the group G. We recall that a pseudo-metric is just a metric without the property assuring that if two points have a null distance then they must coincide. We endow $C^0(X, \mathbb{R})$ with the L^∞ norm and G with the distance $D_G(g_1, g_2) := \max_{x \in X} d(g_1(x), g_2(x))$, so that G becomes a topological group acting on $C^0(X, \mathbb{R})$ by composition on the right. We observe that the action of G on $C^0(X, \mathbb{R})$ is continuous [3].

If G is the trivial group Id, then d_G is the max-norm distance $\|\varphi_1 - \varphi_2\|_\infty$. Moreover, if G_1 and G_2 are subgroups of Homeo(X) and $G_1 \subseteq G_2$, then

$$d_{\text{Homeo}(X)}(\varphi_1, \varphi_2) \le d_{G_2}(\varphi_1, \varphi_2) \le d_{G_1}(\varphi_1, \varphi_2) \le \|\varphi_1 - \varphi_2\|_\infty$$

for every $\varphi_1, \varphi_2 \in C^0(X, \mathbb{R})$.

The direct computation of d_G is usually difficult, due to the size of G. As an example, if $X = \mathbb{R}^3$ and G is the group of all isometries of $\mathbb{R}^3$, a direct computation of d_G would require to evaluate $\|\varphi_1 - \varphi_2 \circ g\|_\infty$ for every isometry $g : \mathbb{R}^3 \to \mathbb{R}^3$. The reader could think of approximating $d_G(\varphi_1, \varphi_2)$ by the value $\mu_S(\varphi_1, \varphi_2) := \inf_{g \in S} \|\varphi_1 - \varphi_2 \circ g\|_\infty$, where S is a sufficiently dense subset S of G. Unfortunately, the use of μ_S would be impractical for data retrieval for two reasons. First of all, in many cases, S should be a very large set in order to obtain a good approximation of d_G, so implying a large computational cost. Secondly, S could not be assumed to be a subgroup of G, even if G is compact (cf. Sect. 3.1 in [3]). For example, this happens when G is the group $SO(3)$ of all orientation-preserving isometries of $\mathbb{R}^3$ that take the point $(0, 0, 0)$ to itself. As a consequence, the function $\mu_S(\varphi_1, \varphi_2)$ would not be a pseudo-metric. This would make the use of μ_S unsuitable for several applications. In Sect. 4, we will see that this difficulty can be worked around by means of persistent homology and the concept of group equivariant non-expansive operator (Theorem 10).

We conclude this section by observing that in many cases we are not interested in every function in $C^0(X, \mathbb{R})$, but in a bounded topological subspace Φ of $C^0(X, \mathbb{R})$. This is due to the fact that the choice of each measuring device restricts the set of functions that can be obtained as data produced by the measurement. From now on, we will assume that a bounded topological subspace Φ of $C^0(X, \mathbb{R})$ has been chosen.

2.1 The Role of d_G in Topological Data Analysis

The comparison of data is usually a process depending on an observer. We could indeed say that data comparison consists in the study of the relationship between an observer and the reality he/she can measure. In this framework, data coincide with measurements. Observers receive and transform data and are, in some sense, defined by the way they perform this transformation. It follows that observers can be defined as collections of suitable operators acting on measurements [16].

According to the dictionary, a "measurement is the assignment of a number to a characteristic of an object or event, which can be compared with other objects or events" [17]. This definition implies that measurements (and hence data) can be seen as functions φ associating a real number $\varphi(x)$ with each point x of a set X of characteristics. (This definition admits a natural extension to vector-valued functions, but for the sake of simplicity, we will treat here only the case of scalar-valued functions). If we wish to develop a theory that can be applied in real situations, we need stability with respect to noise. This justifies the use of topologies on X and on the set Φ of possible measurements on X, as illustrated in the previous section. Furthermore, observers are often endowed with some kind of equivariance, represented by a suitable group G of homeomorphisms. Therefore, we are interested in models where this equivariance can be represented. For example, we usually look for pseudo-metrics that do not distinguish between the shapes of the same object in different spatial positions. The natural pseudo-distance d_G has this property, since it vanishes when the measurements φ, $\varphi \circ g$ are considered, with $\varphi \in \Phi$ and $g \in G$. For this reason, the pseudo-metric d_G can be considered as a ground truth for data comparison in our theoretical setting. This justifies our interest in its study.

3 Theoretical Results About d_G

When the filtering functions are defined on a regular closed manifold, some results restrict the range of values that can be taken by the natural pseudo-distance d_G.

Theorem 1 *([10]) Assume that $\mathcal{M}$ is a closed manifold of class C^1 and that $\varphi_1, \varphi_2 : \mathcal{M} \to \mathbb{R}$ are two functions of class C^1. Set $d := d_{\mathrm{Homeo}(\mathcal{M})}(\varphi_1, \varphi_2)$. Then, a positive integer k exists for which one of the following properties holds:*

(i) k is odd, and kd is the distance between a critical value of φ_1 and a critical value of φ_2;
(ii) k is even, and kd is either the distance between two critical values of φ_1 or the distance between two critical values of φ_2.

Theorem 2 *([11]) Assume that $\mathcal{S}$ is a closed surface of class C^1 and that $\varphi_1, \varphi_2 : \mathcal{S} \to \mathbb{R}$ are two functions of class C^1. Set $d := d_{\mathrm{Homeo}(\mathcal{S})}(\varphi_1, \varphi_2)$. Then, at least one of the following properties holds:*

(i) d is the distance between a critical value of φ_1 and a critical value of φ_2;
(ii) d is half the distance between two critical values of φ_1;
(iii) d is half the distance between two critical values of φ_2;
(iv) d is one third of the distance between a critical value of φ_1 and a critical value of φ_2.

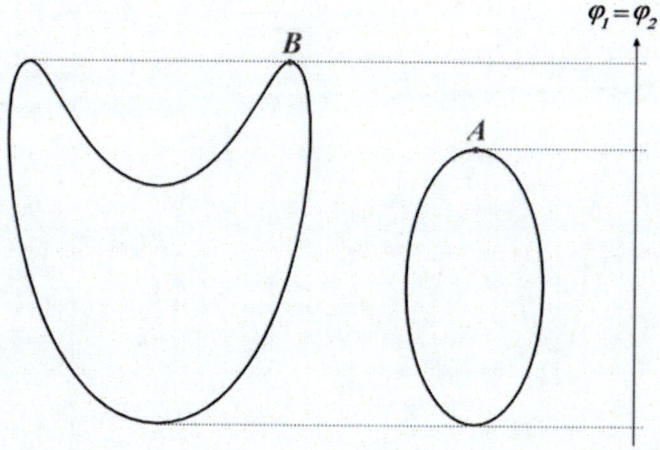

Fig. 1 In this case, the natural pseudo-distance is equal to the distance between two critical values of the filtering functions

Theorem 3 *([12]) Assume that $\mathcal{C}$ is a closed curve of class C^1 and that $\varphi_1, \varphi_2 : \mathcal{C} \to \mathbb{R}$ are two functions of class C^1. Set $d := d_{\text{Homeo}(\mathcal{C})}(\varphi_1, \varphi_2)$. Then, at least one of the following properties holds:*

(i) d is the distance between a critical value of φ_1 and a critical value of φ_2;
(ii) d is half the distance between two critical values of φ_1;
(iii) d is half the distance between two critical values of φ_2.

The statement in the last theorem is sharp, as shown by the following examples.

Example 1 Let us consider the two embeddings of S^1 into $\mathbb{R}^2$ represented in Fig. 1. The ordinate y defines two filtering functions φ_1, φ_2 on S^1. In this case, $d_{\text{Homeo}(S^1)}(\varphi_1, \varphi_2) = |\varphi_1(A) - \varphi(B)|$, i.e., it is the distance between a critical value of φ_1 and a critical value of φ_2.

Example 2 Let us consider the two embeddings of S^1 into $\mathbb{R}^2$ represented in Fig. 2. The ordinate y defines two filtering functions φ_1, φ_2 on S^1. In this case, $d_{\text{Homeo}(S^1)}(\varphi_1, \varphi_2) = \frac{1}{2}|\varphi_1(A) - \varphi_1(B)|$, i.e., it is half the distance between two critical values of φ_1. In Fig. 2, a homeomorphism $g_\varepsilon : S^1 \to S^1$ is displayed, such that $\|\varphi_1 - \varphi_2 \circ g_\varepsilon\|_\infty \leq \frac{1}{2}|\varphi_1(A) - \varphi_1(B)\| + \varepsilon$ (we set $g_\varepsilon(D_\varepsilon) = H_\varepsilon$, $g_\varepsilon(C) = G$ and $g_\varepsilon(E_\varepsilon) = F_\varepsilon$; the first red arc is taken to the second red arc). The equality $d_{\text{Homeo}(S^1)}(\varphi_1, \varphi_2) = \frac{1}{2}|\varphi_1(A) - \varphi_1(B)|$ follows from Theorem 8 in Section 4.

The research concerning the case that G is a proper subgroup of $\text{Homeo}(\mathcal{M})$ is still at its very beginning. As an example of the results concerning this line of research, we cite the following theorem.

Theorem 4 *([13]) Let φ_1, φ_2 be Morse functions from the Lie group S^1 to $\mathbb{R}$ and set $d = d_{S^1}(\varphi_1, \varphi_2)$. At least one of the following statements holds:*

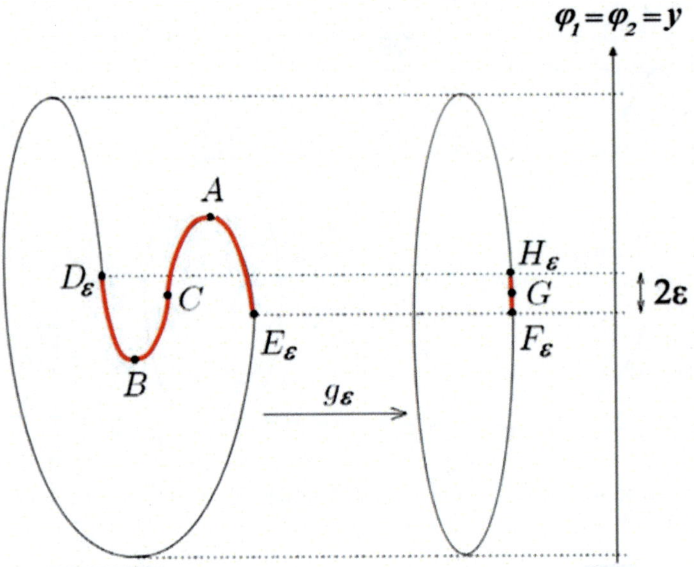

Fig. 2 In this case, the natural pseudo-distance is equal to half the distance between two critical values of the filtering function φ_1

(1) There exist a critical point θ_1 *for* φ_1 *and a critical point* θ_2 *for* φ_2 *such that* $d = |\varphi_1(\theta_1) - \varphi_2(\theta_2)|$;

(2) There exist $\theta_1, \theta_2, \tilde{\theta}_1, \tilde{\theta}_2 \in S^1$ *such that*

- $d = |\varphi_1(\theta_1) - \varphi_2(\theta_2)| = |\varphi_1(\tilde{\theta}_1) - \varphi_2(\tilde{\theta}_2)|$;
- $\frac{d\varphi_1}{d\theta}(\theta_1) = \frac{d\varphi_2}{d\theta}(\theta_2)$ *and* $\frac{d\varphi_1}{d\theta}(\tilde{\theta}_1) = \frac{d\varphi_2}{d\theta}(\tilde{\theta}_2)$;
- $\theta_1 - \theta_2 = \tilde{\theta}_1 - \tilde{\theta}_2$;
- $\frac{d\varphi_1}{d\theta}(\theta_1) \cdot \frac{d\varphi_1}{d\theta}(\tilde{\theta}_1) \cdot (\varphi_1(\theta_1) - \varphi_2(\theta_2)) \cdot (\varphi_1(\tilde{\theta}_1) - \varphi_2(\tilde{\theta}_2)) < 0$.

3.1 Optimal Homeomorphisms

Assume that $\varphi_1, \varphi_2 : X \to \mathbb{R}$ are continuous functions. Let G be a subgroup of $\mathrm{Homeo}(X)$. We say that a homeomorphism $g \in G$ is *optimal* in G for (φ_1, φ_2) if $\|\varphi_1 - \varphi_2 \circ g\|_\infty = d_G(\varphi_1, \varphi_2)$. The following results hold for optimal homeomorphisms.

Theorem 5 *([10]) Assume that* $\mathcal{M}$ *is a* C^1 *closed manifold and that* $\varphi_1, \varphi_2 : \mathcal{M} \to \mathbb{R}$ *are of class* C^1. *If an optimal homeomorphism* $g \in \mathrm{Homeo}(\mathcal{M})$ *for* (φ_1, φ_2) *exists, then* $d_{\mathrm{Homeo}(\mathcal{M})}(\varphi_1, \varphi_2)$ *is the distance between a critical value of* φ_1 *and a critical value of* φ_2.

Theorem 6 *([14]) If* $\varphi_1, \varphi_2 : S^1 \to \mathbb{R}$ *are Morse functions and* $d_{\mathrm{Homeo}(S^1)}(\varphi_1, \varphi_2)$ *vanishes, then an optimal* C^2*-diffeomorphism exists in* $\mathrm{Homeo}(S^1)$ *for* (φ_1, φ_2).

Theorem 7 *([13]) The number of optimal homeomorphisms in the Lie group* S^1 *for a pair* (φ_1, φ_2) *of Morse functions from* S^1 *to* $\mathbb{R}$ *is finite.*

4 A Link Between d_G and Persistent Homology

In this section, we will show that the natural pseudo-distance d_G can be studied by combining persistent homology with the concept of group equivariant non-expansive operator.

4.1 Persistent Homology

Persistent homology can be seen as an efficient method to compute lower bounds and good approximations for the natural pseudo-distance. We recall here some basic definitions and facts concerning persistent homology. The interested reader can find a more detailed and formal treatment in [18–21]. In plain words, persistent homology is a mathematical theory describing the changes of the homology groups of the sub-level sets $X_t = \varphi^{-1}((-\infty, t])$ varying t in $\mathbb{R}$, where φ is a real-valued continuous function defined on a topological space X. We can look at the parameter t as an increasing time, whose change produces the birth and death of k-dimensional holes in the sub-level set X_t. For $k = 0, 1, 2$, the expression "k-dimensional holes" refers to gaps between connected components, tunnels and voids, respectively. The distance between the birthdate and deathdate of a hole is called its *persistence*. The more persistent is a hole, the more important it is for data comparison, since holes with small persistence are usually produced by noise.

As happens for homology, persistent homology can be introduced in several different settings. In this paper, we will use the definition based on Čech homology (cf. [22]).

We start from the following definition.

Definition 1 Let $\varphi : X \to \mathbb{R}$ be a continuous function. If $u, v \in \mathbb{R}$ and $u < v$, we can consider the inclusion i of X_u into X_v. Such an inclusion induces a homomorphism $i^* : H_k(X_u) \to H_k(X_v)$ between the homology groups of X_u and X_v in degree k. The group $PH_k^{\varphi}(u, v) := i^*(H_k(X_u))$ is called the *k-th persistent homology group with respect to the function* $\varphi : X \to \mathbb{R}$, *computed at the point* (u, v). The rank $r_k(\varphi)(u, v)$ of this group is said *the k- th persistent Betti numbers function with respect to the function* $\varphi : X \to \mathbb{R}$, *computed at the point* (u, v).

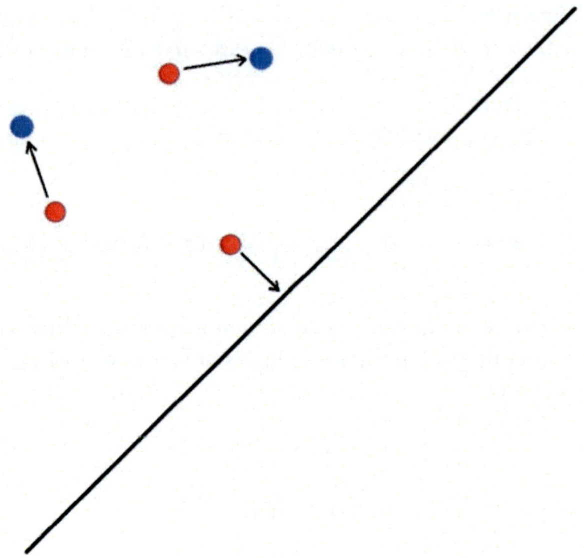

Fig. 3 Example of matching between two persistence diagrams

It can be easily proved that if $g \in \text{Homeo}(X)$, the groups $PH_k^{\varphi}(u, v)$, $PH_k^{\varphi \circ g}(u, v)$ are isomorphic to each other for every $(u, v) \in \mathbb{R}$ with $u < v$ and every $k \in \mathbb{Z}$.

A classical way to describe persistent Betti numbers functions is given by *persistence diagrams*. The k-th persistence diagram $\text{Dgm}_k(\varphi)$ of the function φ is the set of all pairs (b_j, d_j), where b_j and d_j are the birthdate and the deathdate of the j-th k-dimensional hole, respectively, with reference to the filtration $X_t = \varphi^{-1}((-\infty, t])$ varying t in $\mathbb{R}$. When a hole never dies, we set its deathdate equal to ∞. For technical reasons, the points (t, t) are added to each persistence diagram. Two persistence diagrams $\text{Dgm}_k(\varphi_1)$, $\text{Dgm}_k(\varphi_2)$ can be compared by means of the *bottleneck distance* $d_{BN}\left(\text{Dgm}_k(\varphi_1), \text{Dgm}_k(\varphi_2)\right)$. It is defined as the maximum movement of the points of $\text{Dgm}_k(\varphi_1)$ that is necessary to change $\text{Dgm}_k(\varphi_1)$ into $\text{Dgm}_k(\varphi_2)$, measured with respect to the maximum norm (see Fig. 3). If Čech homology is used, each persistent Betti numbers function $r_k(\varphi)$ is equivalent to the corresponding persistence diagram $\text{Dgm}_k(\varphi)$. Therefore, the bottleneck distance induces a metric d_{match} on the set of the persistent Betti numbers functions, so that $d_{\text{match}}\left(r_k(\varphi_1), r_k(\varphi_2)\right) = d_{BN}\left(\text{Dgm}_k(\varphi_1), \text{Dgm}_k(\varphi_2)\right)$. The interested reader can find the formal definitions of persistence diagram and bottleneck distance in [20].

An important property of the metric d_{match} is its stability, as stated in the following result.

Theorem 8 *If k is a natural number and $\varphi_1, \varphi_2 \in C^0(X, \mathbb{R})$, then*

$$d_{\text{match}}(r_k(\varphi_1), r_k(\varphi_2)) \leq d_{\text{Homeo}(X)}(\varphi_1, \varphi_2) \leq \|\varphi_1 - \varphi_2\|_\infty .$$

4.2 Group Equivariant Non-expansive Operators

Let us consider the set $\mathcal{F}(\Phi, G)$ of all maps F from Φ to Φ that verify the following two properties:

(1) $F(\varphi \circ g) = F(\varphi) \circ g$ for every $\varphi \in \Phi$ and every $g \in G$ (i.e., F is equivariant with respect to G);
(2) $\|F(\varphi_1) - F(\varphi_2)\|_\infty \le \|\varphi_1 - \varphi_2\|_\infty$ for every $\varphi_1, \varphi_2 \in \Phi$ (i.e., F is non-expansive).

Obviously, $\mathcal{F}(\Phi, G)$ is not empty, since it contains at least the identity map.

The maps in $\mathcal{F}(\Phi, G)$ are called group equivariant non-expansive operators (GENEOs). In $\mathcal{F}(\Phi, G)$, we define the metric $D_{\text{GENEO}}(F_1, F_2) := \sup_{\varphi\in\Phi} \|F_1(\varphi) - F_2(\varphi)\|_\infty$.

4.3 Persistent Homology as a Tool to Get Lower Bounds for d_G

If $\mathcal{F}$ is a nonempty subset of $\mathcal{F}(\Phi, G)$, then for every fixed k, we can define the following pseudo-metric $D^{\mathcal{F},k}_{\text{match}}$ on Φ:

$$D^{\mathcal{F},k}_{\text{match}}(\varphi_1, \varphi_2) := \sup_{F\in\mathcal{F}} d_{\text{match}}(r_k(F(\varphi_1)), r_k(F(\varphi_2)))$$

for every $\varphi_1, \varphi_2 \in \Phi$, where $r_k(\varphi)$ denotes the k-th persistent Betti numbers function with respect to the function $\varphi : X \to \mathbb{R}$. We will usually omit the index k, when its value is clear from the context or not influential.

We observe that $D^{\mathcal{F}}_{\text{match}}(\varphi_1, \varphi_2 \circ g) = D^{\mathcal{F}}_{\text{match}}(\varphi_1 \circ g, \varphi_2) = D^{\mathcal{F}}_{\text{match}}(\varphi_1, \varphi_2)$ for every $\varphi_1, \varphi_2 \in \Phi$ and every $g \in \text{Homeo}(X)$.

The importance of $D^{\mathcal{F}}_{\text{match}}$ lies in the following two results, showing that it can be used to get information about the natural pseudo-distance d_G.

Theorem 9 *([3]) If $\emptyset \neq \mathcal{F} \subseteq \mathcal{F}(\Phi, G)$, then $D^{\mathcal{F}}_{\text{match}} \le d_G$.*

Theorem 10 *([3]) Let us assume that every function in Φ is non-negative, the k-th Betti number of X does not vanish, and Φ contains each constant function c for which a function $\varphi \in \Phi$ exists such that $0 \le c \le \|\varphi\|_\infty$. Then $D^{\mathcal{F}(\Phi,G)}_{\text{match}} = d_G$.*

As a consequence, the topological and geometrical study of $\mathcal{F}(\Phi, G)$ is important in the research concerning the natural pseudo-distance. Theorem 10 allows us to approximate d_G by approximating $D^{\mathcal{F}(\Phi,G)}_{\text{match}}$.

Two relevant properties of $\mathcal{F}(\Phi, G)$ are expressed by the following results.

Theorem 11 *([3]) If Φ is compact, then $\mathcal{F}(\Phi, G)$ is compact.*

Theorem 12 *([23]) If Φ is convex, then $\mathcal{F}(\Phi, G)$ is convex.*

5 An Open Problem

Let us consider a closed C^1 surface $\mathcal{S}$ and two C^1 filtering functions $\varphi_1, \varphi_2 : \mathcal{S} \to \mathbb{R}$. Let Homeo($\mathcal{S}$) be the group of all self-homeomorphisms of $\mathcal{S}$. We know that $d_{\mathrm{Homeo}(\mathcal{S})}(\varphi_1, \varphi_2) := \inf_{g \in \mathrm{Homeo}(\mathcal{S})} \|\varphi_1 - \varphi_2 \circ g\|_\infty$ is the natural pseudo-distance between φ_1 and φ_2, with respect to the group Homeo($\mathcal{S}$). As we have previously seen, it has been proved in [11] that at least one of the following statements holds:

(1) $d_{\mathrm{Homeo}(\mathcal{S})}(\varphi_1, \varphi_2)$ is the distance between a critical value of φ_1 and a critical value of φ_2;
(2) $d_{\mathrm{Homeo}(\mathcal{S})}(\varphi_1, \varphi_2)$ is half the distance between two critical values of φ_1;
(3) $d_{\mathrm{Homeo}(\mathcal{S})}(\varphi_1, \varphi_2)$ is half the distance between two critical values of φ_2;
(4) $d_{\mathrm{Homeo}(\mathcal{S})}(\varphi_1, \varphi_2)$ is one third of the distance between a critical value of φ_1 and a critical value of φ_2.

Interestingly, no example of two functions $\varphi_1, \varphi_2 : \mathcal{S} \to \mathbb{R}$ is known, such that (4) holds but (1), (2), (3) do not hold. A natural question arises: Can we find an example of two such functions or prove that such an example cannot exist (so improving Theorem 5.7 in [11])?

We recall that the usual technique to compute the natural pseudo-distance $d_{\mathrm{Homeo}(\mathcal{S})}$ consists in

- finding a lower bound for $d_{\mathrm{Homeo}(\mathcal{S})}(\varphi_1, \varphi_2)$ by computing the bottleneck distance $d_{BN}\left(\mathrm{Dgm}_k(\varphi_1), \mathrm{Dgm}_k(\varphi_2)\right)$ between the persistence diagrams in degree k of the functions φ_1 and φ_2 (cf. Theorem 8);
- looking for a sequence (g_i) in Homeo($\mathcal{S}$), such that $\lim_{i\to\infty} \|\varphi_1 - \varphi_2 \circ g_i\|_\infty = d_{BN}\left(\mathrm{Dgm}_k(\varphi_1), \mathrm{Dgm}_k(\varphi_2)\right)$.

If such a sequence (g_i) exists, then the definition of natural pseudo-distance implies that $d_{\mathrm{Homeo}(\mathcal{S})}(\varphi_1, \varphi_2)$ is equal to $d_{BN}\left(\mathrm{Dgm}_k(\varphi_1), \mathrm{Dgm}_k(\varphi_2)\right)$.

Unfortunately, at least one of the following statements holds (cf. [5]):

(a) $d_{BN}\left(\mathrm{Dgm}_k(\varphi_1), \mathrm{Dgm}_k(\varphi_2)\right)$ is the distance between a critical value of φ_1 and a critical value of φ_2;
(b) $d_{BN}\left(\mathrm{Dgm}_k(\varphi_1), \mathrm{Dgm}_k(\varphi_2)\right)$ is half the distance between two critical values of φ_1;
(c) $d_{BN}\left(\mathrm{Dgm}_k(\varphi_1), \mathrm{Dgm}_k(\varphi_2)\right)$ is half the distance between two critical values of φ_2.

Therefore, if (1), (2), (3) do not hold for $\varphi_1, \varphi_2 : \mathcal{S} \to \mathbb{R}$, then $d_{\mathrm{Homeo}(\mathcal{S})}(\varphi_1, \varphi_2)$ cannot be equal to $d_{BN}\left(\mathrm{Dgm}_k(\varphi_1), \mathrm{Dgm}_k(\varphi_2)\right)$. This means that if there exist two C^1 functions $\varphi_1, \varphi_2 : \mathcal{S} \to \mathbb{R}$ verifying (4) but not (1), (2), (3), then we need new methods to compute $d_{\mathrm{Homeo}(\mathcal{S})}(\varphi_1, \varphi_2)$ and to recognize the pair (φ_1, φ_2) as the right example. As a consequence, the answer to the question asked in this section is still unknown.

Acknowledgements Work carried out under the auspices of INdAM-GNSAGA.

References

1. P. Frosini, Measuring shapes by size functions. Proc. SPIE **1607**, 122–133 (1992)
2. P. Frosini, A distance for similarity classes of submanifolds of a Euclidean space. Bull. Austral. Math. Soc. **42**(3), 407–416 (1990)
3. P. Frosini, G. Jabłoński, Combining persistent homology and invariance groups for shape comparison. Discre. Comput. Geometry **55**(2), 373–409 (2016)
4. M. d'Amico, P. Frosini, C. Landi, Natural pseudo-distance and optimal matching between reduced size functions. Acta Appl. Math. **109**(2), 527–554 (2010)
5. P. Donatini, P. Frosini, Lower bounds for natural pseudodistances via size functions. Arch. Inequal. Appl. **2**(1), 1–12 (2004)
6. P. Frosini, C. Landi, Size theory as a topological tool for computer vision. Pattern Recognit Image Anal **9**(4), 596–603 (1999)
7. P. Frosini, M. Mulazzani, Size homotopy groups for computation of natural size distances. B. Belg. Math. Soc.-Sim. **6**(3), 455–464 (1999)
8. P. Frosini, G-invariant persistent homology. Math. Methods Appl. Sci. **38**(6), 1190–1199 (2015)
9. F. Cagliari, B. Di Fabio, C. Landi, The natural pseudo-distance as a quotient pseudo-metric, and applications. Forum Math. **27**(3), 1729–1742 (2015)
10. P. Donatini, P. Frosini, Natural pseudodistances between closed manifolds. Forum Math. **16**(5), 695–715 (2004)
11. P. Donatini, P. Frosini, Natural pseudodistances between closed surfaces. J. Eur. Math. Soc. (JEMS) **9**(2), 331–353 (2007)
12. P. Donatini, P. Frosini, Natural pseudo-distances between closed curves. Forum Math. **21**(6), 981–999 (2009)
13. A. De Gregorio, On the set of optimal homeomorphisms for the natural pseudo-distance associated with the Lie group S^1. Topology Appl. **229**, 187–195 (2017)
14. A. Cerri, B. Di Fabio, On certain optimal diffeomorphisms between closed curves. Forum Math. **26**(6), 1611–1628 (2014)
15. P. Frosini, Una rapida escursione fra le distanze naturali di taglia (Italian), Atti Accademia Peloritana dei Pericolanti Cl. Sci. Fis. Mat. Natur. LXXII **I**, 25–40 (1995)
16. P. Frosini, *Towards an observer-oriented theory of shape comparison*, Eurographics Workshop on 3D Object Retrieval, ed by. A. Ferreira, A. Giachetti, D. Giorgi (The Eurographics Association, 2016), pp. 5–8
17. Wikipedia contributors, *Measurement—Wikipedia, the free encyclopedia*, https://en.wikipedia.org/w/index.php?title=Measurement&oldid=870827913. Online; Accessed Nov 27 2018
18. S. Biasotti, L. De Floriani, B. Falcidieno, P. Frosini, D. Giorgi, C. Landi, L. Papaleo, M. Spagnuolo, *Describing shapes by geometrical-topological properties of real functions*, ACM Comput. Surv. **40**(4), 12:1–12:87 (2008)
19. G. Carlsson, A. Zomorodian, The theory of multidimensional persistence. Discrete Comput. Geom. **42**(1), 71–93 (2009)
20. H. Edelsbrunner, J. Harer, *Persistent homology—a survey*, in Surveys on Discrete and Computational Geometry, Contemporary Mathematics, vol. 453 (American Mathematical Society, Providence, RI, 2008), pp. 257–282
21. Y.S. Oudot, *Persistence theory: from quiver representations to data analysis*. Mathematical Surveys and Monographs, vol. 209 (American Mathematical Society, Providence, RI, 2015), pp. viii+218
22. A. Cerri, B. Di Fabio, M. Ferri, P. Frosini, C. Landi, Betti numbers in multidimensional persistent homology are stable functions. Math. Methods Appl. Sci. **36**(12), 1543–1557 (2013)
23. P. Frosini, N. Quercioli, *Some remarks on the algebraic properties of group invariant operators in persistent homology*, in Machine Learning and Knowledge Extraction ed. by A. Holzinger, P. Kieseberg, A. M. Tjoa, E. Weipp (Springer International Publishing, Cham, 2017), pp. 14–24

A Brief Introduction to Multidimensional Persistent Betti Numbers

Andrea Cerri and Patrizio Frosini

Abstract In this paper, we propose a brief overview about multidimensional persistent Betti numbers (PBNs) and the metric that is usually used to compare them, i.e., the multidimensional matching distance. We recall the main definitions and results, mainly focusing on the 2-dimensional case. An algorithm to approximate n-dimensional PBNs with arbitrary precision is described.

2020 Mathematics Subject Classification. Primary: 55N31; Secondary: 57R19 · 65D18 · 68U05 · 62R40

1 Introduction

Persistent topology and homology are the main tools in topological data analysis. They study how the topology and homology of the sublevel set X_u of a continuous function $f : X \to \mathbb{R}^n$ change when u varies in $\mathbb{R}^n$. The case $n = 1$ has been considered in many papers, starting from the beginning of the '90s (see [1] for historical notes). The case $n > 1$ (i.e., multidimensional persistence) was firstly investigated in [2] as regards homotopy groups, while multidimensional persistence modules were considered in [3, 4] and subsequently studied in other papers including [5–7]. In particular, the *interleaving distance* between multidimensional persistence modules has been formally introduced and discussed in [5]. Another useful tool in persistence theory is given by *multidimensional persistent Betti number functions* (briefly, n-dimensional PBNs) [8], also called *rank invariants* [4]. They have been studied in [9] by means of the so-called *foliation method*. Focusing on the 0th homology, that paper proved that for $n > 1$ a foliation in half-planes can be given, such that the restric-

A. Cerri
FST—Fom Software Technology, Cattolica, Italy
e-mail: acerri@fomsoftware.com

P. Frosini (✉)
Dipartimento di Matematica, Università di Bologna, Bologna, Italy
e-mail: patrizio.frosini@unibo.it

R. L. Devaney et al. (eds.), *Topological Dynamics and Topological Data Analysis*, Springer Proceedings in Mathematics & Statistics 350,
https://doi.org/10.1007/978-981-16-0174-3_18

tion of the n-dimensional PBNs to these half-planes turns out to be 1-dimensional. Each plane in the foliation corresponds to a positive slope line r in $\mathbb{R}^n$ and to the 1-dimensional filtration X_p of X, where X_p is the set of points of X whose images by f are both under and on the left of the point $p \in r$. This approach leads to an algorithm to approximate with arbitrary precision the multidimensional persistent Betti number functions. Furthermore, a stable matching distance between n-dimensional PBNs is available, namely the *n-dimensional matching distance* ([8–10]). The interest in the n-dimensional matching distance between PBNs derives from the fact that, while its computation is pretty simple, the computation of the interleaving distance between persistence modules is NP-hard [11]. This survey paper illustrates the main results concerning n-dimensional PBNs and the n-dimensional matching distance, with particular reference to the case $n = 2$. Finally, we present a recent variant of this last metric, called *coherent matching distance* [12]. For each result, the paper, where the interested reader can find the corresponding proof and further details, is reported.

2 PBNs: Definitions and First Properties

In this section, we recall some basic definitions and properties in persistent homology and topology. For further information, we refer the interested reader to the surveys [1, 13–15]. We will assume that the considered filtering functions are *continuous* and make use of Čech homology. Although different from the more usual setting of tame functions and simplicial or singular homology, our choice is motivated by the following facts:

- the reduction of multidimensional persistence to the 1-dimensional setting is not possible in the setting of tame functions, as observed in [10], but it luckily does in the wider setting of continuous functions;
- using the continuity axiom of Čech homology, it is possible to prove the Representation Theorem 2.5, stating that the PBNs of a scalar-valued filtering function can be completely described by a persistence diagram.

Hereafter, X is a finitely triangulable topological space. The symbol Δ^+ denotes the half-plane $\{(u, v) \in \mathbb{R}^2 : u < v\}$, while Δ^* is the set $\Delta^+ \cup \{(u, \infty) : u \in \mathbb{R}\}$.

2.1 1-Dimensional PBNs

We first consider the case when the filtering function f is real-valued. Indeed, our approach to the multidimensional setting of PBNs is based on a reduction to the 1-dimensional situation. We can consider the sublevel sets of f to define a family of subspaces $X_u = f^{-1}((-\infty, u])$, $u \in \mathbb{R}$, nested by inclusion, i.e., a *filtration of* X. Homology may be applied to derive some topological information about the filtration of X induced by f. The first step is to define persistent homology groups as follows. For $u < v \in \mathbb{R}$, we consider the inclusion of X_u into X_v, which induces a homomorphism of homology groups $H_k(X_u) \to H_k(X_v)$ for every $k \in \mathbb{Z}$. Its image consists of the k-homology classes that live at least from $H_k(X_u)$ to $H_k(X_v)$: It is called the *kth persistent homology group of* (X, f) *at* (u, v), denoted by $H_k^{(u,v)}(X, f)$. By assuming that coefficients are chosen in a field $\mathbb{K}$, we get that homology groups are vector spaces. Therefore, they can be completely described by their dimension, leading to the following definition [16].

Definition 2.1 (Persistent Betti Numbers) The *persistent Betti numbers function* of f in degree k, briefly PBN, is the function $\beta_f : \Delta^+ \to \mathbb{N}$ defined as

$$\beta_f(u, v) = \dim H_k^{(u,v)}(X, f).$$

Since X is finitely triangulable, we have that $\beta_f(u, v) < \infty$ for every $(u, v) \in \Delta^+$. Hereafter, we will assume that a degree $k \in \mathbb{Z}$ has been chosen.

2.1.1 Persistence Diagrams and Representation Theorem.

One of the main properties of 1-dimensional PBNs is that they admit a very simple and compact representation. Precisely, under our assumptions on X and f, and making use of Čech homology, it is possible to prove that each 1-dimensional PBNs can be compactly described by a multiset of points, proper and at infinity, of the real plane. We call them *proper cornerpoints* and *cornerpoints at infinity (or cornerlines)*, respectively.

Definition 2.2 (Proper cornerpoint) For every point $p = (u, v) \in \Delta^+$, the number $\mu(p)$ is the minimum over all the positive real numbers ε, with $u + \varepsilon < v - \varepsilon$, of

$$\beta_f(u + \varepsilon, v - \varepsilon) - \beta_f(u - \varepsilon, v - \varepsilon) - \beta_f(u + \varepsilon, v + \varepsilon) + \beta_f(u - \varepsilon, v + \varepsilon).$$

The number $\mu(p)$ will be called the *multiplicity* of p for β_f. Any point $p \in \Delta^+$ such that the number $\mu(p)$ is strictly positive is said to be a *proper cornerpoint for* β_f.

Definition 2.3 (Cornerpoint at infinity) For every vertical line r, with equation $u = \bar{u}$, $\bar{u} \in \mathbb{R}$, we identify r with $(\bar{u}, \infty) \in \Delta^*$, and define the number $\mu(r)$ as the minimum over all the positive real numbers ε, with $\bar{u} + \varepsilon < 1/\varepsilon$, of

$$\beta_f(\bar{u} + \varepsilon, 1/\varepsilon) - \beta_f(\bar{u} - \varepsilon, 1/\varepsilon).$$

The number $\mu(r)$ will be called the *multiplicity* of r for β_f. When this finite number is strictly positive, r is said to be a *cornerpoint at infinity for* β_f.

The concept of cornerpoint finds application in providing a representation of PBNs [8, 17]. Set $\bar{\Delta}^* = \Delta^* \cup \partial\Delta^+$.

Definition 2.4 (Persistence diagram) The *persistence diagram* $\mathrm{Dgm}(f) \subset \bar{\Delta}^*$ is the multiset of all cornerpoints (both proper and at infinity) for β_f, counted with their multiplicity, union the points of $\Delta := \partial\Delta^+$, counted with infinite multiplicity.

The key role of persistence diagrams is shown in the following Representation Theorem 2.5 [8, 17], claiming that they uniquely determine 1-dimensional PBNs (the converse also holds by definition of persistence diagram).

Theorem 2.5 *(Representation Theorem) For every* $(\bar{u}, \bar{v}) \in \Delta^+$*, we have*

$$\beta_f(\bar{u}, \bar{v}) = \sum_{\substack{(u,v)\in\Delta^* \\ u\leq\bar{u},\, v>\bar{v}}} \mu((u, v)).$$

In practice, Theorem 2.5 states that the value assumed by β_f at a point $(\bar{u}, \bar{v}) \in \Delta^+$ equals the number of cornerpoints lying above and on the left of $(\bar{u}, \bar{v})$. By means of this theorem, 1-dimensional PBNs can be compactly represented as multisets of cornerpoints and cornerpoints at infinity, i.e., as persistence diagrams.

2.1.2 Stability of 1-Dimensional PBNs.

The Representation Theorem 2.5 implies that any distance between persistence diagrams induces a distance between 1-dimensional PBNs. This justifies the following definition 2.6 [8, 17, 18]. Before proceeding, we need to introduce the extended metric $\tilde{d}(p, q) := \|p - q\|_{\tilde{\infty}}$ on Δ^*. For every $p = (u, v), q = (u', v') \in \Delta^*$, we define

$$\|p - q\|_{\tilde{\infty}} = \min\left\{\max\left\{|u - u'|, |v - v'|\right\}, \max\left\{(v - u)/2, (v' - u')/2\right\}\right\}, \quad (1)$$

with the convention about points at infinity that $\infty - c = \infty$ and $c - \infty = -\infty$ when $c \neq \infty$, $\infty - \infty = 0$, $\frac{\infty}{2} = \infty$, $|\pm\infty| = \infty$, $\min\{c, \infty\} = c$ and $\max\{c, \infty\} = \infty$. In plain words, $\tilde{d}(p, q)$ measures the pseudo-distance between two points p and q as the minimum between the cost of moving one point onto the other and the cost of moving both points onto the diagonal Δ, with respect to the max-norm and under the assumption that any two points of the diagonal have vanishing pseudo-distance (we recall that a pseudo-distance d is just a distance missing the condition $d(X, Y) = 0 \Rightarrow X = Y$, i.e., two distinct elements may have vanishing distance with respect to d). When the number of cornerpoints is finite, the matching of persistence diagrams is related to the bottleneck transportation problem, and the matching distance reduces to the bottleneck distance [17]. However, this is not always the case when working with continuous filtering functions, as the number of cornerpoints may be countably infinite.

Definition 2.6 (Matching distance) Let $f, g : X \to \mathbb{R}$ be two continuous functions. For any bijection σ between $\mathrm{Dgm}(f)$ and $\mathrm{Dgm}(g)$, set $\mathrm{cost}(\sigma) := \max_{p \in \mathrm{Dgm}(f)} \|p - \sigma(p)\|_{\widetilde{\infty}}$. The *matching distance* d_{match} between β_f and β_g is defined as

$$d_{\mathrm{match}}\left(\beta_f, \beta_g\right) = \min_{\sigma} \mathrm{cost}(\sigma), \tag{2}$$

where σ ranges over all bijections between $\mathrm{Dgm}(f)$ and $\mathrm{Dgm}(g)$.

We remark that the matching distance is stable with respect to perturbations of the filtering functions, as the following matching stability theorem states:

Theorem 2.7 *(1-Dimensional Stability Theorem) If $f, g : X \to \mathbb{R}$ are two continuous functions, then $d_{\mathrm{match}}(\beta_f, \beta_g) \le \|f - g\|_\infty$.*

For a proof of the previous theorem and more details about the matching distance, the reader is referred to [8, 18] (see also [17, 19] for the bottleneck distance).

2.2 The Foliation Method

We now review the so-called *foliation method*, leading to the definition of a stable distance for multidimensional PBNs [8].

If the considered filtering function is vector-valued, i.e., $f : X \to \mathbb{R}^n$, providing the multidimensional analog of PBNs is straightforward. For $u, v \in \mathbb{R}^n$, with $u = (u_1, \dots, u_n)$ and $v = (v_1, \dots, v_n)$, we say $u \preceq v$ (resp. $u \prec v$) if and only if $u_i \le v_i$ (resp. $u_i < v_i$) for every index $i = 1, \dots, n$. We also endow $\mathbb{R}^n$ with the max-norm $\|(u_1, u_2, \dots, u_n)\|_\infty = \max_{1 \le i \le n} |u_i|$ and use the symbol Δ_n^+ to denote the open set $\{(u, v) \in \mathbb{R}^n \times \mathbb{R}^n : u \prec v\}$.

Given $u \prec v$, the *multidimensional kth persistent homology group of (X, f) at (u, v)* is defined as the image $H_k^{(u,v)}(X, f)$ of the homomorphism $H_k(X_u) \to H_k(X_v)$ induced in homology by the inclusion of $H_k(X_u)$ into $H_k(X_v)$, with $X_u = \{x \in X : f(x) \preceq u\}$.

Definition 2.8 (Persistent Betti Numbers) The *multidimensional persistent Betti numbers function* of $f : X \to \mathbb{R}^n$ in degree k, briefly PBN, is the function $\beta_f : \Delta_n^+ \to \mathbb{N} \cup \{\infty\}$ defined as

$$\beta_f(u, v) = \dim H_k^{(u,v)}(X, f).$$

Since X is finitely triangulable, we have that $\beta_f(u, v) < \infty$ for every $(u, v) \in \Delta_n^+$ (cf. [8, 20]). The key idea underlying the foliation method is that a collection of half-planes in Δ_n^+ can be given, such that the restriction of the multidimensional PBNs to these half-planes turns out to be a 1-dimensional PBNs function in two scalar variables. This approach implies that the comparison of two multidimensional

PBNs can be performed half-plane by half-plane by measuring the distance of appropriate 1-dimensional PBNs. Therefore, the stability of multidimensional PBNs is a consequence of the 1-dimensional PBNs' stability.

We start by recalling that the following parameterized family of half-planes in $\mathbb{R}^n \times \mathbb{R}^n$ is a *foliation* of Δ_n^+ (cf. [9][Prop. 1], [21] and [22]).

Definition 2.9 (Linearly admissible pairs) For every $m = (m_1, \ldots, m_n)$ of $\mathbb{R}^n$ such that $m_i > 0$ for $i = 1, \ldots, n$, and $\sum_{i=1}^n m_i = 1$, and for every $b = (b_1, \ldots, b_n)$ of $\mathbb{R}^n$ such that $\sum_{i=1}^n b_i = 0$, we shall say that the pair (m, b) is *linearly admissible*. We denote the set of all linearly admissible pairs in $\mathbb{R}^n \times \mathbb{R}^n$ by $Ladm_n$. Given a linearly admissible pair (m, b), we define the half-plane $\pi_{(m,b)}$ of $\mathbb{R}^n \times \mathbb{R}^n$ by the following parametric equations:

$$\begin{cases} u = s \cdot m + b \\ v = t \cdot m + b \end{cases}$$

for $s, t \in \mathbb{R}$, with $s < t$.

The set $Ladm_n$ is a set whose closure is $(2n-2)$-dimensional submanifold of $\mathbb{R}^n \times \mathbb{R}^n$ with boundary. The collection of half-planes $\pi_{(m,b)}$ constitute a foliation of Δ_n^+, implying that for each $(u, v) \in \Delta_n^+$ there exists one and only one $(m, b) \in Ladm_n$ such that $(u, v) \in \pi_{(m,b)}$. Observe that m and b only depend on (u, v).

A first property of this foliation is that the restriction of β_f to each leaf can be seen as a particular 1-dimensional PBNs. Intuitively, on each half-plane $\pi_{(m,b)}$ one can find the PBNs corresponding to the filtration of X obtained by sweeping the line through u and v parameterized by $\gamma_{(m,b)} : \mathbb{R} \to \mathbb{R}^n$, with $\gamma_{(m,b)}(\tau) = \tau \cdot m + b$. Each set X_τ in this filtration is given by the points of X that are taken by f into the quadrant $\{u \in \mathbb{R}^n : u \preceq \gamma_{(m,b)}(\tau)\}$.

A second property is that this filtration is equivalent to the one given by the lower level sets of a certain real-valued continuous function. Both these properties are stated in the next theorem, proved in [8, Thm. 4.2], and are intuitively shown in Fig. 1.

Theorem 2.10 *(Reduction Theorem) For every $(u, v) \in \Delta_n^+$, let (m, b) be the only linear admissible pair such that $(u, v) = (s \cdot m + b, t \cdot m + b) \in \pi_{(m,b)}$. Setting $m_* = \min_i m_i$, let moreover $f_{(m,b)} : X \to \mathbb{R}$ be the continuous filtering function defined by setting*

$$f_{(m,b)}(x) = m_* \cdot \max_i \left\{ \frac{f_i(x) - b_i}{m_i} \right\}.$$

Then it holds that

$$\beta_f(u, v) = \beta_{\frac{f_{(m,b)}}{m_*}}(s, t).$$

The Reduction Theorem 2.10 implies that in the multidimensional case, we can obtain an analog D_{match} of the distance d_{match}. The metric D_{match} has a particularly simple form, but yet yields the desired stability properties [8].

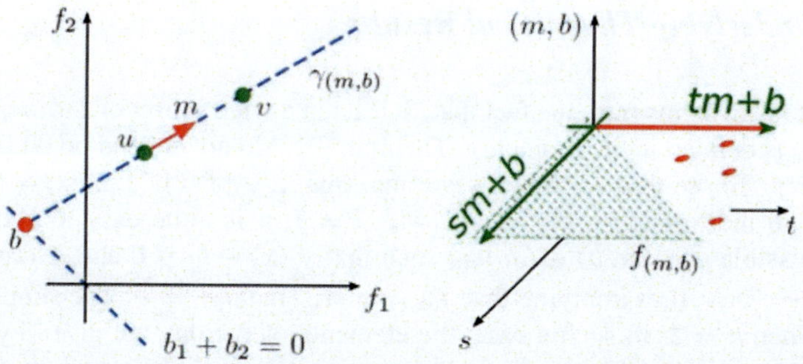

Fig. 1 1-dimensional reduction of 2-dimensional PBNs. Left: a 1-dimensional filtration is constructed sweeping the line through u and v. A unit vector m and a point b are used to parameterize this line as $\gamma_{(m,b)}(\tau) = \tau \cdot m + b$. Right: the persistence diagram of this filtration can be found on the half-plane $\pi_{(m,b)}$

Definition 2.11 (Multidimensional matching distance) Let $f, g : X \to \mathbb{R}^n$ be continuous functions. If $(m, b) \in Ladm_n$, set $d_{(m,b)}\left(\beta_f, \beta_b\right) = d_{\text{match}}\left(\beta_{f_{(m,b)}}, \beta_{g_{(m,b)}}\right)$. The *multidimensional matching distance* D_{match} between β_f and β_g is defined as

$$D_{\text{match}}\left(\beta_f, \beta_g\right) = \sup_{(m,b)\in Ladm_n} d_{(m,b)}\left(\beta_f, \beta_g\right).$$

3 Evaluating the Distance Between Multidimensional PBNs

Definition 2.11 implies that, in general, a direct computation of $D_{\text{match}}\left(\beta_f, \beta_g\right)$ is not feasible, as we should compute the value $d_{(m,b)}\left(\beta_f, \beta_g\right)$ for an infinite number of pairs $(m, b) \in Ladm_n$. On the other hand, taking a non-empty, finite subset $A \subseteq Ladm_n$ and replacing $\sup_{(m,b)\in Ladm_n}$ by $\max_{(m,b)\in A}$ in Definition 2.11, we get a stable and computable pseudo-distance between multidimensional PBNs, say $\tilde{D}_{\text{match}}\left(\beta_f, \beta_g\right)$, which is an approximation of D_{match} to be used in applications.

Computing $\tilde{D}_{\text{match}}\left(\beta_f, \beta_g\right)$ requires the definition of a subset $A \subseteq Ladm_n$ striking a balance between computational cost and approximation accuracy. In fact, it is reasonable that the larger the set A, the smaller the approximation error. On the other hand, the smaller the set A, the faster the computation of $\tilde{D}_{\text{match}}\left(\beta_f, \beta_g\right)$. In this perspective, the goal is to find a set A representing a compromise between these two situations. Additionally, given an arbitrary real value $\varepsilon > 0$ as an error threshold, we might want A depending on ε in a way that $\tilde{D}_{\text{match}}\left(\beta_f, \beta_g\right)$ accomplishes the inequality $\left|D_{\text{match}}\left(\beta_f, \beta_g\right) - \tilde{D}_{\text{match}}\left(\beta_f, \beta_g\right)\right| \le \varepsilon$.

In what follows we review the procedure proposed in [21, 23] to develop an algorithm resulting in an approximation $\tilde{D}_{\text{match}}\left(\beta_f, \beta_g\right)$ of the multidimensional matching distance $D_{\text{match}}\left(\beta_f, \beta_g\right)$, up to an input error threshold ε.

3.1 Underlying Theoretical Results

The first result stems from the fact that, at least in a wide subset of $Ladm_n$, the functions $f_{(m,b)}$ defined in the Reduction Theorem 2.10 do not depend on all the components of f. To see this, we first fix $c = \max\{\max_{x\in X} \|f(x)\|_\infty, \max_{x\in X} \|g(x)\|_\infty\}$. Given two indexes $\bar{\imath}, \bar{\jmath} \in \{1, \ldots, n\}$, with $\bar{\imath} \neq \bar{\jmath}$, it is quite easy to choose a linear admissible pair $(m, b) \in Ladm_n$ such that $f_{\bar{\imath}}(x) - b_{\bar{\imath}} \leq 0$ and $f_{\bar{\jmath}}(x) - b_{\bar{\jmath}} \geq 0$ for every $x \in X$, thus implying that $f_{(m,b)} = m_* \cdot \max_{i\neq\bar{\imath}} \frac{f_i - b_i}{m_i}$. The simplest example is when $n = 2$: In such a case, the elements of $Ladm_2$ are given by $(m, b) = ((m_1, 1 - m_1), (b_1, -b_1))$, with $0 < m_1 < 1$ and $b_1 \in \mathbb{R}$. It is easy to check that, whenever $b_1 \geq c$ (respectively, $b_1 \leq -c$) it holds that $f_{(m,b)}(x) = m_* \cdot \frac{f_2(x)+b_1}{1-m_1}$ (resp. $f_{(m,b)}(x) = m_* \cdot \frac{f_1(x)-b_1}{m_1}$) for every $x \in X$. Similar arguments hold for $g_{(m,b)}$, so that we can write

$$d_{(m,b)}(\beta_f, \beta_g) = \begin{cases} \frac{m_*}{m_1} \cdot d_{\text{match}}(\beta_{f_1}, \beta_{g_1}), & \text{if } \ b_1 \leq -c; \\ \frac{m_*}{1-m_1} \cdot d_{\text{match}}(\beta_{f_2}, \beta_{g_2}), & \text{if } \ b_1 \geq c, \end{cases} \tag{3}$$

the equality in (3) coming from the properties of the matching distance d_{match} (see also [22, Prop. 2.3]).

Based on the above reasonings, the next result [21] states how and when it is possible to reduce the computation of $d_{(m,b)}\left(\beta_f, \beta_g\right)$ to a $(n-1)$-dimensional setting. Set $Ladm_n^+ = \{(m, b) \in Ladm_n : \|b\|_\infty \geq (n-1) \cdot c\}$. For every index $i \in \{1, \ldots, n\}$, we denote by f^i (respectively, g^i) the $\mathbb{R}^{n-1}$-valued function obtained from f (resp. g) by removing the i-th component. Similarly, the symbol m^i (resp. b^i) will be used for the element of $\mathbb{R}^{n-1}$ obtained from m (resp. b) by removing the i-th component.

Theorem 3.1 *Assume that $(m, b) \in Ladm_n^+$. Then an index $\bar{\imath} \in \{1, \ldots, n\}$ exists such that*

$$d_{(m,b)}\left(\beta_f, \beta_g\right) = \frac{m_*}{\min_{i\neq\bar{\imath}} m_i} \cdot d_{(\hat{m},\hat{b})}\left(\beta_{f^{\bar{\imath}}}, \beta_{g^{\bar{\imath}}}\right), \tag{4}$$

with $\left(\hat{m}, \hat{b}\right) \in Ladm_{n-1}$ given by $\hat{m} = \frac{m^{\bar{\imath}}}{(1-m_{\bar{\imath}})}$ and $\hat{b} = b^{\bar{\imath}} + \hat{m} \cdot b_{\bar{\imath}}$.

It is also possible to bound the variation of $d_{(m,b)}\left(\beta_f, \beta_g\right)$ when moving from one half-plane to another in $Ladm_n \setminus Ladm_n^+$. To do this, it is useful to introduce a distance $d : Ladm_n \times Ladm_n \to \mathbb{R}^+$ on the set of admissible pairs [21]. For $(m, b), (m', b') \in Ladm_n$, we set

$$d\left((m, b), (m', b')\right) = \max\left\{\max_{i=1,\ldots,n}\left|\frac{m_*}{m_i} - \frac{m'_*}{m'_i}\right|, \|b - b'\|_\infty\right\}. \tag{5}$$

Based on the above distance, it is possible to prove the following result [21].

Theorem 3.2 *Let* $(m, b) \in Ladm_n \setminus Ladm_n^+$ *and* $(m', b') \in Ladm_n$, *and assume that* $d\left((m, b), (m', b')\right) \leq \delta$. *Then* $\left|d_{(m,b)}\left(\beta_f, \beta_g\right) - d_{(m',b')}\left(\beta_f, \beta_g\right)\right| \leq 2\delta(n \cdot c + 1)$.

Remark 3.3 Note that $d_{(m,b)}\left(\beta_f, \beta_g\right) \leq 2c$ for every $(m, b) \in Ladm_n$ (this is a trivial consequence of Theorem 2.7); thus we have $\left|d_{(m,b)}\left(\beta_f, \beta_g\right) - d_{(m',b')}\left(\beta_f, \beta_g\right)\right| \leq 2c$. Now, if $\delta \geq \frac{1}{n}$ then $2c \leq 2\delta\,(nc + 1)$. Consequently, the inequality claimed by Theorem 3.2 is trivial when $\delta \geq \frac{1}{n}$.

4 An Algorithm for Approximating D_{match}

The above Theorems 3.1 and 3.2 can be used to derive an algorithm for approximating the multidimensional matching distance $D_{\text{match}}\left(\beta_f, \beta_g\right)$.

4.1 The 2-Dimensional Case

We start by providing a detailed treatment of the case $n = 2$, since our approach for higher dimensions is based on a reduction to the 2-dimensional situation. We list the steps in the algorithm described in [21]. For a previous version of the algorithm in the case $n = 2$, the reader is referred to [23].

(a) Fix a threshold error ε. By rescaling appropriately both f and g (and consequently ε), we can assume without loss of generality that $c = 1$. For every $\delta > 0$, we can consider the concept of *regular* δ*-grid over a subset* L *of* $Ladm_2$, i.e., a collection of points $G = \{p = (m, b) \in Ladm_2\}$ such that, denoting by $B_\delta(p)$ the open ball centered at p having radius δ according to the distance d introduced by equality (5), the following statements hold:

   (1) $B_\delta(p) \cap B_\delta(p') = \emptyset$ for every $p, p' \in G$;
   (2) $L \subseteq \cup_{p \in G} \bar{B}_\delta(p)$, with $\bar{B}_\delta(p)$ the closure of $B_\delta(p)$.

(b) We need to fix δ. Because of Remark 3.3 we take δ smaller than $\frac{1}{2}$, say $\delta = \frac{1}{4}$. We also define a finite, regular δ-grid G on $L = Ladm_2 \setminus Ladm_2^+$, see Fig. 2 for some examples. To display the grid, we use the fact that $Ladm_2$ can be identified with the product space $M_2 \times N_2$, with $M_2 = \{m = (m_1, 1 - m_1),\ 0 < m_1 < 1\}$ and $N_2 = \{b = (b_1, -b_1),\ b_1 \in \mathbb{R}\}$. Therefore, we can represent $Ladm_2$ as the subset of the real plane given by $I \times \mathbb{R}$, I the open interval $\{m_1 \in \mathbb{R} : 0 < m_1 < 1\}$. In this perspective, the set $Ladm_2 \setminus Ladm_2^+ = \{(m, b) : \|b\|_\infty < 1\}$ is displayed as $I \times \{b \in \mathbb{R} : |b| < 1\}$. We refer the reader to [21] for a practical construction of G.

(c) Our goal is to compute the largest value for $d_{(m,b)}(\beta_f, \beta_g)$ on $Ladm_2^+$ and on $Ladm_2 \setminus Ladm_2^+$. Equality (2) allows us to simplify the computation of

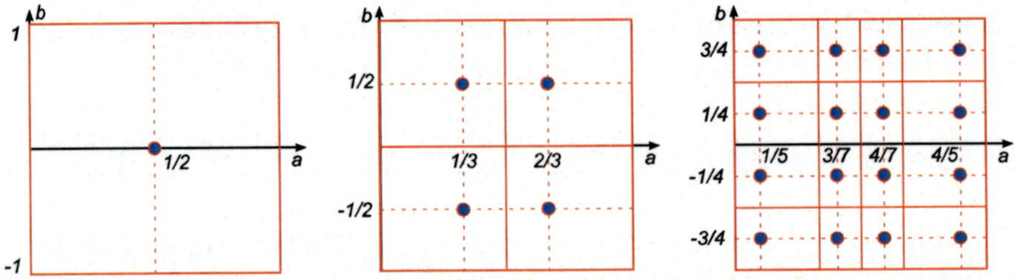

Fig. 2 Regular grids on $Ladm_2 \setminus Ladm_2^+$ for $\delta = 1$ (left), $\delta = 1/2$ (center), and $\delta = 1/4$ (right). The grids are regular with respect to the distance d defined by the equality (5)

$d_{(m,b)}(\beta_f, \beta_g)$ on $Ladm_2^+$. Indeed, it implies that $d_{(m,b)}(\beta_f, \beta_g) \le d_{\text{match}}(\beta_{f_1}, \beta_{g_1})$ if $b = (b_1, -b_1)$ is such that $b_1 \le -c$, while $d_{(m,b)}(\beta_f, \beta_g) \le d_{\text{match}}(\beta_{f_2}, \beta_{g_2})$ if $b_1 \ge c$. Moreover, in the first case, the value $d_{\text{match}}(\beta_{f_1}, \beta_{g_1})$ is achieved when $m = (m_1, 1 - m_1)$ is such that $m_1 \le \frac{1}{2}$; while in the second case, the value $d_{\text{match}}(\beta_{f_2}, \beta_{g_2})$ is achieved when $m_1 \ge \frac{1}{2}$. Thus, it is sufficient to consider the maximum between $d_{\text{match}}(\beta_{f_1}, \beta_{g_1})$ and $d_{\text{match}}(\beta_{f_2}, \beta_{g_2})$ in order to know the value $\max_{Ladm_2^+} d_{(m,b)}(\beta_f, \beta_g)$. We denote such a maximum by D_{ext}.

(d) Theorem 3.2 allows us to control the variation of $d_{(m,b)}(\beta_f, \beta_g)$ in each set $(Ladm_2 \setminus Ladm_2^+) \cap \bar{B}_\delta(p)$, and hence in $Ladm_2 \setminus Ladm_2^+$. For every $p = (m, b) \in G$, we compute the value $d_{(m,b)}(\beta_f, \beta_g)$ and set $D_{int} = \max_{p \in G} d_{(m,b)}(\beta_f, \beta_g)$.

(e) The number $D_{\text{tot}} = \max\{D_{ext}, D_{int}\}$ is then a first approximation of the matching distance $D_{\text{match}}(\beta_f, \beta_g)$. We briefly describe how to refine the value D_{tot} to obtain an approximation of $D_{\text{match}}(\beta_f, \beta_g)$ up to the error threshold ε.

- If the inequality $2\delta \cdot (2c + 1) \le \varepsilon$ holds, by Definition 2.11 and by applying Theorem 3.2, it follows that $|D_{\text{match}}(\beta_f, \beta_g) - D_{\text{tot}}| \le \varepsilon$. Therefore, we stop having as output D_{tot};
- Otherwise, we delete each point $p = (m, b) \in G$ such that the inequality $D_{\text{tot}} - d_{(m,b)}(\beta_f, \beta_g) > 2\delta \cdot (2c + 1)$ holds. Indeed, Theorem 3.2 ensures that D_{tot} will not be achieved (or exceeded) by computing the values $d_{(m,b)}(\beta_f, \beta_g)$ over the sets $\bar{B}_\delta(p)$. Moreover, the grid G is refined as follows: Each p still in G is replaced by four suitable points $p_1, \dots, p_4$, such that $\{p_j, j = 1, \dots, 4\}$ is a regular $\frac{\delta}{2}$-grid on $B_\delta(p)$ based on the four balls $B_{\frac{\delta}{2}}(p_j)$. Finally, D_{int} and D_{tot} are updated according to the new grid G', δ is replaced by $\frac{\delta}{2}$, and the algorithm restarts by checking if the inequality $2\delta \cdot (2c + 1) \le \varepsilon$ holds.

4.1.1 The n-dimensional Case

We can now show how the above procedure can be generalized to the n-dimensional setting, with $n > 2$. Such an extension is partially based on a reduction to the case $n = 2$.

Similarly to what happens in the case $n = 2$, we need to compute the largest value for $d_{(m,b)}(\beta_f, \beta_g)$ on $Ladm_n^+$ and on $Ladm_n \setminus Ladm_n^+$. We fix a threshold error ε. By appropriately rescaling both f and g (and consequently ε), we can assume without loss of generality that $c = 1$, so that $Ladm_n^+ = \{(m, b) \in Ladm_n : \|b\|_\infty \geq n - 1\}$. In $Ladm_n^+$, Theorem 3.1 allows us to reduce the computation of $d_{(m,b)}(\beta_f, \beta_g)$ to a $(n - 1)$-dimensional situation. Indeed, it implies that, for every $(m, b) \in Ladm_n^+$, there exists $(\hat{m}, \hat{b}) \in Ladm_{n-1}$ such that $d_{(m,b)}\left(\beta_f, \beta_g\right) \leq d_{(\hat{m},\hat{b})}\left(\beta_{f^{\bar{\imath}}}, \beta_{g^{\bar{\imath}}}\right)$ for a suitable index $\bar{\imath} \in \{1, \ldots, n\}$. On the other hand, it is possible to prove that, for every $\bar{\imath} \in \{1, \ldots, n\}$ and every $(\hat{m}, \hat{b}) \in Ladm_{n-1}$, there always exists $(m, b) \in Ladm_n^+$ such that $d_{(m,b)}\left(\beta_f, \beta_g\right) = d_{(\hat{m},\hat{b})}\left(\beta_{f^{\bar{\imath}}}, \beta_{g^{\bar{\imath}}}\right)$. As a consequence, the computation of $d_{(m,b)}\left(\beta_f, \beta_g\right)$ over the set $Ladm_n^+$ can be reduced to the one of the $(n - 1)$-dimensional matching distances $D_{\text{match}}\left(\beta_{f^i}, \beta_{g^i}\right)$, for $i = 1, \ldots, n$.

Obviously, we can recursively repeat the same reasonings to progressively decrease the dimensionality of the problem. It turns out that computing the largest value for $d_{(m,b)}(\beta_f, \beta_g)$ on $Ladm_n^+$ can be reduced to the 2-dimensional case, by considering the $\binom{n}{2}$ 2-dimensional matching distances $D_{\text{match}}\left(\beta_{f_{ij}}, \beta_{g_{ij}}\right)$, with $f_{ij} = (f_i, f_j)$ and $g_{ij} = (g_i, g_j)$ for every $i \neq j$.

Similarly to what happens in the 2-dimensional case, Theorem 3.2 allows us to control the variation of $d_{(m,b)}(\beta_f, \beta_g)$ on the set $Ladm_n \setminus Ladm_n^+$. Also in this case, we can define a regular grid G on $Ladm_n \setminus Ladm_n^+$ by extending the above reasonings for the 2-dimensional setting, see [21] for more details.

5 Beyond the Multidimensional Matching Distance D_{match}

In Definition 2.11, we have seen that the multidimensional matching distance $D_{\text{match}}(\beta_f, \beta_g)$ depends on the comparison of the two collections $\{\text{Dgm}(f_{(m,b)})\}$ and $\{\text{Dgm}(g_{(m,b)})\}$, with (m, b) varying in $Ladm_n$. This is done by computing the supremum of the 1-dimensional matching distances $d_{(m,b)}(\beta_f, \beta_g)$ over (m, b). Note that, in principle, a small change of the pair (m, b) can cause a large change in the "optimal" matching, that is, the matching $\sigma : \text{Dgm}(f_{(m,b)}) \to \text{Dgm}(g_{(m,b)})$ whose cost is equal to the distance $d_{(m,b)}(\beta_f, \beta_g)$. In other words, the definition of $D_{\text{match}}(\beta_f, \beta_g)$ is based on a family of optimal matchings that is not required to change continuously with respect to the pair (m, b). This is due to the intrinsically discontinuous definition of $D_{\text{match}}(\beta_f, \beta_g)$, which in turn makes studying its properties difficult.

5.1 The Coherent Matching Distance for 2-Dimensional Persistent Betti Numbers

For these reasons, in [12, 24], a new matching distance between multidimensional PBNs has been introduced, called *coherent matching distance* and initially investigated in the 2-dimensional setting. The definition of the coherent matching distance is based on matchings that change "coherently" with the persistence diagrams of the 1-dimensional filtering functions that we take into account. In other words, the basic idea consists of considering only matchings between the persistence diagrams $\mathrm{Dgm}(f_{(m,b)})$ and $\mathrm{Dgm}(g_{(m,b)})$ that change continuously with respect to the pair (m, b).

The idea of "coherent matching" leads to the discovery of an interesting phenomenon of *monodromy*. While requiring that the matchings change continuously, one has to avoid the pairs (m, b) at which the persistence diagram contains double points, called *singular pairs*. This is done by choosing a connected open set $U \subseteq Ladm_2$ of regular (i.e., non-singular) pairs and assuming that $(m, b) \in U$. In doing this, it is possible to preserve the "identity" of points in the persistence diagram and follow them when moving in U. From this identity, the concept of a family of continuously changing matchings easily arises. Interestingly, turning around a singular pair can produce a permutation in the considered persistence diagram, so that the considered filtering function is associated with a monodromy group. An example of this phenomenon can be found in [12].

Therefore, the definition of "coherent matching" must take a monodromy group into account. In [12], this is done by defining a transport operator $T_\gamma^{(f,g)}$, which continuously transports each matching $\sigma_{(m,b)}$ between the persistence diagrams $\mathrm{Dgm}(f_{(m,b)})$, $\mathrm{Dgm}(g_{(m,b)})$ to a matching $\sigma_{(m',b')}$ between the persistence diagrams $\mathrm{Dgm}(f_{(m',b')})$, $\mathrm{Dgm}(g_{(m',b')})$ along a path γ from (m, b) to (m', b') in the set U. The existence of monodromy implies that the transport of $\sigma_{(m,b)}$ depends not only on the pairs (m, b), (m', b'), but also on the path γ.

Having introduced the transport operator $T_\gamma^{(f,g)}$, we can define the *coherent cost of a matching* $\sigma_{(m,b)}$ by considering the usual cost of all the matchings that are obtained by transportation of $\sigma_{(m,b)}$:

$$\mathrm{cohcost}_U(\sigma_{(m,b)}) = \sup_\gamma \mathrm{cost}\left(T_\gamma^{(f,g)}\left(\sigma_{(m,b)}\right)\right), \tag{6}$$

where γ ranges over the set of all continuous paths from $[0, 1]$ to U starting at (m, b), while $\mathrm{cost}(\sigma)$ is the cost of a matching σ between persistence diagrams (see Definition 2.6).

This done the definition of the coherent matching distance CD_U is straightforward: If two filtering functions $f, g : X \to \mathbb{R}^2$ are given and U does not contain singular pairs neither for f nor for g, then $CD_U(\beta_f, \beta_g)$ is the infimum of the coherent costs of the matchings between the persistence diagrams associated with an admissible pair $(m, b) \in U$ arbitrarily fixed:

$$CD_U(\beta_f, \beta_g) = \inf_{\sigma_{(m,b)}} \text{cohcost}_U\left(\sigma_{(m,b)}\right), \tag{7}$$

with $\sigma_{(m,b)}$ varying in the set of all matchings from $\text{Dgm}(f_{(m,b)})$ to $\text{Dgm}(g_{(m,b)})$.

It is important to remark that the definition of CD_U does not depend on the choice of the pair (m, b) [12, Prop. 12]. Moreover, under suitable conditions for the functions f and g, it is possible to prove that, if $\|f - g\|_\infty < c$ for a non-negative real value c sufficiently small, then $CD_U(\beta_f, \beta_g) \leq |f - g\|_\infty$ [12, Thm. 3].

Another key point here is that the function $\text{cost}(T_\gamma^{(f,g)}(\sigma_{(m,b)}))$ takes its global maximum over γ when the endpoint of γ belongs either to the vertical line $m = \frac{1}{2}$ or to the boundary of U [12, Thm. 6]. This result casts new light on the abundance of examples where the supremum defining the usual matching distance D_{match} is taken for the pairs $(m, b) \in Ladm_2$ with $m \approx \frac{1}{2}$ [12, 23]. Nevertheless, it suggests that the coherent matching distance CD_U could be used in place of the matching distance D_{match} both in theory and applications, as it allows one to manage the parameter space $Ladm_2$ more efficiently.

Acknowledgements Work carried out under the auspices of INdAM-GNSAGA.

References

1. S. Biasotti, L. De Floriani, B. Falcidieno, P. Frosini, D. Giorgi, C. Landi, L. Papaleo, M. Spagnuolo, Describing shapes by geometrical-topological properties of real functions. ACM Comput. Surv. **40**(4), 1–87 (2008)
2. P. Frosini, M. Mulazzani, Size homotopy groups for computation of natural size distances. B. Belg. Math. Soc.-Sim. **6**(3), 455–464 (1999)
3. G. Carlsson, A.J. Zomorodian, *The theory of multidimensional persistence, SCG '07: Proceedings of the twenty-third annual Symposium on Computational Geometry (New York* (USA), ACM, NY, 2007), pp. 184–193
4. G. Carlsson, A.J. Zomorodian, The theory of multidimensional persistence. Discrete Comput. Geom. **42**(1), 71–93 (2009)
5. M. Lesnick, The theory of the interleaving distance on multidimensional persistence modules. Found. Comput. Math. **15**(3), 613–650 (2015)
6. M. Lesnick, M. Wright, *Interactive Visualization of 2-D Persistence Modules*, ArXiv e-prints (2015), arXiv:1512.00180
7. M. Scolamiero, W. Chachólski, A. Lundman, R. Ramanujam, S. Öberg, Multidimensional persistence and noise. Found. Comput. Math. **17**(6), 1367–1406 (2017)
8. A. Cerri, B. Di Fabio, M. Ferri, P. Frosini, C. Landi, Betti numbers in multidimensional persistent homology are stable functions. Math. Method. Appl. Sci. **36**(12), 1543–1557 (2013)
9. S. Biasotti, A. Cerri, P. Frosini, D. Giorgi, C. Landi, Multidimensional size functions for shape comparison. J. Math. Imaging Vis. **32**(2), 161–179 (2008)
10. F. Cagliari, B. Di Fabio, M. Ferri, One-dimensional reduction of multidimensional persistent homology. Proc. Amer. Math. Soc. **138**(8), 3003–3017 (2010)
11. H.B. Bjerkevik, M.B. Botnan, *Computational complexity of the interleaving distance*, in 34th International Symposium on Computational Geometry (SoCG 2018), ed. by B. Speckmann, C.D. Tóth, (Dagstuhl, Germany), Leibniz International Proceedings in Informatics (LIPIcs), Schloss Dagstuhl–Leibniz-Zentrum fuer Informatik, 2018, pp. 13:1–13:15

12. A. Cerri, M. Ethier, P. Frosini, *On the geometrical properties of the coherent matching distance in 2D persistent homology*, J. Appl. and Comput. Topology, **138**(8), 3003–3017 (2010). https://doi.org/10.1007/s41468-019-00041-y
13. H. Edelsbrunner, J. Harer, *Computational Topology: An Introduction*, American Mathematical Society, Providence (R.I.) (2010)
14. R. Ghrist, Barcodes: the persistent topology of data. Bull. Amer. Math. Soc. **45**(1), 61–75 (2008)
15. A.J. Zomorodian, *Topology for computing*. Cambridge Monographs on Applied and Computational Mathematics, vol. 16 (Cambridge University Press, Cambridge, 2005)
16. H. Edelsbrunner, D. Letscher, A.J. Zomorodian, Topological persistence and simplification. Discrete Comput. Geom. **28**(4), 511–533 (2002)
17. D. Cohen-Steiner, H. Edelsbrunner, J. Harer, Stability of persistence diagrams. Discrete Comput. Geom. **37**(1), 103–120 (2007)
18. M. d'Amico, P. Frosini, C. Landi, Natural pseudo-distance and optimal matching between reduced size functions. Acta Appl. Math. **109**, 527–554 (2010)
19. F. Chazal, D. Cohen-Steiner, M. Glisse, L.J. Guibas, S.Y. Oudot, *Proximity of persistence modules and their diagrams, SCG '09: Proceedings of the 25th annual Symposium on Computational Geometry (New York* (USA), ACM, NY, 2009), pp. 237–246
20. F. Cagliari, C. Landi, Finiteness of rank invariants of multidimensional persistent homology groups. Appl. Math. Lett. **24**(4), 516–518 (2011)
21. A. Cerri, P. Frosini, *A new approximation algorithm for the matching distance in multidimensional persistence*, J. Comp. Math., **38**(2), 291–309. https://doi.org/10.4208/jcm.1809-m2018-0043. http://global-sci.org/intro/article_detail/jcm/14518.html
22. A. Cerri, P. Frosini, *Invariance properties of the multidimensional matching distance in Persistent Topology and Homology*, Tech. Report no. 2765, Università di Bologna, 2010, Available at http://amsacta.cib.unibo.it/2765/
23. S. Biasotti, A. Cerri, P. Frosini, D. Giorgi, A new algorithm for computing the 2-dimensional matching distance between size functions. Pattern Recogn. Lett. **32**(14), 1735–1746 (2011)
24. A. Cerri, M. Ethier, P. Frosini, *The coherent matching distance in 2d persistent homology*, Lecture Notes in Computer Science, Proceedings of the 6th International Workshop on Computational Topology in Image Context, Marseille, France, ed. by A. Bac, J.-L. Mari, Springer International Publishing Switzerland, **9667** (2016), 216–227

Some New Methods to Build Group Equivariant Non-expansive Operators in TDA

Nicola Quercioli

Abstract Group equivariant operators are playing a more and more relevant role in machine learning and topological data analysis. In this paper, we present some new results concerning the construction of G-equivariant non-expansive operators (GENEOs) from a space Φ of real-valued bounded continuous functions on a topological space X to Φ itself. The space Φ represents our set of data, while G is a subgroup of the group of all self-homeomorphisms of X, representing the invariance we are interested in.

Keywords Natural pseudo-distance · Filtering function · Group action · Group equivariant non-expansive operator · Persistent homology · Persistence diagram · Topological data analysis

2000 Mathematics Subject Classification. Primary: 55N31; Secondary: 62R40

1 Introduction

In the recent years, topological data analysis (TDA) has imposed itself as a useful tool in order to manage huge amount of data of the present digital world [1]. In particular, persistent homology has assumed a relevant role as an efficient tool for qualitative and topological comparison of data [2]; since in several applications, we can express the acts of measurement by $\mathbb{R}^m$-valued functions defined on a topological space, so inducing filtrations on such a space [3]. These filtrations can be analyzed by means of the standard methods used in persistent homology. For further and detailed information about persistent homology, we refer the reader to [4].

The importance of group equivariance in machine learning is well known (see, e.g., [5–8]). Our work on group equivariant non-expansive operators (GENEOs) is devoted to possibly establish a link between persistence theory and machine learning.

N. Quercioli (✉)
Department of Mathematics, University of Bologna, Bologna, Italy
e-mail: nicola.quercioli2@unibo.it

R. L. Devaney et al. (eds.), *Topological Dynamics and Topological Data Analysis*, Springer Proceedings in Mathematics & Statistics 350,
https://doi.org/10.1007/978-981-16-0174-3_19

Our basic idea is that acts of measurement are directly influenced by the observer, and we should mostly focus on well approximating the observer, rather than precisely describing the data (see, e.g., [9]). In some sense, we could see the observer as a collection of GENEOs acting on a suitable space of data and encode in the choice of these operators the invariance we are interested in.

The concept of invariance group leads us to consider the *natural pseudo-distance* as our main tool to compare data. Let us consider two real-valued functions φ, ψ on a topological space X, representing the data we want to compare, and a group G of self-homeomorphisms of X. Roughly speaking, the computation of the natural pseudo-distance d_G between φ and ψ is the attempt of finding the best correspondence between these two functions with respect to the invariance group G.

Unfortunately, d_G is difficult to compute, but [10] illustrates a possible path to approximate the natural pseudo-distance by means of a dual approach involving persistent homology and GENEOs. In particular, one can see that a good approximation of the space $\mathcal{F}(\Phi, G)$ of all GENEOs corresponds to a good approximation of the pseudo-distance d_G. In order to extend our knowledge about $\mathcal{F}(\Phi, G)$, we devote this paper to introduce some new methods to construct new GENEOs from a given set of GENEOs.

The outline of our paper follows. In Sect. 2, we briefly present our mathematical framework. In Sect. 3, we give a new result about building GENEOs by power means and show some examples to explain why this method is useful and meaningful. In Sect. 4, we illustrate a new procedure to build new GENEOs by means of series of GENEOs. In particular, this is a first example of costruction of an operator starting from an infinite set of GENEOs.

2 Our Mathematical Model

In this section, the mathematical model illustrated in [10] will be briefly recalled. Let X be a (non-empty) topological space and Φ be a topological subspace of the topological space $C_b^0(X, \mathbb{R})$ of the continuous bounded functions from X to $\mathbb{R}$, endowed with the topology induced by the sup-norm $\|\cdot\|_\infty$. The elements of Φ represent our data and are called *admissible filtering functions* on the space X. We also assume that Φ contains at least the constant functions c such that $|c| \leq \sup_{\varphi\in\Phi} \|\varphi\|_\infty$. The invariance of the space Φ is represented by the action of a subgroup G of the group $\mathrm{Homeo}(X)$ of all homeomorphisms from X to itself. The group G is used to act on Φ by composition on the right, i.e., we suppose that $\varphi \circ g$ is still an element of Φ for any $\varphi \in \Phi$ and any $g \in G$. In other words, the functions φ and $\varphi \circ g$, elements of Φ, are considered equivalent to each other for every $g \in G$.

In this theoretical framework, we use the *natural pseudo-distance* d_G to compare functions.

Definition 1 For every $\varphi_1, \varphi_2 \in \Phi$, we can define the function $d_G(\varphi_1, \varphi_2) := \inf_{g\in G} \sup_{x\in X} |\varphi_1(x) - \varphi_2(g(x))|$ from $\Phi \times \Phi$ to $\mathbb{R}$. The function d_G is called the *natural pseudo-distance* associated with the group G acting on Φ.

We can consider this (extended) pseudo-metric as the ground truth for the comparison of functions in Φ with respect to the action of the group G. Unfortunately, d_G is usually difficult to compute. However, the natural pseudo-distance can be studied and approximated by a method involving *G-equivariant non-expansive operators.*

Definition 2 A G-equivariant non-expansive operator (GENEO) for the pair (Φ, G) is a function

$$F : \Phi \longrightarrow \Phi$$

that satisfies the following properties:

1. F is G-equivariant: $F(\varphi \circ g) = F(\varphi) \circ g, \quad \forall\, \varphi \in \Phi, \quad \forall\, g \in G$;
2. F is non-expansive: $\|F(\varphi_1) - F(\varphi_2)\|_\infty \leq \|\varphi_1 - \varphi_2\|_\infty, \quad \forall\, \varphi_1, \varphi_2 \in \Phi$.

The symbol $\mathcal{F}(\Phi, G)$ is used to denote the set of all G-equivariant non-expansive operators for (Φ, G). Obviously, $\mathcal{F}(\Phi, G)$ is not empty because it contains at least the identity operator.

Remark 1 The non-expansivity property means that the operators in $\mathcal{F}(\Phi, G)$ are 1-Lipschitz functions, and therefore, they are continuous. We underline that GENEOs are not required to be linear.

If X has nontrivial homology in degree k, the following key result holds [10].

Theorem 1 $d_G(\varphi_1, \varphi_2) = \sup_{F\in\mathcal{F}(\Phi,G)} d_{\mathrm{match}}(\mathrm{Dgm}_k(F(\varphi_1)), \mathrm{Dgm}_k(F(\varphi_2)))$, *where* $\mathrm{Dgm}_k(\varphi)$ *denotes the k-th persistence diagram of the function* $\varphi : X \to \mathbb{R}$ *and* d_{match} *is the classical matching distance.*

Persistent homology and the natural pseudo-distance are related to each other by Theorem 1 via GENEOs. This result enables us to approximate d_G by means of G-equivariant non-expansive operators. The construction of new classes of GENEOs is consequently a relevant step in the approximation of the space $\mathcal{F}(\Phi, G)$, and hence in the computation of the natural pseudo-distance, so justifying the interest for the results shown in Sects. 3 and 4.

3 Building New GENEOs by Means of Power Means

In this section, we introduce a new method to build GENEOs, concerning the concept of power mean. Now we recall a proposition that enables us to find new GENEOs, based on the use of 1-Lipschitz functions (see [11]).

Proposition 1 *Let L be a 1-Lipschitz function from $\mathbb{R}^n$ to $\mathbb{R}$, where $\mathbb{R}^n$ is endowed with the norm $\|(x_1, \dots, x_n)\|_\infty = \max\{|x_1|, \dots, |x_n|\}$. Assume also that $F_1, \dots, F_n$ are GENEOs for (Φ, G). Let us define the function $L^*(F_1, \dots, F_n) : \Phi \longrightarrow C_b^0(X, \mathbb{R})$ by setting*

$$L^*(F_1, \dots, F_n)(\varphi)(x) := L(F_1(\varphi)(x), \dots, F_n(\varphi)(x)).$$

If $L^*(F_1, \dots, F_n)(\Phi) \subseteq \Phi$, the operator $L^*(F_1, \dots, F_n)$ is a GENEO for (Φ, G).

In order to apply this proposition, we recall some definitions and properties about power means and p-norms. Let us consider a sample of real numbers $x_1, \dots, x_n$ and a real number $p > 0$. As well known, the power mean $M_p(x_1, \dots, x_n)$ of $x_1, \dots, x_n$ is defined by setting

$$M_p(x_1, \dots, x_n) := \left(\frac{1}{n}\sum_{i=1}^{n} |x_i|^p\right)^{\frac{1}{p}}.$$

In order to proceed, we consider the function $\|\cdot\|_p : \mathbb{R}^n \longrightarrow \mathbb{R}$ defined by setting

$$\|x\|_p = (|x_1|^p + |x_2|^p + \dots + |x_n|^p)^{\frac{1}{p}}$$

where $x = (x_1, \dots, x_n)$ is a point of $\mathbb{R}^n$. It is well know that, for $p \ge 1$, $\|\cdot\|_p$ is a norm and that for any $x \in \mathbb{R}^n$, we have $\lim_{p\to\infty} \|x\|_p = \|x\|_\infty$. Finally, it is easy to check that if $x \in \mathbb{R}^n$ and $0 < p < q < \infty$, it holds that

$$\|x\|_q \le \|x\|_p \le n^{\frac{1}{p}-\frac{1}{q}} \|x\|_q. \tag{1}$$

For q tending to infinity, we obtain a similar inequality:

$$\|x\|_\infty \le \|x\|_p \le n^{\frac{1}{p}} \|x\|_\infty. \tag{2}$$

Now we can define a new class of GENEOs. Let us consider $F_1, \dots, F_n$ GENEOs for (Φ, G) and $p > 0$. Let us define the operator $M_p(F_1, \dots, F_n) : \Phi \longrightarrow C_b^0(X, \mathbb{R})$ by setting

$$M_p(F_1, \dots, F_n)(\varphi)(x) := M_p(F_1(\varphi)(x), \dots, F_n(\varphi)(x)).$$

Theorem 2 *If $p \ge 1$ and $M_p(F_1, \dots, F_n)(\Phi) \subseteq \Phi$, $M_p(F_1, \dots, F_n)$ is a GENEO for (Φ, G).*

Proof If we show that M_p is a 1-Lipschitz function for $p \ge 1$, Proposition 1 will ensure us that $M_p(F_1, \dots, F_n)$ is a GENEO.

Let $p \ge 1$ and $x, y \in \mathbb{R}^n$. Since $\|\cdot\|_p$ is a norm, the reverse triangle inequality holds. Therefore, because of (2), we have that:

$$\begin{aligned}
\left|\left(\frac{1}{n}\sum_{i=1}^{n}|x_i|^p\right)^{\frac{1}{p}} - \left(\frac{1}{n}\sum_{i=1}^{n}|y_i|^p\right)^{\frac{1}{p}}\right| &= \left(\frac{1}{n}\right)^{\frac{1}{p}} \left|\left(\sum_{i=1}^{n}|x_i|^p\right)^{\frac{1}{p}} - \left(\sum_{i=1}^{n}|y_i|^p\right)^{\frac{1}{p}}\right| \\
&= \left(\frac{1}{n}\right)^{\frac{1}{p}} \left|\|x\|_p - \|y\|_p\right| \\
&\leq \left(\frac{1}{n}\right)^{\frac{1}{p}} \|x-y\|_p \\
&\leq \left(\frac{1}{n}\right)^{\frac{1}{p}} n^{\frac{1}{p}} \|x-y\|_\infty = \|x-y\|_\infty.
\end{aligned}$$

Hence, for $p \geq 1$, M_p is non-expansive (i.e., 1-Lipschitz), and the statement of our theorem is proved.

Remark 2 If $0 < p < 1$ and $n > 1$, M_p is not a 1-Lipschitz function. This can be easily proved by showing that for $x_2 = x_3 = \cdots = x_n = 1$ the derivative $\frac{\partial M_p}{\partial x_1}$ is not bounded.

3.1 Examples

In this subsection, we want to justify the use of the operator M_p. In order to make this point clear, let us consider the space Φ of all 1-Lipschitz functions from the unit circle S^1 to $[0, 1]$ and the invariance group G of all rotations of S^1. Now, we can take into consideration the following operators:

- the identity operator $F_1 : \Phi \longrightarrow \Phi$;
- the operator $F_2 : \Phi \longrightarrow \Phi$ defined by setting $F_2(\varphi) := \varphi \circ \rho_{\frac{\pi}{2}}$ for any $\varphi \in \Phi$, where $\rho_{\frac{\pi}{2}}$ is the rotation through a $\frac{\pi}{2}$ angle.

Let us set $\bar{\varphi} = |\sin x|$ and $\bar{\psi} = \sin^2 x$. As we can see in Figs. 1 and 2, the functions $F_i(\bar{\varphi})$ and $F_i(\bar{\psi})$ have the same persistence diagrams for $i = 1, 2$. In order to distinguish $\bar{\varphi}$ and $\bar{\psi}$, we define the operator $F : \Phi \longrightarrow \Phi$ by setting $F(\varphi) := M_1(F_1, F_2)(\varphi) = \frac{F_1(\varphi)+F_2(\varphi)}{2}$. In particular,

$$F(\bar{\varphi}) := M_1(F_1, F_2)(\bar{\varphi}) = \frac{F_1(\bar{\varphi}) + F_2(\bar{\varphi})}{2} = \frac{|\sin x| + |\cos x|}{2} \tag{3}$$

and

$$F(\bar{\psi}) := M_1(F_1, F_2)(\bar{\psi}) = \frac{F_1(\bar{\psi}) + F_2(\bar{\psi})}{2} = \frac{\sin^2 x + \cos^2 x}{2} = \frac{1}{2}. \tag{4}$$

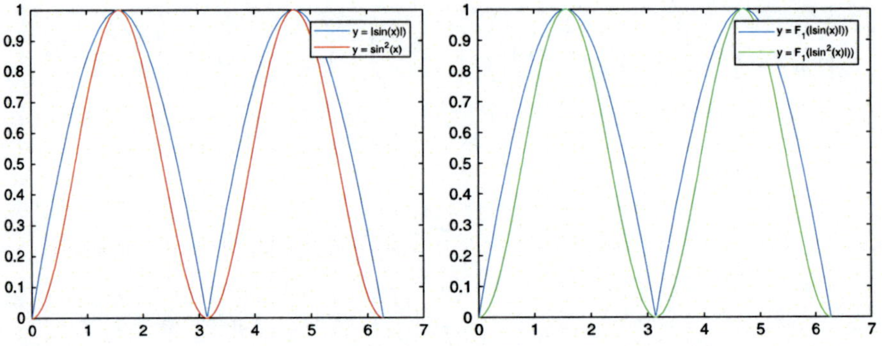

Fig. 1 On the left: $\bar{\varphi}$ and $\bar{\psi}$ have the same persistence diagrams. On the right: $F_1(\bar{\varphi})$ and $F_1(\bar{\psi})$ have the same persistence diagrams

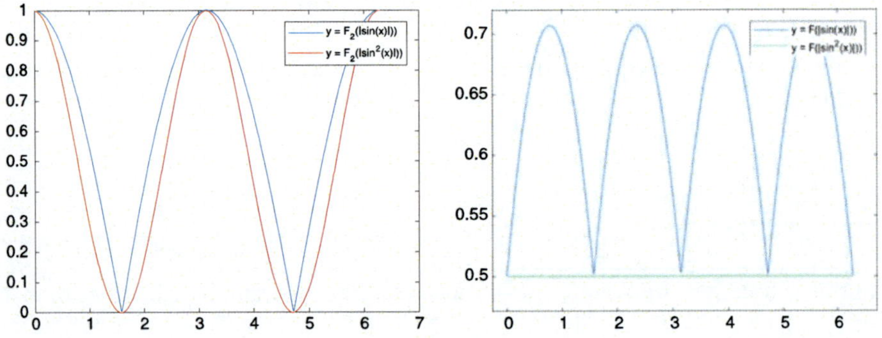

Fig. 2 On the left: $F_2(\bar{\varphi})$ and $F_2(\bar{\psi})$ have the same persistence diagrams. On the right: the persistence diagrams of $F(\bar{\varphi})$ and $F(\bar{\psi})$ are different from each other

We can easily check that $F(\bar{\varphi})$ and $F(\bar{\psi})$ have different persistence diagrams; thus F allows us to distinguish between $\bar{\varphi}$ and $\bar{\psi}$. All this proves that the use of the operator M_1 can increase the information, letting F_1 and F_2 cooperate.

A similar argument still holds for values of p greater than one. Under the same hypotheses about Φ, we can consider the same GENEOs F_1, F_2 and the functions $\bar{\varphi} = |\sin x|$ and $\hat{\psi} = (\sin^2 x)^{\frac{1}{p}}$. For the sake of simplicity, we fixed $p = 3$ in order to represent the following figures. As we can see in Figs. 3 and 4, we cannot distinguish $\bar{\varphi}$ and $\hat{\psi}$ by using persistent homology since their persistence diagrams coincide. Neither applying F_1 nor F_2 can help us, but when we apply $M_p(F_1, F_2)$, we can distinguish $\bar{\varphi}$ from $\hat{\psi}$ by means of their persistence diagrams (see Fig. 4).

These examples justify the use of the previously defined power mean operators $M_p(F_1, \ldots, F_n)$ to combine the information given by the operators $F_1, \ldots, F_n$.

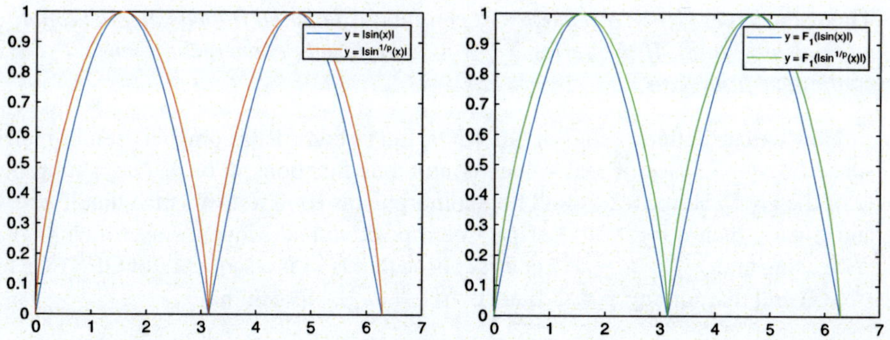

Fig. 3 On the left: $\bar{\varphi}$ and $\hat{\psi}$ have hence the same persistence diagrams. On the right: On the right: $F_1(\bar{\varphi})$ and $F_1(\hat{\psi})$ have the same persistence diagrams

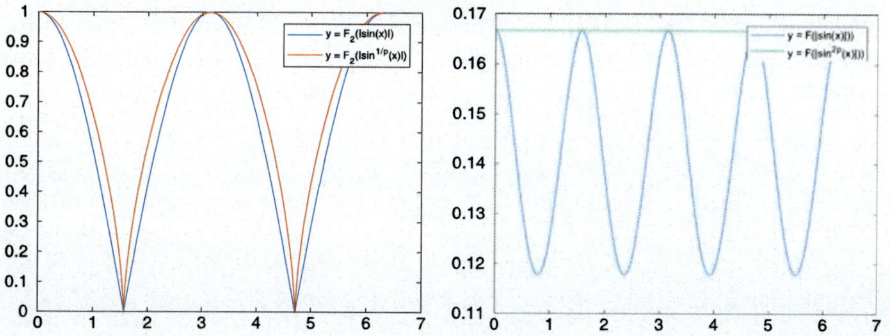

Fig. 4 On the left: $F_2(\bar{\varphi})$ and $F_2(\hat{\psi})$ have the same persistence diagrams. On the right: the persistence diagrams of $F(\bar{\varphi})$ and $F(\hat{\psi})$ are different from each other

4 Series of GENEOs

First we recall some well-known results about series of functions.

Theorem 3 *Let (a_k) be a positive real sequence such that (a_k) is decreasing and $\lim_{k\to\infty} a_k = 0$. Let (g_k) be a sequence of bounded functions from the topological space X to $\mathbb{C}$. If there exists a real number $M > 0$ such that*

$$\left|\sum_{k=1}^{n} g_k(x)\right| \leq M \tag{5}$$

for every $x \in X$ and every $n \in \mathbb{N}$, then the series $\sum_{k=1}^{\infty} a_k g_k$ is uniformly convergent on X.

The second result ensures us that a uniformly convergent series of continuous functions is a continuous function.

Theorem 4 *Let (f_n) be a sequence of continuous function from a compact topological space X to $\mathbb{R}$. If the series $\sum_{k=1}^{\infty} f_k$ is uniformly convergent, then $\sum_{k=1}^{\infty} f_k$ is continuous from X to $\mathbb{R}$.*

Now we can define a series of GENEOs. Let us consider a compact pseudo-metric space (X, d), a space of real-valued continuous functions Φ on X and a subgroup G of the group $\mathrm{Homeo}(X)$ of all homeomorphisms from X to X, such that if $\varphi \in \Phi$ and $g \in G$, then $\varphi \circ g \in \Phi$. Let (a_k) be a positive real sequence such that (a_k) is decreasing and $\sum_{k=1}^{\infty} a_k \leq 1$. Let us suppose that (F_k) is a sequence of GENEOs for (Φ, G) and that for any $\varphi \in \Phi$ there exists $M(\varphi) > 0$ such that

$$\left| \sum_{k=1}^{n} F_k(\varphi)(x) \right| \leq M(\varphi) \tag{6}$$

for every $x \in X$ and every $n \in \mathbb{N}$. These assumptions fulfill the hypotheses of the previous theorems and ensure that the following operator is well-defined. Let us consider the operator $F : C_b^0(X, \mathbb{R}) \longrightarrow C_b^0(X, \mathbb{R})$ defined by setting

$$F(\varphi) := \sum_{k=1}^{\infty} a_k F_k(\varphi). \tag{7}$$

Proposition 2 *If $F(\Phi) \subseteq \Phi$, then F is a GENEO for (Φ, G).*

Proof

- Let $g \in G$. Since F_k is G-equivariant for any k and g is uniformly continuous (because X is compact), F is G-equivariant:

$$\begin{aligned} F(\varphi \circ g) &= \sum_{k=1}^{\infty} a_k F_k(\varphi \circ g) \\ &= \sum_{k=1}^{\infty} a_k (F_k(\varphi) \circ g) \\ &= \left(\sum_{k=1}^{\infty} a_k F_k(\varphi) \right) \circ g \\ &= F(\varphi) \circ g \end{aligned}$$

for any $\varphi \in \Phi$.

- Since F_k is non-expansive for any k and $\sum_{k=1}^{\infty} a_k \leq 1$, F is non-expansive:

$$\begin{aligned}
\|F(\varphi_1) - F(\varphi_2)\|_\infty &= \left\| \sum_{k=1}^{\infty} a_k F_k(\varphi_1) - \sum_{k=1}^{\infty} a_k F_k(\varphi_2) \right\|_\infty \\
&= \left\| \lim_{n\to\infty} \left(\sum_{k=1}^{n} a_k F_k(\varphi_1) - \sum_{k=1}^{n} a_k F_k(\varphi_2) \right) \right\|_\infty \\
&= \lim_{n\to\infty} \left\| \sum_{k=1}^{n} a_k (F_k(\varphi_1) - F_k(\varphi_2)) \right\|_\infty \\
&\leq \lim_{n\to\infty} \sum_{k=1}^{n} (a_k \|F_k(\varphi_1) - F_k(\varphi_2)\|_\infty) \\
&\leq \lim_{n\to\infty} \sum_{k=1}^{n} (a_k \|\varphi_1 - \varphi_2\|_\infty) \\
&= \sum_{k=1}^{\infty} a_k \|\varphi_1 - \varphi_2\|_\infty \\
&\leq \|\varphi_1 - \varphi_2\|_\infty.
\end{aligned}$$

5 Conclusions

In this work, we have illustrated some new methods to build new classes of G-equivariant non-expansive operators (GENEOs) from a given set of operators of this kind. The leading purpose of our work is to expand our knowledge about the topological space $\mathcal{F}(\Phi, G)$ of all GENEOs. If we can well approximate the space $\mathcal{F}(\Phi, G)$, we can obtain a good approximation of the natural pseudo-distance d_G (Theorem 1). Searching new operators is a fundamental step in getting more information about the structure of $\mathcal{F}(\Phi, G)$, and hence, we are asked to find new methods to build GENEOs. Moreover, the approximation of $\mathcal{F}(\Phi, G)$ can be seen as an approximation of the considered observer, represented as a collection of GENEOs. Many questions remain open. In particular, we should study an extended theoretical framework that involves GENEOs from the pair (Φ, G) to a different pair (Ψ, H). A future research about this is planned to be done.

Acknowledgements The research described in this article has been partially supported by GNSAGA-INdAM (Italy).

References

1. G. Carlsson, Topology and data. Bull. Amer. Math. Soc. (N.S.), textbf46(2), 255–308 (2009)
2. H. Edelsbrunner, D. Morozov, Persistent homology: theory and practice. European Congress of Mathematics (2013), 31–50
3. S. Biasotti, L. De Floriani, B. Falcidieno, P. Frosini, D. Giorgi, C. Landi, L. Papaleo, M. Spagnuolo, Describing shapes by geometrical-topological properties of real functions. ACM Comput. Surv. **40**(4), 12:1–12:87 (2008)
4. H. Edelsbrunner, J.L. Harer, Persistent homology—a survey. Contemp. Mathe. **453**, 257–282 (2008)
5. F. Anselmi, L. Rosasco, T. Poggio, On invariance and selectivity in representation learning. Informat. Inf. J. IMA **5**(2), 134–158 (2016)
6. T. Cohen, M. Welling, *Group equivariant convolutional networks*, in Proceedings of the 33rd International Conference on Machine Learning, PMLR 48 (2016), 2990–2999
7. D. Marcos, M. Volpi, N. Komodakis, D. Tuia, *Rotation equivariant vector field networks*, in Proceedings of the 2017 IEEE International Conference on Computer Vision (ICCV) (2017), 5058–5067
8. J. Masci, D. Boscaini, M. M. Bronstein, P. Vandergheynst, *Geodesic convolutional neural networks on Riemannian manifolds*, in Proceedings of the 2015 IEEE International Conference on Computer Vision Workshop (ICCVW), IEEE Computer Society (2015), 832–840
9. P. Frosini, *Towards an observer-oriented theory of shape comparison*, Proceedings of the 8th Eurographics Workshop on 3D Object Retrieval, ed. by A. Ferreira, A. Giachetti, D. Giorgi. Lisbon, Portugal (2016), 5–8
10. P. Frosini, G. Jabłoński, Combining persistent homology and invariance groups for shape comparison. Discrete Comput. Geometry **55**(2), 373–409 (2016)
11. P. Frosini, N. Quercioli, *Some remarks on the algebraic properties of group invariant operators in persistent homology*, in Lecture Notes in Computer Science, Proceedings of the International Cross-Domain Conference, CD-MAKE 2017, Reggio, Italy, August 29-September 1, 2017, ed. by A. Holzinger, P. Kieseberg, A.M. Tjoa, E. Weippl, MAKE Topology, Springer, Cham, LNCS 10410 (2017), 14–24

Topological Stability of the Hippocampal Spatial Map and Synaptic Transience

Yuri Dabaghian

Abstract Spatial awareness in mammals is based on internalized representations of the environment—cognitive maps—encoded by networks of spiking neurons. Although behavioral studies suggest that these maps can remain stable for long periods, it is also well-known that the underlying networks of synaptic connections constantly change their architecture due to various forms of neuronal plasticity. This raises a principal question: how can a dynamic network encode a stable map of space? In the following, we discuss some recent results obtained in this direction using an algebro-topological modeling approach, which demonstrate that emergence of stable cognitive maps produced by networks with transient architectures is not only possible, but also may be a generic phenomenon.

Keywords Spatial map · Topological dynamics · Emergent phenomena · Hippocampal learning

1 Introduction

General background. Spatial awareness in mammals is based on an internalized representation of the environment. Many parts of the brain are contributing to this representation, providing different types of information: cue positions [1], geometry of the navigated paths [2], orientations [3, 4], traveled distances [5, 6], velocities [7], qualitative geometric relationships [8, 9], etc. A principal question addressed by neuroscience is how all these types of data are captured by neuronal activity and what are the computational algorithms employed by various networks for processing this information.

At the current stage, our understanding of the mechanisms of spatial cognition is based mostly on empirical observations. For example, it was found that a major

Y. Dabaghian (✉)
Department of Neurology, McGovern Medical School, The University of Texas at Houston, Houston, TX 77030, USA
e-mail: Yuri.A.Dabaghian@uth.tmc.edu

R. L. Devaney et al. (eds.), *Topological Dynamics and Topological Data Analysis*, Springer Proceedings in Mathematics & Statistics 350,
https://doi.org/10.1007/978-981-16-0174-3_20

role in cognitive representation of the ambient space is played by the hippocampus: a vast number of experiments demonstrate that the hippocampal network contributes a "cognitive map" $\mathcal{C}$ that is crucial for the animal's ability to navigate, to find its nest and food sources, etc. [10, 11]. Experimentally, the properties of the cognitive map are studied by mapping hippocampal activity into the studied environment $\mathcal{E}$,

$$f : \mathcal{C} \to \mathcal{E}.$$

In experiments with rodents (e.g., rats or mice), this mapping is constructed by ascribing the $(x - y)$ coordinates to very spike produced by the hippocampal principal neurons according to the animal's position at the time when the spike was fired [12]. As shown in [13], such mapping produces spatial clusters of spikes, indicating that these neurons, known as the "place cells," fire only in certain places—their respective "place fields." The spatial layout of the place fields in $\mathcal{E}$—the place field map $M_{\mathcal{E}}$ (Fig. 1a)—is therefore viewed as a geometric representation of the cognitive map of that particular environment, $\mathcal{C}(\mathcal{E})$. Electrophysiological recordings in "morphing" arenas demonstrate that $M_{\mathcal{E}}$ is flexible: as the environment is slowly deformed, the place fields shift and change their shapes, but largely preserve their mutual overlaps, adjacency and containment relationships [14–17]. Thus, the order in which the place cells spike during the animal's navigation remains invariant within a certain range of geometric transformations [18–23], which implies that $\mathcal{C}(\mathcal{E})$ may be viewed as a coarse framework of qualitative spatiotemporal relationships rather than precise geometry, i.e., that the hippocampal map is topological in nature.

From the computational perspective, this observation suggests that the information contained in place cell spiking should be interpreted topologically. In [24–29] we proposed an approach for such analyses, based on a schematic representations of the information supplied by place cells (co)activity. Specifically, if groups of coactive place cells, e.g., $c_0, c_1, \ldots, c_n$, are viewed as abstract simplexes, $\sigma = [c_0, c_1, \ldots, c_n]$, then the pool of the coactive place cell combinations observed by a given moment t forms a simplicial "coactivity complex" $\mathcal{T}(t)$ whose topology represents the topological structure of the cognitive map of the underlying environment (see [24–29] and Fig. 1b).

The evolution of $\mathcal{T}(t)$ reflects how the net spatial information accumulates in time: starting from a few simplexes at the beginning of navigation, the complex $\mathcal{T}(t)$ grows and eventually, if the parameters of spiking activity fall within the biological range of values, assumes a shape that is topologically equivalent to the shape of the navigated environment in a biologically plausible period $T_{\min}$—a theoretical estimate of the time required to "learn" the environment [24–29].

Curiously, the key building blocks of this model—the coactive groups of the hippocampal place cells represented by the coactivity simplexes, have physiological counterparts, called "cell assemblies"—functionally interconnected groups of neurons that work as operational units of the hippocampal network [30, 31]. In [32], it was shown that this correspondence can be made accurate: the construction of the coactivity complex may be adjusted so that its maximal simplexes (i.e., the simplexes that are not subsimplexes of any larger simplex) represent place cell assemblies, rather

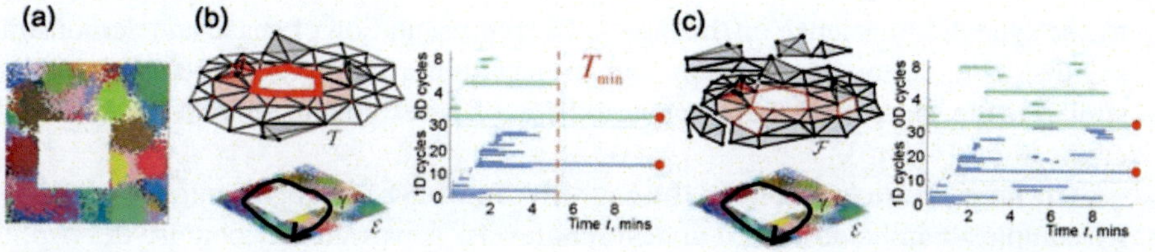

Fig. 1 Place cells and cell assemblies. **a** Simulated place field map $M_{\mathcal{E}}$ in a small ($1m \times 1m$) planar environment $\mathcal{E}$ with a square hole: dots of a particular color, marking the locations where a specific place cells produced spikes, form spatial clusters—the place fields. Shown is a map produced for $N = 300$ place cells with a median maximal firing rate $f = 14$ Hz and place field size 20 cm. **b** The net pool of coactivities is represented by the coactivity complex $\mathcal{T}$ (top), which provides a topological representation of the environment $\mathcal{E}$ (bottom). E.g., the non-contractible simplicial path shown by red chain of simplexes corresponds to a non-contractible physical path γ around the central hole in $\mathcal{E}$. The coactivity complex $\mathcal{T}$ assumes its topological shape as the spatial information provided by the place cells accumulates. The panel on the right shows the timelines of $0D$ (top) and $1D$ (bottom) topological loops in $\mathcal{T}$, computed using Persistent Homology theory methods [49–52]. The minimal time T_{min} required to eliminate the spurious loops and extract the persistent ones (marked by the red bullets) provides an estimate for the time required by a given place cell ensemble to learn the topological structure of the navigated environment [24–29]. **c** If the simplexes may not only appear but also disappear, then the structure of the resulting "flickering" coactivity complex $\mathcal{F}(M_{\mathcal{E}})$ may never saturate, i.e., transient topological defects, described by Zigzag Persistent Homology theory [53–55] may persist indefinitely

than arbitrary combinations of coactive place cells. An important physiological property of the cell assemblies is that these are *dynamic* structures: They may form among the cells that demonstrate repeated coactivity and disband as a result of deterioration of synaptic connections, caused by reduction or cessation of spiking, then reappear due to a subsequent surge of coactivity, then disband again and so forth [30, 31]. In the model, the appearance and disappearance of the cell assemblies are represented by the the appearances and disappearances of the corresponding simplexes, so that the rewiring dynamics of the cell assembly network and the dynamics of the resulting cognitive map is represented by a dynamic—"flickering"—cell assembly complex, denoted below as $\mathcal{F}(t)$. Unlike its "perennial" counterpart $\mathcal{T}(t)$ that can only grow and stabilize with time, the flickering complex $\mathcal{F}(t)$ may inflate, shrink, fragment into pieces, produce transient holes, fractures, gaps and other dynamic "topological defects" (Fig. 1c).

Thus, on the one hand, the dynamics of $\mathcal{F}(t)$ may be viewed as a natural consequence of the network's plasticity: studies show that the lifetime of the hippocampal cell assemblies ranges between minutes [33–35] and hundreds of milliseconds [36, 37], suggesting that the hippocampal network perpetually rewires [38]. On the other hand, behavioral and cognitive studies show that spatial memories in rats can last for days and months [39–41]. This poses a principal question: *how can a large-scale spatial representation of the environment be stable if the neuronal substrate changes at a much shorter timescale*?

A principal answer to this question is suggested by an algebro-topological model of the dynamic cell assembly networks, which allows studying the effect produced by the synaptic transience on the large-scale representation of space and demonstrating that a stable topological map can from within a biologically plausible period, similar to the "perennial" learning period $T_{\min}(\mathcal{T})$, despite the rapid transience of the connections [42–45].

The large-scale topology of the cognitive map $\mathcal{C}(\mathcal{E})$, as represented by a coactivity complex, can be described at different levels. A particularly concise description of a topological shape is given in terms of its topological loops (non-contractible surfaces identified up to topological equivalence) in different dimensions, i.e., by its Betti numbers b_n, $n = 0, 1, \ldots$ [46, 47]. For example, the number of inequivalent topological loops that can be contracted to a zero-dimensional ($0D$) vertex, $b_0(\mathcal{F})$, corresponds to the number of the connected components in $\mathcal{F}(t)$; the number of loops that contract to a one-dimensional ($1D$) chain of links, $b_1(\mathcal{F})$, defines the number of holes and so forth [46, 47]. The full list of the Betti numbers of a space or a complex X is known as its topological barcode, $\mathfrak{b}(X) = (b_0(X), b_1(X), b_2(X), \ldots)$, which captures the topological identity of X [48]. For example, the barcode $\mathfrak{b} = (1, 1, 0, \ldots)$ corresponds to a topological annulus, the barcode $\mathfrak{b} = (1, 0, 1, 0, \ldots)$—to a two-dimensional ($2D$) sphere S^2, the barcode $\mathfrak{b} = (1, 2, 1, 0, \ldots)$—to a torus T^2 and so forth [49]. Thus, by comparing the barcode of the coactivity complex $\mathfrak{b}(\mathcal{F})$ to the barcode of the environment $\mathfrak{b}(\mathcal{E})$ one can establish whether their topological shapes match, $\mathfrak{b}(\mathcal{F}(t)) = \mathfrak{b}(\mathcal{E})$, i.e., whether the coactivity complex provides a faithful representation of the environment at a given moment t. The mathematical methods required for these analyses—Persistent Homology [49–52] and Zigzag Persistent Homology theories [53–55], also outlined in [56, 57], allow building a dynamical model of the cognitive map and addressing the question "*How can a rapidly rewiring network produce and sustain a stable cognitive map*?"

2 Overview of the Results

An efficient implementation of the coactivity complex $\mathcal{F}(t)$ is based on the "cognitive graph" model of the hippocampal network [12, 59], in which each active place cell c_i corresponds to a vertex v_i of a graph $\mathcal{G}$, whose connections $\varsigma_{ij} = [v_i, v_j]$ represent pairs of coactive cells. An assembly of place cells $c_0, c_1, \ldots, c_n$ then corresponds to the fully interconnected subgraph, i.e., to a maximal clique $\varsigma = [v_0, v_1, \ldots, v_n]$ of $\mathcal{G}$ [29, 30, 32]. Since cliques, as combinatorial objects, can be viewed as simplexes spanned by the same sets of vertexes, the collection of $\mathcal{G}$-cliques defines a clique simplicial complex [60] that serves as an instantiation of the coactivity complex [26–29, 32]. The dynamics of the clique coactivity complexes can be modeled based on the dynamics of the links of the corresponding coactivity graph $\mathcal{G}$. In the following, we discuss two such approaches, both of which demonstrate a possibility of encoding stable cognitive maps by transient cell assembly networks.

2.1 Decaying Clique Complexes

Consider the following dynamics of the coactivity graph $\mathcal{G}$.

- The vertexes of $\mathcal{G}$ appear when the corresponding cells become active for the first time and never disappear, since according to the experiments, place cells' spiking in learned environments remains stable [62].
- The connection ς_{ij} between the vertexes v_i and v_j appears with probability $p_+ = 1$ if the cells c_i and c_j become active within a $w = 1/4$ second period (for biological motivations of the w-value see [25, 61]). The exact time t of the link's appearance can be associated with any moment within w.
- An existing link ς_{ij} between cells c_i and c_j disappears with the probability

$$p_-(t) = \frac{1}{\tau} e^{-t/\tau}, \tag{1}$$

  where the time t is counted from the moment of the link's last activation and τ defines the link's *proper* decay time.
- The dynamics of the higher order cliques, e.g., their decay times, are fully determined by the link decay period τ. In the following, the notations $\mathcal{G}_\tau$ and $\mathcal{F}_\tau$ will refer, respectively, to the flickering coactivity graph and the corresponding flickering clique coactivity complex with the connections' proper decay rate $1/\tau$.

Note that the ongoing place cell activity can reinstate some decayed links in $\mathcal{G}_\tau$ and rejuvenate (i.e., reset the decay of) some existent ones, thus producing an *effective* link's mean lifetime $\tau_e > \tau$ and leading to diverse topological dynamics of the coactivity complex $\mathcal{F}_\tau$. As mentioned previously, a key determinant of this dynamics is the sequence in which the rat traverses place fields in a map $M_\mathcal{E}$. Fixing $M_\mathcal{E}$ and the animal's trajectory $\gamma(t)$ settles the times at which place cell combinations become active (notwithstanding the stochasticity of neuronal firing [63, 64]), so that the Betti numbers b_k of $\mathcal{F}_\tau(t)$ become dependent primarily on the parameters of neuronal spiking activity: firing rates, place field sizes, etc., and on the links' decay time τ. In the following, we will review some of these dependencies for the case of the environment shown on Fig. 1a, and discuss how they affect the net topological structure of the corresponding cognitive map. For more details see [42–45].

Dynamics of the decaying flickering coactivity complexes. If τ is too small (e.g., if the coactivity simplexes tend to disappear between two consecutive co-activations of the corresponding cells), then the flickering complex should rapidly deteriorate without assuming the required topological shape. In contrast, if τ is too large, then the effect of the decaying connections should be small, i.e., the flickering complex $\mathcal{F}_\tau(t)$ should follow the dynamics of its "perennial" counterpart $\mathcal{T}(t) \equiv \mathcal{F}_\infty(t)$, computed for the same place cell spiking parameters. In particular, if the firing rates and the place field sizes are such that $\mathcal{T}(t)$ assumes the correct topological shape in a biologically viable time $T_{\min}(\mathcal{T})$, then a similar behavior should be expected from its slowly decomposing counterpart $\mathcal{F}_\tau(t)$. However, for intermediate values of τ that exceed the characteristic interval Δt between two consecutive activations of a typical link

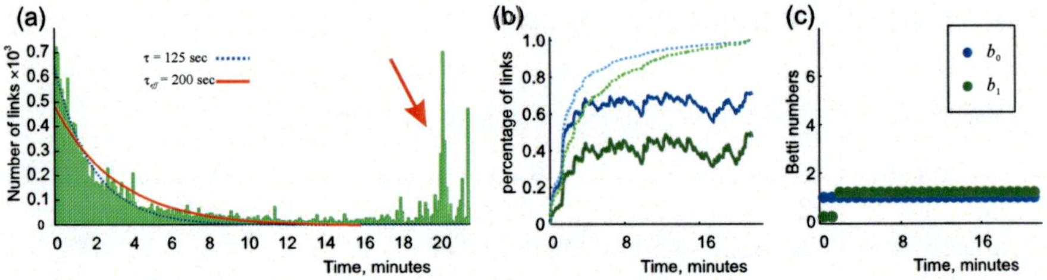

Fig. 2 Topological dynamics of the decaying coactivity complex. a The histogram of the connections' durations between their consecutive appearances and disappearances: The shorter lifetimes are distributed exponentially (the red line fit) and the population of the "survivor" links produces a bulging tail of the distribution (red arrow). The dashed blue line shows the shape of the distribution (1). **b** The population of $1D$ (blue trace) and $2D$ (green trace) simplexes in the decaying "flickering" complex $\mathcal{F}_\tau(t)$, compared to the population of $1D$ and $2D$ simplexes in the perennial complex $\mathcal{T}(t)$ (dashed lines). The size of $\mathcal{F}_\tau(t)$ remains dynamic, whereas $\mathcal{T}(t)$ saturates in about 10 minutes. **c** Betti numbers $b_0(\mathcal{F}_\tau(t))$ (blue) and $b_1(\mathcal{F}_\tau(t))$ (green) remain unchanged after a short initial stabilization period

in $\mathcal{G}$—a the natural timescale defined by the statistics of the rat's movements—the topological dynamics of $\mathcal{F}_\tau(t)$ may exhibit a rich variety of behaviors.

Simulations show that the characteristic inter-activation interval in the environment shown on Fig. 1a is about $\Delta t \approx 30$ seconds. For the proper decay times that generously exceed Δt, e.g., $2.5\Delta t \lesssim \tau \lesssim 4.5\Delta t$, the histogram of the time intervals $\Delta t_{\varsigma,i}$ between the ith consecutive birth and death of a link ς is bimodal: the relatively short lifetimes are exponentially distributed, with the *effective* link lifetimes about twice higher $\tau_e^{(2)} \approx 2\tau$ (higher order simplexes decay more rapidly, e.g., $\tau_e^{(3)} \approx \tau$, etc.). In addition, there emerges a pool of long-living connections that persist throughout the entire navigation period (Fig. 2a). In other words, the flickering coactivity complex $\mathcal{F}_\tau(t)$ acquires a stable "core" formed by a population of "surviving simplexes", enveloped by a population of rapidly recycling, "fluttering," simplexes.

The numbers of d-dimensional simplexes in $\mathcal{F}_\tau(t)$ (its f-numbers in terminology of [65]) rapidly grow at the onset of the navigation, when $\mathcal{F}_\tau(t)$ inflates, but then begin to saturate by the time a typical link makes an appearance (in the case of the environment shown on Fig. 1a, this takes a few minutes). The characteristic size of $\mathcal{F}_\tau(t)$ grows to about a half of the size of its perennial counterpart, $\mathcal{F}_\infty(t) \equiv \mathcal{T}(t)$, with about 15% fluctuations (Fig. 2b). Thus, the population of simplexes in $\mathcal{F}_\tau(t)$ is transient: although the change of the size of $\mathcal{F}_\tau(t)$ from one moment of time to the next are relatively small, the number of simplexes that are present at a given moment of time t, but missing at a later moment t', rapidly grows as a function of temporal separation $|t - t'|$, becoming comparable to the sizes of either $\mathcal{F}_\tau(t)$ or $\mathcal{F}_\tau(t')$ in approximately one effective link-decay span [44, 45].

Meanwhile, the large-scale topology of $\mathcal{F}_\tau(t)$ changes significantly slower: after a brief initial stabilization period that roughly corresponds to the perennial learning time $T_{\min}(\mathcal{T})$, the topological barcode $\mathfrak{b}(\mathcal{F}_\tau)$ remains similar to the barcode of

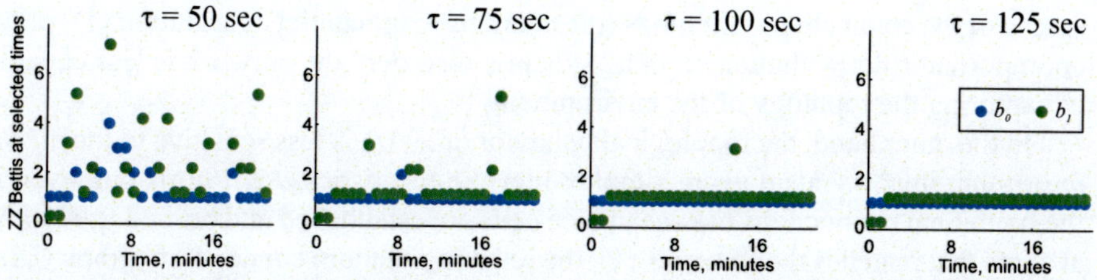

Fig. 3 **Topological stabilization**. As the decay constant τ grows, the topological shape of $\mathcal{F}_\tau(t)$ stabilizes. Shown are the Betti numbers b_0 (blue dots) and b_1 (green dots) at select moments of time, computed for several values of τ

the navigated environment $\mathcal{E}$, exhibiting occasional topological fluctuations at the $T_{\min}$-timescale (Fig. 2c). Thus, the coactivity complex $\mathcal{F}_\tau$ can preserve not only its approximate size, but also its topological shape, despite the ongoing recycling of its simplexes.

As τ reduces, the topological fluctuations intensify (Fig. 3) and vice versa, as τ grows, the effective lifetimes $\tau_e^{(2)}$ and $\tau_e^{(3)}$, as well as the number of the simplexes actualized at a given moment increase approximately linearly, resulting in a growing "stable core" that stabilizes the overall topological structure of $\mathcal{F}_\tau(t)$. Given the physiological range of parameters (simulated rat speed, place cell firing rates, place field sizes, etc.), a *complete suppression* of topological fluctuations in the coactivity complex is achieved after the decay times exceed a finite threshold τ_p^*, comparable to the time required to revisit a typical spot in the environment. This value gives a theoretical estimate for the rate of physiological transience that permits stable representations of the environment $\mathcal{E}$ [44].

Alternative lifetime statistics may strongly influence the topological dynamics of the cognitive map. For example, if the links' lifetimes are fixed, i.e., if the decay probability is defined by

$$p_-(t) = \begin{cases} 1 & \text{if } t = \tau \\ 0 & \text{if } t \neq \tau, \end{cases} \tag{2}$$

then the topological structure of the resulting "quenched-decay" coactivity complex $\mathcal{F}_\tau^*(t)$ changes dramatically. Even though the rejuvenation effects widen the effective distribution of the links' lifetimes (as before, in addition to a population of short-lived links with lifetimes close to τ, there appears a population of the "survivor" simplexes), the resulting topological dynamics is more unstable: $\mathcal{F}_\tau^*(t)$ may split into dozens of islets containing short-lived, spurious topological defects, even for the values of τ that reliably produce physical Betti numbers for the exponentially distributed lifetimes (1).

As the decay slows down (i.e., as τ grows), the population of survivor links also grows and the topological structure of $\mathcal{F}_\tau^*(t)$ stabilizes; nevertheless, the robust, "physical" Betti numbers are attained at much (twice or more) higher values of τ than with the exponentially decaying links. Physiologically, this implies that the statistical spread of the connections' lifetimes (the tail of the exponential distribution (1)) plays an important role: without a certain "synaptic disorder" the network is less capable of capturing the topology of the environment.

On the other hand, the topological behavior of $\mathcal{F}_\tau(t)$ is less sensitive to the *mechanism* that implements a given simplex-recycling statistics. As it turns out, even if the functional connections between place cells are established and pruned *randomly*, at a rate that matches the statistics (1), the resulting random connectivity graph $\mathcal{G}_r(t)$ produces a random clique complex $\mathcal{F}_r(t)$ with topological properties similar to those of $\mathcal{F}_\tau(t)$. In particular, the Betti numbers of $\mathcal{F}_r(t)$ converge to the Betti numbers of the environment about as quickly as the Betti numbers of its decaying counterpart $\mathcal{F}_\tau(t)$, exhibiting similar pattern of the topological fluctuations. Thus, the model suggests that the dynamic topology of the flickering complex may be controlled by the statistics of the decays and by the sheer number of simplexes present at a given moment, rather than by nature of the network's activity (e.g., random versus driven by the animal's moves).

2.2 Finite Latency Complexes

An alternative model of flickering clique complexes can be built by restricting the period over which the coactivity graph is formed to a shorter time window ϖ [32]. In such approach, the coactivity simplexes that emerge within the starting ϖ-period, ϖ_1, will constitute a coactivity complex $\mathcal{F}(\varpi_1)$; the simplexes appearing within the next window, ϖ_2, obtained by shifting ϖ_1 over a small step $\Delta\varpi$, will form the complex $\mathcal{F}(\varpi_2)$ and so forth. For large consecutive window overlaps ($\Delta\varpi \ll \varpi$), a given clique-simplex ς (as defined by the set of its vertexes) may appear through a chain of consecutive windows, $\varpi_1, \varpi_2, \ldots, \varpi_{k-1}$, then disappear at the kth step ϖ_k (i.e., $\varsigma \in \mathcal{F}(\varpi_{k-1})$, but $\varsigma \notin \mathcal{F}(\varpi_k)$), then reappear in a later window $\varpi_{l\geq k}$, then disappear again, and so forth. One may then use the midpoints t_k of the windows in which ς has (re)appeared (or any other point within ϖ_k) to define the moments of ς's (re)births, and the matching points in the windows where it disappears to define the times of its deaths. By construction, the duration of ς's existence between its k-th consecutive appearance and disappearance, $\delta t_{\varsigma,k}$, can be as short as the shift step $\Delta\varpi$ or as long as the animal's navigation session.

Simulations show that for ϖ exceeding the perennial learning time $T_{min}(\mathcal{T})$ and $\Delta\varpi \approx 0.01\varpi$, the intervals $\delta t_{\varsigma,k}$ (as well as their means averaged over k, $t_\varsigma = \langle \delta t_{\varsigma,k} \rangle_k$ and of their net existence times $\Delta T_\varsigma = \sum_k \delta t_{\varsigma,k}$) are exponentially distributed, which allows characterizing the simulated cell assemblies by a half-life, τ_ϖ. Specifically, for the physiological range of parameters of the neuronal activity in the environment shown on Fig. 1 and $\varpi \approx 1.2 T_{\min}(\mathcal{T})$, the lifetime of a typical maximal simplex varies within $\tau_\varsigma \approx 3 - 12$ s (depending on the simplex' dimensionality), which is much shorter than the proper decay time in the previous model (1) and closer to the experimentally established range of values [30].

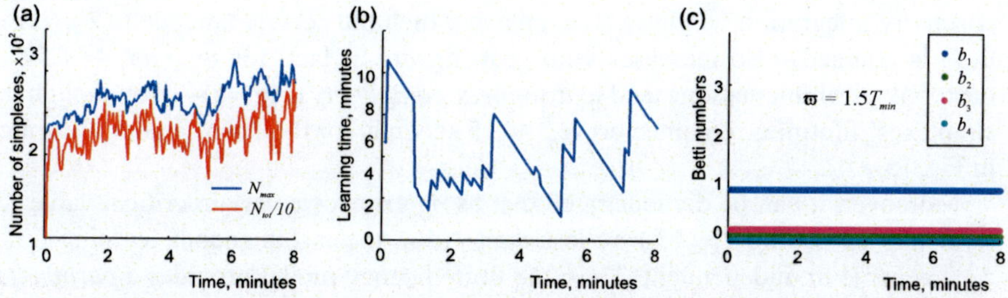

Fig. 4 Topological dynamics in the finite latency flickering complexes. **A**. The number of maximal simplexes ($N_{\max}$, blue trace) and total number of simplexes ($N_{tot}/10$, red trace) in the coactivity complex $\mathcal{F}_{\varpi}(t)$. **B**. The instantaneous learning time $T_{\min}^{(k)}$ as a function of the discrete time t_k, computed for $\varpi = 1.5T_{\min}(\mathcal{T})$. **C**. The low-dimensional Betti numbers, b_1, b_2, b_3 and b_4 as a function of the discrete time, computed using $\varpi = 1.5T_{\min}(\mathcal{T})$ remain stable, demonstrating full topological stabilization of $\mathcal{F}_{\varpi}(t)$

Dynamics of the finite latency flickering coactivity complexes. It is natural to view the individual, "instantaneous" complexes $\mathcal{F}(\varpi_i)$ as instantiations of a single "finite latency" flickering coactivity complex, $\mathcal{F}(\varpi_i) = \mathcal{F}_{\varpi}(t_i)$. As it turns out, such complexes exhibit a number similarities with the decaying complexes $\mathcal{F}_{\tau}(t)$. For example, the complex $\mathcal{F}_{\varpi}(t)$ does not fluctuate significantly: for $\varpi \geq T_{\min}(\mathcal{T})$, the number of simplexes contained in $\mathcal{F}_{\varpi}(t)$ changes within about $5 - 10\%$ of its mean value during the entire navigation period, but the pool of the *actualized* maximal simplexes is renewed at about ϖ timescale (Fig. 4a). Biologically, this implies that the simulated cell assembly network fully rewires over a ϖ period, similar to the effective link decay time $\tau_e^{(2)}$ computed in the previous model.

On the other hand, the large-scale topological properties of $\mathcal{F}_{\varpi}(t)$ are much more stable, similarly to the topological properties of $\mathcal{F}_{\tau}(t)$. For example, for sufficiently long latencies, $\varpi \gtrsim 1.2\,T_{\min}(\mathcal{T})$, the time required to produce the correct barcode $\mathfrak{b}(\mathcal{F}_{\varpi}) = \mathfrak{b}(\mathcal{E})$ within each window ϖ_k is typically finite, $T_{min}^{(k)} = T_{\min}(\mathcal{F}(\varpi_k)) < \infty$ (Fig. 4b). Moreover, the average learning period $\bar{T}_{min} = \langle T_{min}^{(k)} \rangle_k$ is typically similar to the perennial learning time $T_{\min}(\mathcal{T})$, with a variance of about $20 - 40\%$ of the mean. This result shows that the topological dynamics in the cognitive map of a semi-randomly foraging animal is largely time-invariant, i.e., the accumulation of the topological information can start at any point (e.g., at the onset of the navigation) and produce the result in an approximately the same time period. In effect, this justifies using perennial coactivity complexes for estimating T_{min} in [24–29]. It should also be mentioned, however, that there also exists a number of differences between the topological dynamics of $\mathcal{F}_{\varpi}(t)$ and $\mathcal{F}_{\tau}(t)$, e.g., the topological fluctuations in $\mathcal{F}_{\varpi}(t)$ are mostly limited to $1D$ loops, $2D$ surfaces and $3D$ bubbles ($b_0(t) = 1, b_{n>4}(t) = 0$), whereas the fluctuations in $\mathcal{F}_{\tau}(t)$ also affect higher dimensions.

As ϖ widens, the mean lifetime t_{ς} of maximal simplexes grows, suppressing the topological fluctuations in $\mathcal{F}_{\varpi}(t)$ and vice versa, as the memory window shrinks, the fluctuations of the topological loops intensify. The proportion of the "successful" coactivity integration windows (i.e., ϖ_ks in which the correct barcode $\mathfrak{b}(\mathcal{F}_{\varpi}(t)) = \mathfrak{b}(\mathcal{E})$ is attained) also increases with growing ϖ. In fact, for $\varpi \geq \varpi_* \approx 1.5T_{\min}$ the topological fluctuations tend to *disappear completely* (Fig. 4c)—even though the simplexes' lifetimes remain short ($\tau_{\varpi}^* \approx 15$ seconds for the environment illustrated in Fig. 1a).

Moreover, it can be demonstrated that as ϖ exceeds a certain critical value ϖ_c (typically exceeding $T_{\min}(\mathcal{T})$ by less than 40%), the instantaneous learning times $T_{min}^{(k)}$ become ϖ-independent. Thus, the finite latency model provides a *parameter-free* characterization of the time required by a network of place cell assemblies to represent the topology of the environment and establishes the timescale of the topological fluctuations in the simulated cognitive map.

3 Discussion

The topological model of the hippocampal cognitive map offers a connection between the spatial information processed by the individual place cells and the resulting global map emerging at the neuronal ensemble level, for both stable [24–29, 32] and transient [42–44] cell assembly networks. The elements of the model are embedded into the framework of simplicial topology: The groups of coactive cells are represented by abstract coactivity simplexes, whereas the spatial map encoded by the activity of neuronal populations is represented by the corresponding simplicial complexes. In particular, the formation and the disbanding of the cell groups is represented by the appearing and the disappearing coactivity simplexes, which combine into flickering coactivity complexes with nontrivial topological dynamics.

Generically, these dynamics occur at three principal timescales. The fastest timescale corresponds to the rapid recycling of the local connections—the starting point of the model. The large-scale topological loops, described by the time-dependent Betti numbers, unfold at a timescale that is by about an order of magnitude slower than the fluctuations at the simplex-level. Lastly, the topological fluctuations occur over certain robust base values that provide lasting, qualitative information about the environment.

The model demonstrates that for sufficiently slow simplex-recycling rates, the topological fluctuations at the intermediate timescale freeze out, i.e., the simulated cognitive map may transition into a topologically stable state, with static (or nearly static) Betti numbers. Physiologically, this implies that if the hippocampal place cell assemblies rewire sufficiently slowly, then the hippocampal map may remain stable despite the recycling of the connections in its neuronal substrate. Thus, the model suggests that plasticity of neuronal connections, which is ultimately responsible for the network's ability to incorporate new information [66–68], does not necessarily degrade the large-scale, qualitative information acquired by the system. Quite the opposite: renewing the connections allows correcting errors, e.g., removing some spurious, accidental topological obstructions fortuitously incorporated into the cognitive map. In other words, a network capable of not only accumulating, but also disposing information, exhibits better learning capacity, suggesting that physiological learning should involve a balanced contribution of both "learning" and "forgetting" components [69–71].

Remarkably, the three dynamic timescales suggested by the model have their direct biological counterparts: the *short-term memory*, which refers to temporary maintenance of ongoing (working) associations [72, 73], the *intermediate-term memory* that is acquired and updated at the "operational" timescale [74, 75], and the *long-term memory* that captures more persistent, qualitative information are broadly recognized in the literature. Physiologically, these types of memory are associated with different parts of the brain (hippocampal and cortical networks); thus, the model reaffirms functional importance of the complementary learning systems for processing spatial information at different levels of spatiotemporal granularity, from a theoretical viewpoint [76–78].

The model allows exploring the effects produced on the cognitive map by the parameters of neuronal activity and the synaptic structure. For example, it can be shown, e.g., that the deterioration caused by an overly rapid decay of the network's connections may be compensated by increasing neuronal activity, e.g., boosting the place cell firing rates [44] or via contributions of the "off-line", endogenous activity of the hippocampal network—the so-called "place cell replays" [79, 80]. The latter are commonly viewed as manifestations of the animal's "mental explorations" of its cognitive map [81–84] and are believed to help learning and to reinforce the map's stability [85, 86]. This belief is largely validated by the model, which shows that sufficiently frequent, broadly distributed place cell replays, produced without temporal clustering, significantly reduce the topological fluctuations in the cognitive map $\mathcal{C}$, thus helping to separate the fast and the slow timescales and to extract stable topological information for the long-term, qualitative representation of the environment [45]. Physiologically, these results suggest that indiscriminate, repetitive reactivations of memory sequences prevent deterioration of cognitive frameworks.

As a closing comment, it can be mentioned that dynamical simplicial complexes previously appeared in physical literature as discrete models of quantum space-time fluctuations, in the context of Simplicial Quantum Gravity theories [87, 88]. It was shown that such complexes exhibit rich geometrical and topological dynamics, e.g., they can exist in different geometric phases, experience phase transitions between ordered and disordered states, etc., yielding regular behavior in the thermodynamic "classical" limit. Here, the dynamical simplicial complexes appear in a very different context—as schematic models of the cognitive map's topological structure [12, 89], which is naturally discrete (being encoded by finite neuronal populations) and transient due to the plasticity of the underlying network. Nevertheless, the statistical mechanics of these "neuronal" complexes also points at a variety of geometric and topological states developing at several timescales. In particular, using the instantaneous Betti numbers as intensive (size independent) statistical variables allows describing these complexes' temporal architecture and identifying the emergent topological stability phenomena.

References

1. M. Leathers, C. Olson, In monkeys making value-based decisions. LIP Neurons Encode Cue Salience Action Value Sci **338**, 132–135 (2012)
2. D. Nitz, Tracking route progression in the posterior parietal cortex. Neuron **49**(5), 747–56 (2006)
3. R. Muller, J. Ranck Jr., J. Taube, Head direction cells: properties and functional significance. Current Opinions Neurobiol. **6**, 196–206 (1996)
4. J. Taube, Head direction cells and the neurophysiological basis for a sense of direction. Prog. Neurobiol. **55**, 225–256 (1998)
5. A. Terrazas, M. Krause, P. Lipa, K. Gothard, C. Barnes, B. McNaughton, Self-motion and the hippocampal spatial metric. J. Neurosci. **25**, 8085–8096 (2005)
6. E. Moser, M.-B. Moser, A metric for space. Hippocampus **18**, 1142–1156 (2008)
7. E. Kropff, J. Carmichael, M.-B. Moser, E. Moser, Speed cells in the medial entorhinal cortex. Nature **523**, 419–424 (2015)
8. D. Nitz, Spaces within spaces: rat parietal cortex neurons register position across three reference frames. Nat. Neurosci. **15**, 1365–1367 (2012)
9. F. Sargolini, M. Fyhn, T. Hafting, B. McNaughton, M. Witter, M. Moser, E. Moser, Conjunctive representation of position, direction, and velocity in entorhinal cortex. Science **312**, 758–762 (2006)
10. J. O'Keefe, L. Nadel, *The hippocampus as a cognitive map* (Oxford University Press, Oxford, 1978)
11. P. Best, A. White, A. Minai, Spatial processing in the brain: the activity of hippocampal place cells. Ann. Rev. Neurosci. **24**, 459–486 (2001)
12. A. Babichev, S. Cheng, Y. Dabaghian, Topological schemas of cognitive maps and spatial learning. Front. Comput. Neurosci. **10**, 18 (2016)
13. J. O'Keefe, J. Dostrovsky, The hippocampus as a spatial map. Preliminary evidence from unit activity in the freely-moving rat. Brain Res. **34**(1): 171–5 (1971)
14. K. Gothard, W. Skaggs, B. McNaughton, Dynamics of mismatch correction in the hippocampal ensemble code for space: interaction between path integration and environmental cues. J Neurosci. **16**, 8027–8040 (1996)
15. J. Leutgeb, S. Leutgeb, A. Treves, R. Meyer, C. Barnes et al., Progressive transformation of hippocampal neuronal representations in "morphed" environments. Neuron **48**, 345–358 (2005)
16. T. Wills, C. Lever, F. Cacucci, N. Burgess, J. O'Keefe, Attractor dynamics in the hippocampal representation of the local environment. Science **308**, 873–876 (2005)
17. D. Touretzky, W. Weisman, M. Fuhs, W. Skaggs, A. Fenton et al., Deforming the hippocampal map. Hippocampus **15**, 41–55 (2005)
18. B. Poucet, T. Herrmann, Exploratory patterns of rats on a complex maze provide evidence for topological coding. Behav Processes **53**, 155–162 (2001)
19. K. Diba, G. Buzsaki, Hippocampal network dynamics constrain the time lag between pyramidal cells across modified environments. J Neurosci. **28**, 13448–13456 (2008)
20. A. Alvernhe, F. Sargolini, B. Poucet, Rats build and update topological representations through exploration. Anim. Cogn. **15**, 359–368 (2012)
21. X. Wu, D. Foster, Hippocampal replay captures the unique topological structure of a novel environment. J Neurosci. **34**, 6459–6469 (2014)
22. Z. Chen, S.N. Gomperts, J. Yamamoto, M.A. Wilson, Neural representation of spatial topology in the rodent hippocampus. Neural computation **26**, 1–39 (2014)
23. Y. Dabaghian, V. Brandt, L. Frank, Reconceiving the hippocampal map as a topological template. eLife, 1–17 (2014), https://doi.org/10.7554/eLife.03476
24. Y. Dabaghian, F. Mémoli, L. Frank, G. Carlsson, A Topological Paradigm for Hippocampal Spatial Map Formation Using Persistent Homology. PLoS Comput. Biol. **8**(8), e1002581 (2012)
25. M. Arai, V. Brandt, Y. Dabaghian, The effects of theta precession on spatial learning and simplicial complex dynamics in a topological model of the hippocampal spatial map. PLoS Comput. Biol. **10**(6), e1003651 (2014)

26. E. Basso, M. Arai, Y. Dabaghian, The effects of gamma synchronization on spatial learning in a topological model of the hippocampal spatial map. PloS Comput. Biol. **12**, 9 (2016)
27. K. Hoffman, A. Babichev, Y. Dabaghian, A model of topological mapping of space in bat hippocampus. Hippocampus **26**, 1345–1353 (2016)
28. Y. Dabaghian, Through synapses to spatial memory maps: a topological model. Sci. Reports **9**, 572 (2018)
29. Y. Dabaghian, From topological analyses to functional modeling: the case of hippocampus. In submission (2019)
30. G. Buzsaki, Neural syntax: cell assemblies, synapsembles, and readers. Neuron **68**, 362–385 (2010)
31. K. Harris, J. Csicsvari, H. Hirase, G. Dragoi, G. Buzsaki, Organization of cell assemblies in the hippocampus. Nature **424**, 552–556 (2003)
32. A. Babichev, F. Mémoli, D. Ji, Y. Dabaghian, A topological model of the hippocampal cell assembly network. Frontiers in Comput. Neurosci. **10**, 50 (2016)
33. Y. Billeh, M. Schaub, C. Anastassiou, M. Barahona, C. Koch, Revealing cell assemblies at multiple levels of granularity. J. Neurosci. Methods **236**, 92–106 (2014)
34. P. Goldman-Rakic, Cellular basis of working memory. Neuron **14**, 477–485 (1995)
35. N. Hiratani, T. Fukai, Interplay between Short- and Long-Term Plasticity in Cell-Assembly Formation. PLoS One **9**, e101535 (2014)
36. M. Whittington, R. Traub, N. Kopell, B. Ermentrout, E. Buhl, Inhibition-based rhythms: experimental and mathematical observations on network dynamics. Int J Psychophysiol. **38**, 315–336 (2000)
37. G. Bi, M. Poo, Synaptic modification by correlated activity: Hebb's postulate revisited. Annu. Rev. Neurosci. **24**, 139–166 (2001)
38. S. Bennett, A. Kirby, G. Finnerty, Rewiring the connectome: evidence and effects. Neurosci. Biobehav. Rev. **88**, 51–62 (2018)
39. W. Meck, R. Church, D. Olton, Hippocampus, time, and memory. Behav. Neurosci. **127**, 655–668 (2013)
40. N. Clayton, T. Bussey, A. Dickinson, Can animals recall the past and plan for the future? Nat. Rev. Neurosci. **4**, 685–691 (2003)
41. M. Brown, R. Farley, E. Lorek, Remembrance of places you passed: Social spatial working memory in rats. J. Exper. Psych.: Animal Behav. Processes **33**, 213–224 (2007)
42. A. Babichev, Y. Dabaghian, Persistent Memories in Transient Networks. Springer Proc. Phys. **191**, 179–188 (2017)
43. A. Babichev, Y. Dabaghian, Transient cell assembly networks encode stable spatial memories. Sci. Rep. **7**, 3959 (2017)
44. A. Babichev, D. Morozov, Y. Dabaghian, Robust spatial memory maps encoded by networks with transient connections. PLoS Comput. Bio. **14**(9), e1006433 (2018)
45. A. Babichev, D. Morozov, Y. Dabaghian, Replays of spatial memories suppress topological fluctuations in cognitive map. Netw Neurosci Special Issue Topolog Neurosci **3**(3), 707–724 (2019)
46. A. Hatcher, *Algebraic topology* (Cambridge University Press, Cambridge, 2002)
47. P. Aleksandrov, *Elementary concepts of topology* (Ungar Pub Co, New York, 1965)
48. R. Ghrist, Barcodes: The persistent topology of data. Bull. Amer. Math. Soc. **45**, 61–75 (2008)
49. G. Singh, F. Mémoli, T. Ishkhanov, G. Sapiro, G. Carlsson, D. Ringach, Topological analysis of population activity in visual cortex. J. Vis. **8**(11), 1–18 (2008)
50. G. Carlsson, Topology and data. Bull. Amer. Math. Soc. **46**, 255–308 (2009)
51. P. Lum, G. Singh, A. Lehman, T. Ishkanov, M. Vejdemo-Johansson, M. Alagappan, J. Carlsson, G. Carlsson, Extracting insights from the shape of complex data using topology. Sci. Rep. **3**, 1236 (2013)
52. A. Zomorodian, G. Carlsson, Computing persistent homology. Dis. Comput. Geometry **33**, 249–274 (2005)
53. G. Carlsson, Vd Silva, Zigzag Persistence. Found. Comput. Math. **10**, 367–405 (2010)

54. G. Carlsson, V.D. Silva, D. Morozov, Zigzag persistent homology and real-valued functions. *Proceedings of the 25th annual Symposium on Computational Geometry*. (ACM,Aarhus, Denmark, 2009), pp. 247–256
55. H. Edelsbrunner, D. Letscher, A. Zomorodian, Topological persistence and simplification. Discrete Comput. Geometry **28**, 511–533 (2002)
56. A. Zomorodian, *Topology for computing* (Cambridge University Press, Cambridge, 2005)
57. H. Edelsbrunner, J. Harer, *Computational topology: an introduction* (AMS, USA, 2010)
58. N. Burgess, J. O'Keefe, Cognitive graphs, resistive grids, and the hippocampal representation of space. J Gen. Physiol. **107**, 659–662 (1996)
59. R. Muller, M. Stead, J. Pach, The hippocampus as a cognitive graph. J Gen. Physiol. **107**, 663–694 (1996)
60. J. Jonsson, *Simplicial complexes of graphs* (Springer, Berlin, Germany, 2008)
61. K. Mizuseki, A. Sirota, E. Pastalkova, G. Buzsaki, Theta oscillations provide temporal windows for local circuit computation in the entorhinal-hippocampal loop. Neuron **64**, 267–280 (2009)
62. L. Thompson, P. Best, Long-term stability of the place-field activity of single units recorded from the dorsal hippocampus of freely behaving rats. Brain. Res. **509**, 299–308 (1990)
63. M. Shapiro, Plasticity, hippocampal place cells, and cognitive maps. Arch. Neurol. **58**, 874–881 (2001)
64. A.A. Fenton, R.U. Muller, Place cell discharge is extremely variable during individual passes of the rat through the firing field. Proc. Natl. Acad. Sci. **95**(6), 3182–3187 (1998)
65. M. Gromov, On the number of simplexes of subdivisions of finite complexes. Mathe. Notes Acad. Sci. USSR **3**, 326–332 (1968)
66. T. McHugh, S. Tonegawa, CA3 NMDA receptors are required for the rapid formation of a salient contextual representation. Hippocampus **19**, 1153–1158 (2009)
67. B. Leuner, E. Gould, Structural plasticity and hippocampal function. Annu. Rev. Psychol. **61**, 111–140 (2010)
68. A. Schaefers, K. Grafen, G. Teuchert-Noodt, Y. Winter, Synaptic remodeling in the dentate gyrus, CA3, CA1, subiculum, and entorhinal cortex of mice: effects of deprived rearing and voluntary running. Neural Plast. **2010**, 11 (2010)
69. D. Dupret, A. Fabre, M. Döbrössy, A. Panatier, J. Rodriguez, S. Lamarque, V. Lemaire, S. Oliet, P. Piazza, D. Abrous, Spatial learning depends on both the addition and removal of new hippocampal neurons. PLoS Biol. **5**, e214 (2007)
70. B. Kuhl, A. Shah, S. DuBrow, A. Wagner, Resistance to forgetting associated with hippocampus-mediated reactivation during new learning. Nat. Neurosci. **13**, 501–506 (2010)
71. J. Murre, A. Chessa, M. Meeter, A mathematical model of forgetting and amnesia. Frontiers Psych. **4**, 76 (2013)
72. N. Cowan, What are the differences between long-term, short-term, and working memory? Prog. Brain Res. **169**, 323–338 (2008)
73. D. Hebb, *The organization of behavior; a neuropsychological theory* (Wiley, Hoboken, New Jersey, United State, 1949)
74. H. Eichenbaum, T. Otto, N. Cohen, Two functional components of the hippocampal memory system. Behav. Brain Sci. **17**, 449–472 (1994)
75. R. Kesner, M. Hunsaker, The temporal attributes of episodic memory. Behav. Brain Res. **215**, 299–309 (2010)
76. J. McClelland, B. McNaughton, R. O'Reilly, Why there are complementary learning systems in the hippocampus and neocortex: insights from the successes and failures of connectionist models of learning and memory. Psychol. Rev. **102**, 419–457 (1995)
77. S. Fusi, P. Drew, L. Abbott, Cascade Models of Synaptically Stored Memories. Neuron **45**, 599–611 (2005)
78. R. O'Reilly, J. McClelland, Hippocampal conjunctive encoding, storage, and recall: avoiding a trade-off. Hippocampus **4**, 661–682 (1994)
79. M. Karlsson, L. Frank, Awake replay of remote experiences in the hippocampus. Nat. Neurosci. **12**, 913–918 (2009)

80. G. Dragoi, S. Tonegawa, Preplay of future place cell sequences by hippocampal cellular assemblies. Nature **469**, 397–401 (2011)
81. D. Foster, M. Wilson, Reverse replay of behavioural sequences in hippocampal place cells during the awake state. Nature **440**, 680–683 (2006)
82. A. Johnson, A. Redish, Neural Ensembles in CA3 Transiently Encode Paths Forward of the Animal at a Decision Point. J. Neurosci. **27**, 12176–12189 (2007)
83. J. Hopfield, Neurodynamics of mental exploration. Proc. Natl. Acad. Sci. **107**, 1648–1653 (2010)
84. Y. Dabaghian, Maintaining Consistency of Spatial Information in the Hippocampal Network: A Combinatorial Geometry Model. Neural Comput. **28**, 1051–1071 (2016)
85. L. Roux, B. Hu, R. Eichler, E. Stark, G. Buzsaki, Sharp wave ripples during learning stabilize the hippocampal spatial map. Nat. Neurosci. **20**, 845–853 (2017)
86. G. Girardeau, K. Benchenane, S. Wiener, G. Buzsaki, M. Zugaro, Selective suppression of hippocampal ripples impairs spatial memory. Nat. Neurosci. **12**, 1222–1223 (2010)
87. J. Ambjørn, M. Carfora, A. Marzuoli, *The geometry of dynamical triangulations* (Springer, Berlin, New York, 1997)
88. H. Hamber, *Quantum gravitation: the Feynman path integral approach* (Springer, Berlin, 2009)
89. A. Babichev, Y. Dabaghian, Topological schemas of memory spaces. Frontiers Comput. Neurosci. **12**, (2018). https://doi.org/10.3389/fncom.2018.00027

Intuitionistic Fuzzy Graph Morphological Topology

Abraham Jacob and P. B. Ramkumar

Abstract In this paper, we define morphological topology ($\mathcal{M}$-topology) on intuitionistic fuzzy graph (IFG). We also define neighbourhood graph, continuity and isomorphism between $\mathcal{M}$-topologies.

Keywords $\mathcal{M}$-topology · Neighbourhood graph · Continuity · Weak neighbourhood graph · Continuous function · Isomorphism

1 Introduction

Mathematical morphology (MM) is a set theoretic tool for image analysis in digital image processing. Accuracy in image analysis has great importance in any applications like medical imaging. Process of converting an image into digital form involves sampling and quantization. A digital image in an array of squares, called pixels which represents intensity values corresponding to sampling points. It is considered as a grid-shaped graph with vertices as sampling points and edges determined by adjacency relation. After thresholding, a planner graph is obtained with different connected components as vertices.

Vincent [21] introduced graph morphology Laurent Najman and Fernand Meyer [13] did their work on Mathematical morphology on edge and vertex weighted graphs based on lattice structure. Fuzziness helps to handle uncertain situations. Ramkumar and Abraham [16] defined dilation and erosion on intuitionistic fuzzy soft graphs (IFSG).

A. Jacob (✉)
Rajagiri School of Engineering & Technology, A P J Abdul Kalam Technological University, Kerala, India

P. B. Ramkumar
Department of Mathematics Centre for Topology and Applications, Rajagiri School of Engineering & Technology, Kerala, India
e-mail: ramkumar_pb@rajagiritech.edu.in

R. L. Devaney et al. (eds.), *Topological Dynamics and Topological Data Analysis*, Springer Proceedings in Mathematics & Statistics 350,
https://doi.org/10.1007/978-981-16-0174-3_21

We proved dilation and erosion of IFSG of an image is a member of a complete lattice with respect to the operations union and intersection with partial order '$\subseteq$' as IFSG subgraph. Topology is a study of properties of some objects that are invariant under certain invertible transformation. In this paper, we define p_n adjacency vertices of a vertex in IFG as a neighbourhood vertices of this vertex in Sect. 3. We also defined morphological topology ($\mathcal{M}$-topology), neighbourhood graph, continuity and isomorphism of two $\mathcal{M}$-topologies in Sect. 3.

2 Preliminaries

Definition 1 An intuitionistic fuzzy graph (IFG) is of the form $G = (G^*, G^\times, \mu_1, \gamma_1, \mu_2, \gamma_2)$, where

1. $G^* = \{v_1, v_2, \ldots v_n\}$ such that $\mu_1 : G^* \to [0, 1]$ and $\gamma_1 : G^* \to [0, 1]$, the membership and non-membership grades of the element $v_i \in G^*$, respectively, and $0 \leq \mu_1(v_i) + \gamma_1(v_i) \leq 1$ for every $v_i \in G^*$; $i = 1, 2, \ldots, n$.
2. $G^\times \subseteq G^* \times G^*$ where $\mu_2 : G^\times \to [0, 1]$ and $\gamma_2 : G^\times \to [0, 1]$, the membership and non-membership grades of the element $e_{v_i v_j}$ in G^*, respectively, are such that

   (a) $\mu_2(e_{v_i v_j}) \leq \min\{\mu_1(v_i), \mu_1(v_j)\}$
   (b) $\gamma_2(e_{v_i v_j}) \leq \max\{\gamma_1(v_i), \gamma_1(v_j)\}$
   (c) $0 \leq \mu_2(e_{v_i v_j}) + \gamma_2(e_{v_i v_j}) \leq 1$ for every edges $e_{v_i v_j}$ in $G^\times, i = 1, 2, 3, \ldots, n$, $j = 1, 2, 3, \ldots, n$.

Definition 2 Let u_i and u_j be two vertices in IFG

$$G_i = (G^*, G^\times, \mu_{1i}, \gamma_{1i}, \mu_{2i}, \gamma_{2i}).$$

Then, u_j is said to be n-path adjacency vertex (p_n adjacency vertex) to u_i if they are connected by at most n edges. It is denoted by $u_i\ p_n-\text{adj}\ u_j$.

Definition 3 Let $e_{u_i u_j}$ and $e_{u_k u_l}$ be two edges in the IFG $G_i = (G^*, G^\times, \mu_{1i}, \gamma_{1i}, \mu_{2i}, \gamma_{2i})$. Then $e_{u_k u_l}$ is said to be n-path adjacency edge (p_n-adjacency edge) to $e_{u_i u_j}$ if either u_i or u_j is connected to u_k or u_l by almost n edges. It is denoted by $e_{u_i u_j}\ p_n-\text{adj}\ e_{u_k u_l}$.

Now we define the dilation and erosion on intuitionistic fuzzy graphs.

Definition 4 Let $G_i = (G^*, G^\times, \mu_{1i}, \gamma_{1i}, \mu_{2i}, \gamma_{2i})$ be IFG. Let $\mathcal{G}$ be set all intuitionistic fuzzy graphs $G_i = (G^*, G^\times, \mu_{1i}, \gamma_{1i}, \mu_{2i}, \gamma_{2i})$ defined on $G = (G^*, G^\times)$ where each pair in $\mathcal{G}$ satisfies the property intuitionistic fuzzy subgraph with G_i. We define a partial order $\subseteq$ as IF subgraph. Let 0 be an IF graph with all vertices and edges of membership grade 0 end non-membership grade 1 and 1 be an IFG with all vertices and edges of membership grade 1 and non-membership 0. Suprimum end infimum of two IFG G_1 and G_2 in $\mathcal{G}$ is defined as follows.

$$G_1 \vee G_2 = G_1 \cup G_2 = (G^*, G^\times, \mu_{11} \vee \mu_{12}, \gamma_{11} \wedge \gamma_{12}, \mu_{21} \vee \mu_{22}, \gamma_{21} \wedge \gamma_{22})$$
$$G_1 \wedge G_2 = G_1 \cap G_2 = (G^*, G^\times, \mu_{11} \wedge \mu_{12}, \gamma_{11} \vee \gamma_{12}, \mu_{21} \wedge \mu_{22}, \gamma_{21} \vee \gamma_{22})$$

Then, $(\mathcal{G}, \wedge, \vee, 0, 1)$ is a complete lattice. Now define dilation and erosion of vertices and edges in IFG G_i in the following:

1. For each elements u_k in G^*, $\delta_{1i} = G^* \to [0, 1]$ and

$$\in_{1i}\ G^* \to [0, 1] \text{ by}$$
$$\delta_{1i}(u_k) = \left(\sup_{uj}(\mu_{1i}(u_j), \inf_{uj} \gamma_{1i}(u_j) \right)$$
$$\in_{1i}\ (u_k) = \left(\inf_{uj}(\mu_{1i}(u_j), \sup_{uj} \gamma_{1i}(u_j) \right),$$

   where u_j is either u_k or u_j p_1-adj u_k.
2. For each elements $e_{u_k u_l}$ in $G^\times$,

$$\delta_{2i} : G^\times \to [0, 1] \text{ and}$$
$$\in_{2i} : G^\times \to [0, 1] \text{ by}$$
$$\delta_{2i}(e_{u_k u_l}) = \left(\sup_{e_{u_i uj}} \mu_{2i}(e_{u_i u_j}), \inf_{e_{u_i u_j}} \gamma_{2i}(e_{u_i u_j}) \right)$$
$$\in_{2i}\ (e_{u_k u_l}) = \left(\inf_{e_{u_i u_j}} (\mu_{2i}(e_{u_i u_j}), \sup_{e_{u_i u_j}} \gamma_{2i}(e_{u_i u_j}) \right)$$

   where $e_{u_i u_j}$ is $e_{u_k u_l}$ or $e_{u_i u_j}$ p_n-adj $e_{u_k u_l}$.

Then $G_{iE} = (\in_{1i}, \in_{2i})$ is called p_n adjacency eroded IFG and $G_{iD} = (\delta_{1i}, \delta_{2i})$ is called p_n adjacency dilated IFG.

A theorem [16] states the p_n adjacency dilated IFG G_{iD} and p_n adjacency eroded IFG are again IFG. Therefore, $\mathcal{G}$ is closed under dilation and erosion on IFG G_i. This motivates us to define morphological topology ($\mathcal{M}$-topology).

3 Morphological Topology

Now we take $G_i = (G^*, G^\times, \mu_1, \gamma_1, \mu_2, \gamma_2)$ as IFG corresponding to an image obtained for analysis. We proved [] that p_n adjacency dilated IFG G_{iD} and p_n adjacency eroded IFG G_{iE} are IF graphs, motivates us to take $\mathcal{M}$ as the collection of

IFG in which each pair satisfy the IF subgraph property with G_i, for defining morphological topology ($\mathcal{M}$-topology).

Definition 5 Let G_i be any IFG. Let $\mathcal{M}$ be the collection of IFG in which each pair in $\mathcal{M}$ satisfy IF subgraph property with G_i. Then, $\mathcal{M}$ is called $\mathcal{M}$-topology if the following axioms are satisfied.

1. $0, 1 \in \mathcal{M}$ and $G_i \in \mathcal{M}$.
2. $\mathcal{M}$ is closed under arbitrary union of IF graphs.
3. $\mathcal{M}$ is closed under finite intersection of IF graphs.

where 0 is the IFG with vertices and edges of membership grade 0 and non membership grade 1, and 1 on the IFG with vertices and edges of membership grade 1 and non membership grade 0.

Then, the pair $(G_i, \mathcal{M})$ is called $\mathcal{M}$-topological space. Members of $\mathcal{M}$-topology $\mathcal{M}$ are called open IFG.

Similar images should have similar IF graphs. Therefore, these IF graphs are topologically isomorphic. Before defining isomorphism, neighbourhood graph and continuity of a vertex are defined below.

Definition 6 Let $G_i = (G^*, G^\times, \mu_1, \gamma_1, \mu_2, \gamma_2)$ be IFG. Let v_i be a vertex in G_i. Then, neighbourhood vertices of the vertex v_i are defined as the set of all P_n-adjacency vertices of V_i and it is denoted by $n(v_i)$.

Definition 7 Let $G_i = (G^*, G^\times, \mu_1, \gamma_1, \mu_2, \gamma_2)$ be IFG. Let v_i be a vertex in G_i. Let N_i be a subgraph of G_i. This N_i is said to be neighbourhood graph of the vertex v_i if there is an open IF subgraph μ_i of N_i containing neighbourhood vertices of v_i.

Every open subgraph is $\mathcal{M}$-topology containing neighbourhood vertices of v_i is a neighbourhood of v_i.

Definition 8 Let $G_i = (G^*, G^\times, \mu_1, \gamma_1, \mu_2, \gamma_2)$ be IFG. Let v_i be a vertex in G_i. Then, the smallest open IF subgraph of G_i containing neighbourhood vertices of v_i is called weak neighbourhood graph of v_i.. Then, s Let N_i be a subgraph of G_i. This N_i is said to be neighbourhood graph of the vertex v_i if there is an open IF subgraph μ_i of N_i containing neighbourhood vertices of v_i.

Definition 9 Let $(G_i, \mathcal{M}_i)$ and $(G_j, \mathcal{M}_j)$ be two $\mathcal{M}$-topological spaces. Let v_i be a vertex in G_i. Let $f : (G_i, \mathcal{M}_i) \rightarrow (G_j, \mathcal{M}_j)$ be a function. Then, f is said to be continuous at the vertex v_i if for every neighbourhood graph N_j of the vertex $f(v_i)$ in G_j, there is a neighbourhood graph N_i of v_i in G_i such that $f(N_i)$ is a IF subgraph of N_j and $f(N_i) \subseteq N_j$.

Example 1 Let $G_i = (G^*, G^\times, \mu_1, \gamma_1, \mu_2, \gamma_2)$ be a IFG where $G^* = \{v_1, v_2, v_3, v_4\}$ be the vertex set $G^\times = \{e_{v_1v_2}, e_{v_2v_3}, e_{v_3v_4}, e_{v_4v_1}\}$ (See Fig. 1).

Then neighbourhood vertices of $v_i = n(v_1) = \{v_1, v_2, v_4\}$.

Let $\mathcal{M} = \{0, 1, G_i, G_i'\}$. Then three axioms of $\mathcal{M}$-topology are satisfied. Thus, $\mathcal{M}$ is $\mathcal{M}$-topology on IFG G_i.

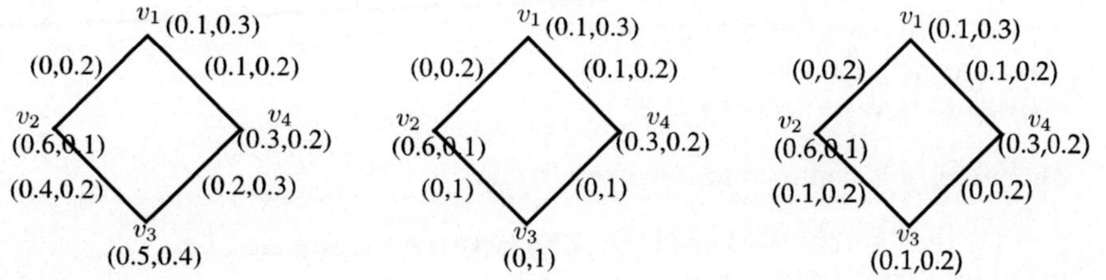

(a) IFG G_i (b) IF subgraph G_i' of G_i (c) Neighbourhood graph G_i'' of vertex v_i

Fig. 1 .

Let G_i'' be a IF subgraph of G_i. Since IFG G_i' is an open subgraph of G_i'', G_i' is a neighbourhood graph of the vertex v_1 in G_i.

Example 2 Let $G_i = (G^*, G^\times, \mu_1, \gamma_1, \mu_2, \gamma_2)$ be IFG.

(a) If $\mathcal{M} = \{0, 1, G_i\}$ then $\mathcal{M}$ is called discrete $\mathcal{M}$-topology on G_i
(b) If $\mathcal{M}$ in the collection of all IFG with IF subgraph property with G_i, 0 and 1, then $\mathcal{M}$ is called indiscrete $\mathcal{M}$-topology on G_i.

Theorem 1 *Let $(G_i, \mathcal{M}_i)$ and $(G_j, \mathcal{M}_j)$ be topological spaces. Let v_i be a vertex in G_i.*

Let $f : (G_i, \mathcal{M}_i) \to (G_j, \mathcal{M}_j)$ be a function. Then, the following are equivalent.

1. *f is continuous at v_i*
2. *The inverse image of every neighbourhood graph of $f(v_i)$ in G_j is a neighbourhood graph of v_i in G_i.*

Proof (1) $\Rightarrow$ (2) Let N_j be a neighbourhood graph of a vertex $f(v_i)$ in G_j.

By definition of neighbourhood graph, there is an open neighbourhood subgraph M_j of N_j containing neighbourhood vertices of $f(v_i)$.

Since f is continuous at v_i in G_i for each neighbourhood IF graph N_j of the vertex $f(v_i)$ in G_j, there is a neighbourhood IF graph N_i of the vertex v_i in G_i such that $f(N_i)$ is a IF subgraph of N_j

$$\therefore\ f(N_i) \subseteq N_j \quad N_i \subseteq f^{-1}(N_j).$$

Since N_i is a neighbourhood IF graph of the vertex v_i, there is an open IF subgraph of N_i containing neighbourhood vertices of v_i.

Therefore, $f^{-1}(N_j)$ is a neighbourhood graph of the vertex v_i in G_i since $N_i \subseteq f^{-1}(N_j)$.

(2) $\Rightarrow$ (1)

Let N_j be a neighbourhood graph of $f(v_i)$ in G_j.

$\Rightarrow f^{-1}(N_j)$ is a neighbourhood graph of v_i in G_i

$\Rightarrow$ there is an open IF subgraph of M_i of $f^{-1}(N_j)$
containing neighbourhood vertices of v_i in G_i
$\therefore M_i \subseteq f^{-1}(N_j) \Rightarrow f(M_i) \subseteq N_j$

Therefore, f is continuous at the vertex v_i. □

Definition 10 Let $(G_i, \mathcal{M}_i)$ and $(G_j, \mathcal{M}_j)$ be two $\mathcal{M}$-topologies. Let v_i be a vertex in G_i. Let $f : (G_i, \mathcal{M}_i) \to (G_j, \mathcal{M}_j)$ be a function. Then, if f is continuous on G_i then f is continuous at each vertices of G_i.

Theorem 2 *Let $(G_i, \mathcal{M}_i)$ and $(G_j, \mathcal{M}_j)$ be two $\mathcal{M}$-topologies. Let $f : (G_i, \mathcal{M}_i) \to (G_j, \mathcal{M}_j)$ be a function. Then, the following are equivalent.*

1. *f is continuous*
2. *Every inverse image of an open IF subgraph if G_j is an open IF subgraph of G_i.*

Proof (1) $\Rightarrow$ (2)
Let v_i be any vertex in G_i. Let M_j be an IF subgraph in $\mathcal{M}_j$ containing $f(v_i)$. Since f is continuous, there is a neighbourhood graph N_i of the vertex v_i such that $f(N_i)$ is a IF subgraph of M_j

$$\begin{aligned} \therefore\ & f(N_i) \subseteq M_j \\ & \Rightarrow N_i \subseteq f^{-1}(M_j) \\ & \Rightarrow f^{-1}(M_j) \text{ is an open subgraph in } \mathcal{M}_i. \end{aligned}$$

(2) $\Rightarrow$ (1)
Let v_i be any vertex in G_i.
Let M_j be open IF subgraph of G_j
$\Rightarrow$ by assumption, $f^{-1}(M_j)$ is open IF subgraph of G_i
$\Rightarrow f^{-1}(M_j)$ is a neighbourhood of a vertex v_i
$\Rightarrow$ there is an open IF subgraph M_i in G_i containing v_i and which IF subgraph of $f^{-1}(M_j)$.
Therefore, $M_i \subseteq f^{-1}(M_j) \Rightarrow f(M_i) \subseteq M_j$.
Therefore, f is continuous. □

Now, we define isomorphism below.

Definition 11 Let $G_i = (G_i^*, G_i^\times, \mu_{1i}, \gamma_{1i}, , \mu_{2i}, \gamma_{2i})$ and $G_j = (G_j^*, G_j^\times, \mu_{1j}, \gamma_{1j}, \mu_{2j}, \gamma_{2j})$ be two IF graphs. Let $(G_i, \mathcal{M}_i)$ and $(G_j, \mathcal{M}_j)$ be two $\mathcal{M}$-topologies on G_i and G_j, respectively. An isomorphism $f : G_i^* \to G_j^*$ which satisfies the following conditions.

1. $\mu_{1i}(v_i) = \mu_{1j}(f(v_i))$, $\gamma_{1i}(v_i) = \gamma_{1j}(f(v_i))$,
2. $\mu_{2i}(e_{v_iv_j}) = \mu_{2j}(f(e_{v_iv_j}))$, $\gamma_{2i}(e_{v_iv_j}) = \gamma_{2j}(f(e_{v_iv_j}))$

for each vertices v_i and edges $e_{v_iv_j}$ in G_i.

Observation 1 Isomorphism is an equivalent relation.

Observation 2 Let $(G_i, \mathcal{M}_i)$ and $(G_j, \mathcal{M}_j)$ be two $\mathcal{M}$-topologies. Let $f : (G_i, \mathcal{M}_i) \to (G_j, \mathcal{M}_j)$ be a function. Then, $\mathcal{M}_i$ and $\mathcal{M}_j$ are isomorphic if and only if f is a bijective mapping and f and f^{-1} are continuous.

4 Conclusion

As a foundation theory of topology in intuitionistic fuzzy graph morphology, we have defined morphological topology ($\mathcal{M}$-topology) associated with neighbourhood graph and continuity of a vertex in IFG with examples. Isomorphism of two $\mathcal{M}$-topologies defined in this paper can be applied for image analysis in future.

References

1. N. Alsheri, M. Akram, Intuitionistic fuzzy planner graphs. Discrete Dyn. Nat. Soc. **2014**, 9 p. Article ID 39782
2. K. Atanassov, *Intuitionistic Fuzzy Sets: Theory and Applications* (Springer-Verlag, Heidelberg, 1999)
3. D. Baets, E. Kerre, M. Gadan, The fundamentals of fuzzy mathematical morphology Part 1: basic concepts. Int. J. General Syst. **23**, 155–171 (1995)
4. N. Cagman, S. Enginoglu, F. Citak, Fuzzy soft set theory and its applications. Iraninan J. Fuzzy Syst. **8**(3), 137–147 (2001)
5. P.M. Dhanya, A. Sreekumar, M. Jathavedan, P.B. Ramkumar, Algebra of morphological dilation on intuitionistic fuzzy hypergraph IJSRSET **4**(1) (2018)
6. P.M. Dhanya, A. Sreekumar, M. Jathavedan, P.B. Ramkumar, Document modeling and clustering using hypergraph, Int. J. Appl. Eng. Res. **12**(10), 2127–2135 (2017), ISSN (0973-4562)
7. H. Heijmans, L. Vincent, Graph morphology in image analysis. *Mathematical Morphology in Image Processing* (1992), pp. 171–203
8. K.D. Joshy, *Introduction to General Topology* (Wiley, Eastern Limited, 1992)
9. K.G. Karunambigai, R. Parvathi, Intuitionistic fuzzy graphs. J. Comput. Intell. Theory Appl. **20**, 139–150 (2006)
10. E. Melin, Digital geometry & Khalimsky space. Uppsala Dissert. Mathe. **54** (2008)
11. J.R. Munkres, Topology, 2nd edn. Pearson
12. A. Nagoor Gani, S. Anupriya, Spilt domination on intuitionistic fuzzy graph, in *Advanced in Computational Mathematics and its Applications (ACMA)* Vol. 2(2) (2012), ISSN 2167-6356
13. L. Najman, F. Meyer, *A short tour of mathematical morphology on edge and vertex weighted graphs*, Image Processing and Analysis with Graphs Theory and Practice, ed. by O. Lezoray, L. Grady, CRC Press (2012), pp. 141–174. Digital Imaging and Computer Vision. 9781439855072
14. L. Najman, J. Cousty, *A Graph-Based Mathematical Morphology Reader* (Elsevier, USA, 2014)
15. L. Najman, H. Talbot, *Mathematical Morphology from theory to Applications* (Wiley, USA, 2008)
16. P.B. Ramkumar, A. Jacob, Morphology on intuitionistic fuzzy soft graphs. Int. J. Adv. Res. Trends Eng. Technol. **5**(Spl. issue)
17. A. Rosenfeld, Digital topology. American Mathe. Monthly **86**(8), 621–630 (1979)
18. J. Serra, *Image Analysis and Mathematical Morphology* (Academic Press, New York, 1982)

19. A.M. Shyla, T.K. Mathew Varkey, Intuitionistic fuzzy soft graph. Int. J. Fuzzy Mathe. Arch. (2016)
20. P. Sunitha, An elementary introduction to intuitionistic fuzzy soft graphs. J. Math. Comput. Sci. **6**(4), 668–681 (2016)
21. L. Vincent, Graphs and mathematical morphology. Signal Process **16**, 365–88 (1989)

Some Properties of the Bitopological Space Associated With the 3-Uniform Semigraph of Cycle Graph

Asha G. Pillai and P. B. Ramkumar

Abstract In this paper,the neighbourhood N_i of the vertex 'i' of the 3 uniform semigraph $C_{m,1}$ is defined as $N_i = V - C_i$ where V is the vertex set and C_i is the set of vertices which are consecutively adjacent to 'i'. Let E denote the collection of end vertices and M denote the collection of middle vertices of $C_{m,1}$.Define $\tau_E = \cap_{i \in E} P(N_i)$ and $\tau_M = \cap_{i \in M} P(N_i)$.$\tau_E$ and τ_M are the discrete topologies on the end vertex set and the middle vertex set respectively. Define $\tau_E^{'} = V \cup \tau_E$ and $\tau_M^{'} = V \cup \tau_M$. $\tau_E^{'}$ and $\tau_M^{'}$ are two different topologies defined on the vertex set and hence $(V, \tau_E^{'}, \tau_M^{'})$ is a bitopological space. Different topological properties of this bitopological space are discussed.

Keywords Semigraph · 3-uniform semigraph · Bitopological space

1 Introduction

A semigraph G is a pair (V, X) where V is a nonempty set whose elements are called vertices of G and X is a set of n-tuples,called the edges of G, of distinct vertices for $n \geq 2$ satisfying the following conditions.

(i) Any two edges have atmost one vertex in common.
(ii) Two edges $(u_1, u_2, ...u_n)$ and $(v_1, v_2, ...v_m)$are equal only if
(a) $n = m$ and (b) either $u_i = v_i$ for $1 \leq i \leq n$ or $u_i = v_{n-i+1}$ for $1 \leq i \leq n$..

A. G. Pillai (✉)
TocH Institute of Science and Technology, Kerala, India

Research Scholar, Rajagiri School of Engineering and Technology, A P J Abdul Kalam Technological University, Kerala, India

P. B. Ramkumar
Department of Mathematics, Centre for Topology and Applications, Rajagiri School of Engineering and Technology, Kerala, India
e-mail: ramkumar_pb@rajagiritech.edu.in

R. L. Devaney et al. (eds.), *Topological Dynamics and Topological Data Analysis*,
Springer Proceedings in Mathematics & Statistics 350,
https://doi.org/10.1007/978-981-16-0174-3_22

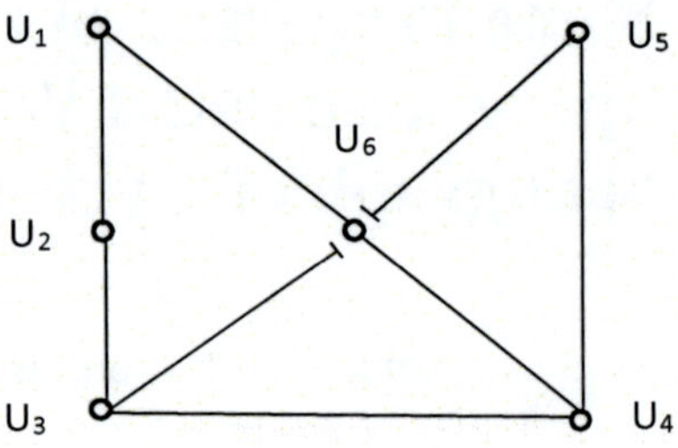

Fig. 1 Example for semigraph

If $e = (u_1, u_2....u_n)$ is an edge of the semigraph G,u_1 and u_n are called the end vertices of e and $u_2, u_3, ...u_{n-1}$ are called the middle vertices of e.

Two vertices of G are adjacent if both of them belong to an edge and two edges are adjacent if they have a common vertex. The pairs of vertices (u_1, u_2), (u_2, u_3)... of the edge e are called consecutively adjacent.

Example

In Fig. 1, the edges are (u_1, u_2, u_3), (u_3, u_4), $(u_4, u_5)(u_1, u_6, u_4)(u_3, u_6)(u_5, u_6)$. For the edge (u_1, u_2, u_3) ,u_1 and u_3 are the end vertices and u_2 is the middle vertex. Also $u_1, u_2 and u_3$ are adjacent as they are part of the same edge. $u_1, u_2 and u_2, u_3$ are consecutively adjacent.The number of vertices in an edge is called the cardinality of the edge (Fig. 1).

2 3-Uniform Semigraph of a Cycle Graph

The semigraph obtained by introducing a middle vertex to each edge of the cycle C_m,where C_m denote the cycle with m vertices, is a 3-uniform semigraph. It is denoted by $C_{m,1}$. In a 3-uniform semigraph, the cardinality of each edge is 3 (Fig. 2).

Examples

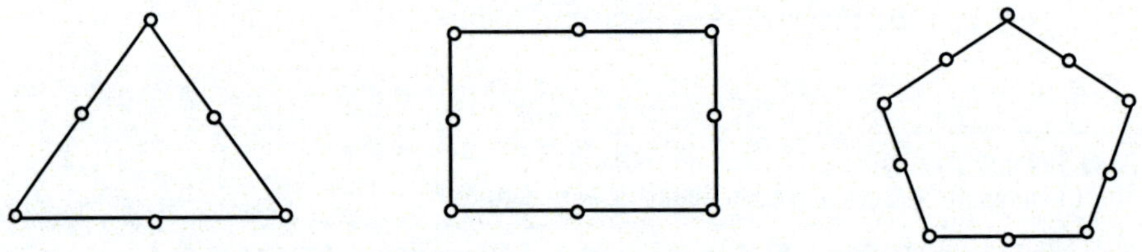

Fig. 2 Examples for 3-uniform semigraph

3 Bitopological Space

A non-empty set X with two distinct topologies $\tau_1 and \tau_2$ defined on X is called a bitopological space, denoted by (X, τ_1, τ_2).

Definition 1 A bitopological space (X, τ_1, τ_2) is said to be weakly pairwise T_0 if for each pair of distinct points, there exists a τ_1-open set or a τ_2 open set containing one but not the other.

Definition 2 A bitopolgical space (X, τ_1, τ_2) is said to be pairwise T_0 if for each pair (x, y) of distinct points of X, there is either a τ_1-open set containing x but not y or a τ_2-open set containing y but not x.

Definition 3 A bitopolgical space (X, τ_1, τ_2) is said to be weakly pairwise T_1 if for each pair (x, y) of distinct points of X, there is a τ_1-open set G and a τ_2-open set H such that $x \in G, y \notin G$ and $y \in H, x \notin H$ or $x \in H, y \notin H and y \in G, y \notin G.$

Definition 4 A bitopolgical space (X, τ_1, τ_2) is said to be pairwise T_1 if for each pair (x, y) of distinct points of X, there is a τ_1-open set G containing x but not y or a τ_2-open set H containing y but not x.

Definition 5 A bitopological space (X, τ_1, τ_2) is said to be weakly pairwise T_2 if for each pair (x, y) of distinct points of X, there is a τ_1-open set G and a τ_2-open set H with $G \cap H = \phi$ such that $x \in G$ and $y \in H$ or $x \in H$ and $y \in G$.

Definition 6 A bitopolgical space (X, τ_1, τ_2) is said to be pairwise T_2 if for each pair (x, y) of distinct points of X, there is a τ_1-open set G and a (τ_2)-open set H with $G \cap H = \phi$ such that $x \in G$ and $y \in H$.

Definition 7 A bitopological space (X, τ_1, τ_2) is said to be double compact if both the spaces (X, τ_1) and (X, τ_2) are compact.

Definition 8 A bitopological space (X, τ_1, τ_2) is said to be pairwise normal if for a τ_1-closed set P and a τ_2-closed set Q with $P \cap Q = \phi$,there is a $G \in \tau_1$ and $H \in \tau_2$ such that $P \subset H$ and $Q \subset G with G \cap H = \phi$.

Definition 9 A bitopological space (X, τ_1, τ_2) is said to be pairwise compact if every τ_1-open cover of X has a finite τ_2-open subcover and every τ_2-open cover of X has a τ_1-open subcover.

Definition 10 A bitopological space (X, τ_1, τ_2) is said to be pairwise connected if X cannot be expressed as the union of two nonempty disjoint open sets G and H such that $G \subset \tau_1$and $H \subset \tau_2$.

4 Bitopological Space on $C_{m,1}$

Let the vertex set of a 3-uniform semigraph of a cycle graph be $V = \{1, 2, 3, ...\}$ where $E = \{1, 3, 5, ...\}$ are the end vertices and $M = \{2, 4, 6, ...\}$ are the middle vertices.

Define a neighbourhood for each vertex as follows.

For each vertex $'i'$, define the neighbourhood $'N_i'$ as $N_i = V - C_i$, where C_i is the set of vertices which are consecutively adjacent to $'i'$.

Consider the collection $\tau_E = \bigcap_{i \in E} P(N_i)$ and $\tau_M = \bigcap_{i \in M} P(N_i)$, where $P(N_i)$ denote the power set of N_i.

τ_E and τ_M are the discrete topologies on E and M respectively.

Illustration 1 In $C_{3,1}$, (Fig. 3)

$N_1 = \{1, 3, 4, 5\}$,$N_2 = \{2, 4, 5, 6\}$
$N_3 = \{1, 3, 5, 6\}$,$N_4 = \{1, 2, 4, 6\}$
$N_5 = \{1, 2, 3, 5\}$,$N_6 = \{2, 3, 4, 6\}$

$\tau_E = \{\phi, \{1\}, \{3\}, \{5\}, \{1, 3\}\{1, 5\}\{3, 5\}\{1, 3, 5\}\}$, which is the discrete topology on E.

$\tau_M = \{\phi, \{2\}, \{4\}, \{6\}, \{2, 4\}, \{2, 6\}, \{4, 6\}, \{2, 4, 6\}\}$, which is the discrete topology on M.

Fig. 3 $C_{3,1}$

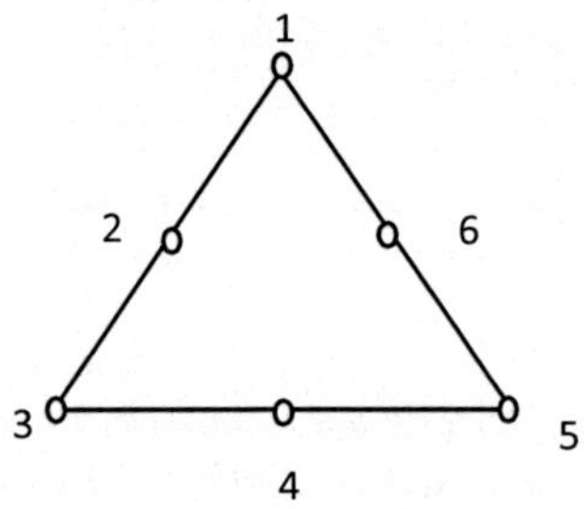

Fig. 4 $C_{4,1}$

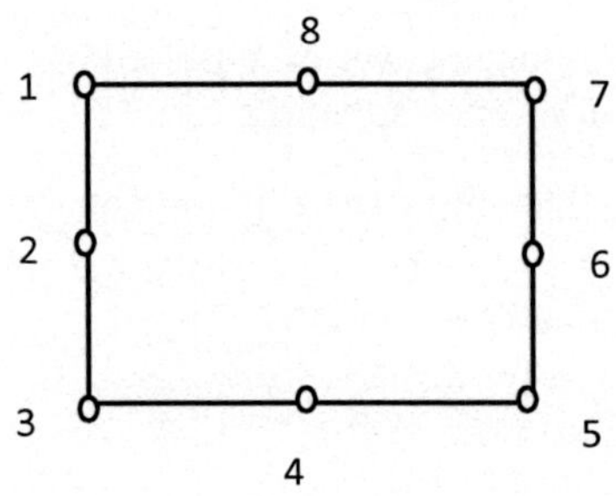

Illustration 2 For $C_{4,1}$ (Fig. 4),

$N_1 = \{1, 3, 4, 5, 6, 7\}$,$N_2 = \{2, 4, 5, 6, 7, 8\}$
$N_3 = \{1, 3, 5, 6, 7, 8\}$,$N_4 = \{1, 2, 4, 6, 7, 8\}$
$N_5 = \{1, 2, 3, 5, 7, 8\}$,$N_6 = \{1, 2, 3, 4, 6, 8\}$
$N_7 = \{1, 2, 3, 4, 5, 7\}$,$N_8 = \{2, 3, 4, 5, 6, 8\}$

$\tau_E = \{\phi, \{1\}, \{3\}, \{5\}, \{7\}, \{1, 3\}, \{1, 5\}, \{1, 7\}, \{3, 5\}, \{3, 7\}, \{5, 7\}, \{1, 3, 5\}, \{1, 3, 7\}, \{1, 5, 7\}, \{3, 5, 7\}, \{1, 3, 5, 7\}\}$ which is the discrete topology on E.

$\tau_M = \{\phi, \{2\}, \{4\}, \{6\}, \{8\}, \{2, 4\}, \{2, 6\}, \{2, 8\}, \{4, 6\}, \{4, 8\}, \{6, 8\}, \{2, 4, 6\}, \{2, 4, 8\}, \{2, 6, 8\}, \{4, 6, 8\}, \{2, 4, 6, 8\}$ which is the discrete topology on M.

Now define $\tau_{E'} = V \cup \tau_E$ and $\tau_{M'} = V \cup \tau_M$.

Both τ_E' and τ_M' are topologies on the vertex set V.

Since τ_E' and τ_M' are defined over the same vertex set V,(V, τ_E', τ_M') is a bitpological space .

5 Some Topological Properties of (V, τ_E', τ_M')

Proposition 1 *The bitopological space* (V, τ_E', τ_M') *is weakly pairwise* T_0.

Proof Let $i, j \in V$ such that $i \neq j$.

If $i \in E$,$\{i\} \in \tau_E'$,which contains i but not j.

If $i \in M$,$\{i\} \in \tau_M'$,which contains i but not j.

Therefore for every pair of vertices $(i, j), i \neq j$, there exists a τ_E' open set or a τ_M' open set containing one but not the other. Hence(V, τ_E', τ_M') is weakly pairwise T_0.

Proposition 2 *The bitopological space* (V, τ_E', τ_M') *is pairwise* T_0

Proof Let $i, j \in V$ such that $i \neq j$.If $i \in E$, $\{i\} \in \tau_E'$ which contains *i* but not j. If $i \in M$, $\{i\} \in \tau_M'$which contains *i* but not *j*. Therefore for every pair of vertices $(i, j), i \neq j$, there exists a τ_E'-open set containing *i* but not *j* or *a* τ_M'-open set containing i but not j and hence (V, τ_E', τ_M') is pairwise T_0.

Proposition 3 *The bitopological space* (V, τ_E', τ_M') *is not pairwise* T_1 *and not weakly pairwise* T_1.

Proof Let $i, j \in V$ such that $i \neq j$.

Let $i, j \in E$. Let $G = \{i\}$. G is τ_E' open which contains *i* but not *j*. Now, *V* is the only τ_M' open set which contains *j*. Since V contains all the vertices,$i \in V$. Therefore there exists no τ_M' open set *H* which contains *j* but not *i*.

$\therefore (V, \tau_E', \tau_M')$ is not pairwise T_1.

Same argument leads to the conclusion that (V, τ_E', τ_M') is not weakly pairwise T_1.

Proposition 4 *The bitopological space* (V, τ_E', τ_M') *is not pairwise* T_2 *and it is not weakly pairwise* T_2.

Proof Let $i, j \in V$ such that $i \neq j$.

Let $i, j \in E$.Let $G = \{i\}$. G is τ_E' open which contains i but not j.Now, V is the only τ_M' open set which contains j.Let $H = V$.Since V contains all the vertices,$i \in V$.$\Rightarrow G \cap H \neq \phi$.We cannot find two open sets G and H with $G \cap H = \phi$ and $i \in G, j \in H$.Hence (V, τ_E', τ_M') is not pairwise T_2.

Same argument leads to the conclusion that (V, τ_E', τ_M') is not weakly pairwise T_2.

Proposition 5 *The bitopological space (V, τ_E', τ_M') is double compact.*

Proof (i).To show that $(V, \tau_E')$is compact.

Let β be an open cover for $(V, \tau_E)'$.

$\Rightarrow \bigcup_{B\in\beta} B = V$ .By the choice of τ_E',every open cover must contain V.$\Rightarrow$Any subcollection of β including V is a finite subcover of β.$\Rightarrow$ $(V, \tau_E')$is compact.

(ii).To show that $(V, \tau_M')$is compact.

β be an open cover for $(V, \tau_M)'$.

$\Rightarrow \bigcup_{B\in\beta} B = V$ .By the choice of τ_M',every open cover must contain V.$\Rightarrow$Any subcollection of β including V is a finite subcover of β.$\Rightarrow$ $(V, \tau_M')$is compact.

Combining $(i)and(ii)$ (V, τ_E', τ_M') is double compact.

Proposition 6 *The bitopological space (V, τ_E', τ_M') is pairwise normal.*

Proof Let P be a τ_E'-closed set and Q be a τ_M'-closed set such that $P \cap Q = \phi$.By the choice of τ_E' and τ_M' ,there exists only one such pair ,namely,P=M and Q=E.

Let G=M and H=E.Clearly,$P \subseteq G$ and $Q \subseteq H$,with $G \cap H = \phi$.

$\therefore$.For a τ_E'-closed set P and a τ_M'-closed set Q,with $P \cap Q = \phi$,there exists $G \in \tau_M'$ and $H \in \tau_E'$ with $G \cap H = \phi$ such that $P \subseteq G$ and $Q \subseteq H$.

$\Rightarrow (V, \tau_E', \tau_M')$ is pairwise normal.

Proposition 7 *The bitopological space (V, τ_E', τ_M') is pairwise disconnected.*

Proof Let G=E and H=M.$G \in \tau_E', H \in \tau_M', G \cap H = \phi and G \cup H = V$.

Hence V can be expressed as the union of two nonempty disjoint open sets G and H such that $G \subset \tau_E'$ and $H \subset \tau_M'$.

$\therefore$. (V, τ_E', τ_M') is pairwise disconnected.

6 Conclusion

In this paper,we have introduced two different topologies on the vertex set of a 3-uniform semigraph of a cycle graph to make it a bitopological space.Some topological properties of the space are verified.Further studies are in progress.

References

1. E. Sampath Kumar, Semigraphs, Combinatorial Optimization. Narosa Publishing House (2004), pp. 125–138
2. J.C. Kelly, Bitopological Spaces. Proc. London Mathe. Soc. **s3-13**(1) 71–89 (1963)
3. N. Murugesan, D. Narmatha, Some properties of Semigraph and its associated graphs. Int. J. Eng. Res. Technol. (IJERT). **3**(5), pp. 898–903 (2014)
4. V. Kaladevi, K. Marimuthu, Distance based indices of bipartite graph associated with a 3-uniform semigraph of a cycle graph. Global J. Pure Appl. Mathe. **13**(9)
5. K.A. Abdu, A. Kilicman, Bitopological Spaces on undirected graphs J. Mathe. Comput. Sci. **18**, 231–232 (2018), www.isr.publications.com/jmcs
6. M.J. Saegrove, On Bitopological spaces, Retrospective Theses and Dissertations, IOWA State University (1971), https://lib.dr.iastate.edu/rtd/4914

Hypergraph Topology

Chandran R. Deepthi and P. B. Ramkumar

Abstract Consider a hypergraph H with vertex set $V = \{v_1, v_2, \dots v_n\}$ and hyperedge set $E = \{e_1, e_2, \dots e_m\}$. Two edges are adjacent if their intersection is non-empty. A neighbourhood of a vertex v_i, denoted by $N(v_i)$ is defined as the collection of vertices in adjacent edges of v_i. Hence, every edge is contained in a neighbourhood. A hypergraph topology is a family T of neighbourhood of vertices in V which satisfies the following conditions

- ϕ ,$V \in T$
- If $N(v_i), N(v_j) \in T$ then $N(v_i) \cap N(v_j) \in T$
- If $N(v_i) \in T$ for each $i \in I$ then $\cup_{i \in I} N(v_i) \in T$

The elements of T are called open sets. Thus, a topology T defined on a hypergraph H is called hypergraph topological space, denoted by (H, T). Also for a subhypergraph, similarly a subhypergraph induced topology is defined. The concept of closed sets, continuity, connectedness, metric and homeomorphism are also discussed

Keywords Hypergraph · Neighbourhood of a vertex · Hypergraph topology · Hypergraph topological space

C. R. Deepthi (✉)
Department of Mathematics, L B S Institute of Technology for Women, Thiruvananthapuram, Kerala, India

Research Scholar, Rajagiri School of Engineering and Technology, A P J Abdul Kalam Technological University, Ernakulam, Kerala, India

P. B. Ramkumar
Department of Mathematics, Centre for Topology and Applications, Rajagiri School of Engineering and Technology, Ernakulam, Kerala, India
e-mail: ramkumar_pb@rajagiritech.edu.in

R. L. Devaney et al. (eds.), *Topological Dynamics and Topological Data Analysis*, Springer Proceedings in Mathematics & Statistics 350,
https://doi.org/10.1007/978-981-16-0174-3_23

1 Introduction

The hypergraph topology is not concerned with the physical layout of the hypergraph, but shows what connections exist between the vertices and hyperedges. Thus, for the same hypergraph we can define different topologies. Hypergraph H is a pair (V, E) where V is the set of vertices and E is the set of edges called hyperedges which is a continuous closed curve containing the vertices. Each hyperedge consists of any finite number of vertices. *Topological Hyper-Graphs* were defined in a paper *Topological Hyper-Graphs by Sarit Buzaglo, Rom Pinchasi and Gunter Rote in 4 December 2007*. They defined topological hypergraphs as vertices enclosed by Jordan Curves. A family of simple closed curves in the plane is a family of *pseudo-circles*, if every two curves in the family are either disjoint or properly cross at precisely two points. They discussed more on graphical properties in topological hypergraphs. In this paper, also hyperedge is defined as pseudo-circles. But here we concentrated only on topological properties. Instead of topological hypergraph, we defined it as hypergraph topological space by defining hypergraph topology. The paper also discusses the concept of closed sets, continuity and connectedness, and their properties. This is further extended to homeomorphism.

2 Preliminaries

Hypergraph H is a pair (V, E) where V is the set of vertices and E is the set of edges called hyperedges which is a continuous closed curve containing the vertices. Each hyperedge consists of any finite number of vertices. Consider a hypergraph H with vertex set $V = \{v_1, v_2, \dots v_n\}$ and hyperedge set $E = \{e_1, e_2, \dots e_m\}$. Two edges are *adjacent* if their intersection is non-empty. To every hypergraph, we define a subhypergraph as follows. A subhypergraph is a hypergraph with some vertices or edges removed. Subhypergraph H_A induced by a subset A of V is defined as $H_A = (A, \{e_i \cap A; e_i \cap A \neq \emptyset\})$ where $e_i \epsilon E$.
For example, consider the above hypergraph H. Define $A \subset V$ as $A = \{v_1, v_3, v_6\}$. Then $H_A = (A, e_i \cap A \neq \emptyset; i = 1, 2, 3) = (A, E_A), where E_A = \{e_{1_A}, e_{3_A}\}$ The subhypergraph H_A is shown below (Fig. 1).

As our interest is on connected hypergraphs, hyperpaths is defined. A *hyperpath* between vertices v_1 and v_k is defined as an alternative sequence of distinct vertices and hyperedges $v_1, e_1, v_2, e_2, \dots, e_{k-1}, v_k$ such that $\{v_i, v_{i+1}\} \subseteq e_i$ for $1 \leq i \leq k-1$.
A hypergraph is *connected* if there is a hyperpath between every pair of vertices. Otherwise, it is disconnected.
A *metric* is a function $d : X \times X \to R$ which satisfies the following conditions

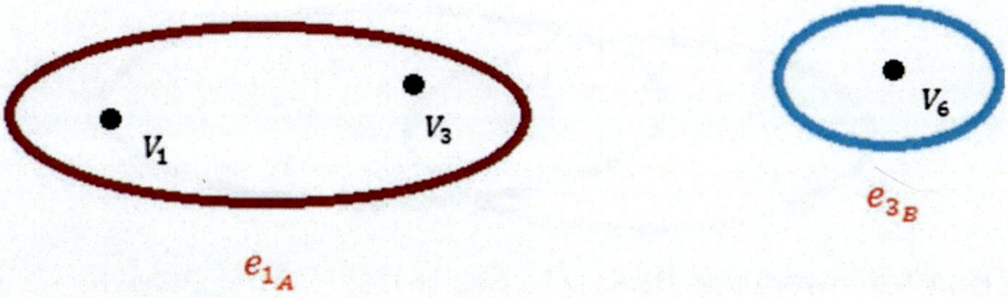

Fig. 1 Subhypergraph

- $d(v_i, v_j) \geq 0$
- $d(v_i, v_j) = 0 \iff v_i = v_j$
- $d(v_i, v_j) = d(v_j, v_i)$
- $d(v_i, v_k) \leq d(v_i, v_j) + d(v_j, v_k)$

3 Hypergraph Topological Space

A *neighbourhood of a vertex* v_i denoted by $N(v_i)$ is defined as the collection of vertices in adjacent edges of v_i. Hence, every edge is contained in a neighbourhood. A *hypergraph topology* is a family T of neighbourhood of vertices in V which satisfies the following conditions

- $\phi, V \in T$
- If $N(v_i), N(v_j) \in T$ then $N(v_i) \cap N(v_j) \in T$
- If $N(v_i) \in T$ for each $i \in I$ then $\cup_{i \in I} N(v_i) \in T$

The elements of T are called *open sets*. Thus, a topology T defined on a hypergraph H is called *hypergraph topological space*, denoted by (H, T). Since neighbourhood of a vertex is an open set, the *closed set* is the complement of a neighbourhood. For example, a hypergraph H is shown below (Fig. 2).

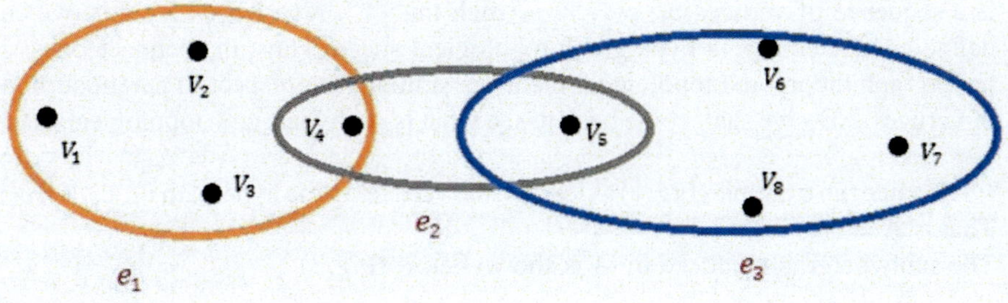

Fig. 2 Hypergraph

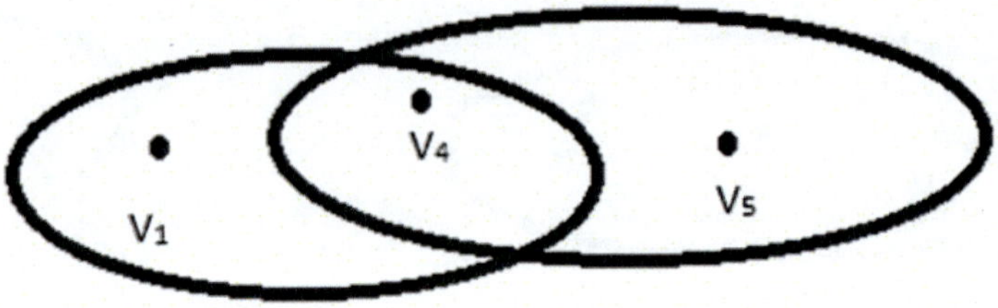

Fig. 3 Subhypergraph induced by A

Here $V = \{v_1, v_2, v_3, v_4, v_5, v_6, v_7, v_8\}$ and $E = \{e_1, e_2, e_3\}$ where $e_1 = \{v_1, v_2, v_3, v_4\}$, $e_2 = \{v_4, v_5\}$ $e_3 = \{v_5, v_6, v_7, v_8\}$
$N(v_1) = \{v_1, v_2, v_3, v_4, v_5\} = N(v_2) = N(v_3)$,
$N(v_4) = \{v_1, v_2, v_3, v_4, v_5, v_6, v_7, v_8\} = V = N(v_5)$,
$N(v_6) = \{v_4, v_5, v_6, v_7, v_8\} = N(v_7) = N(v_8)$
The hypergraph topology that can be defined are

- $T = \{\phi, V, N(v_1)\}$
- $T = \{\phi, V, N(v_5)\}$
- $T = \{\phi, V, N(v_8)\}$
- $T = \{\phi, V, N(v_3)\}$, etc.

(H, T) is the hypergraph topological space. $H = (V, E)$ with $A \subset V$. H_A is the subhypergraph of H.

The subhypergraph topology on H_A is defined by $T_A = \{A \cap N(v_i); N(v_i) \epsilon T\}$.
$N(v_1) = N(v_2) = N(v_3) = \{v_1, v_2, v_3, v_4, v_5\}$,$N(v_4) = N(v_5) = V$,$N(v_6) = N(v_7) = N(v_8) = \{v_4, v_5, v_6, v_7, v_8\}$
$T = \{\phi, V, N(v_1)\}$.
Using the above hypergraph $A = \{v_1, v_3, v_6\}$.$A \cap \phi = \phi$,$A \cap V = A \cap N(v_4) = A$, $A \cap N(v_1) = \{v_1, v_3\}$.
$T_A = \{\phi, A, \{v_1, v_3\}\}$ is not a topology induced by A. This is because the subhypergraph has disjoint hyperedges. So we shall redefine A using hyperpaths .

In the context of hypergraph topology, a hyperpath between vertices v_1 and v_k is a sequence of vertices $v_1, v_2, \ldots, v_k$ such that $\cap_{i=1}^{k} N(v_i) \neq \phi$. By this, we can define connectedness in hypergraph topological space. Thus, the connectedness in hypergraph theory and topological theory is similar with respect to neighbourhood of vertices. Now we shall come back to see what is subhypergraph topological space.

Consider an example (Fig. 1) as let A be the vertices in the hyperpath v_1, e_1, v_4, e_2, v_5. That is $A = \{v_1, v_4, v_5\}$.
The subhypergraph induced by A is shown below (Fig. 3).

Let $T = \{\phi, V, N(v_8)\}$ Then $A \cap \phi = \phi$, $A \cap V = A$, $A \cap N(v_8) = \{v_4, v_5\}$. Thus, $T_A = \{\phi, A, \{v_4, v_5\}\}$ is the induced subhypergraph topology.
As we defined the open set, we can think of closed sets.

4 Closed Set

A set other than V is closed if its complement is a neighbourhood of a vertex. Thus, a set is closed if its complement is open. Thus, ϕ and V are both closed and open.
In the previous example by Fig. 1, $\phi^c = V$,
$V^c = \phi$, $N(v_1)^c = \{v_6, v_7, v_8\} = N(v_2)^c = N(v_3)^c$,
$N(v_4)^c = \phi = N(v_5)^c$
$N(v_6)^c = \{v_1, v_2, v_3\} = N(v_7)^c = N(v_8)^c$ are not neighbourhoods.
The closed set in this topology is ϕ, V, $\{v_6, v_7, v_8\}$, $\{v_1, v_2, v_3\}$
Hence, complement of every set in a hypergraph topology is a closed set.
So similar to simple theorems in general topology, we have these in hypergraph topology also.

Theorem 1 *The union of any collection of open set is open.The intersection of a finite number of open set is open.*

Proof Every neighbourhood is an open set. By the definition of hypergraph topology, any union of elements in a topology, T, is contained in T, and the union of open sets is open in T.
Similarly, by the definition of hypergraph topology, intersection of elements of T is in T. That is, intersection of open sets is open. □

Theorem 2 *The intersection of a collection of closed set is closed. The union of a finite number of closed set is closed.*

Proof The complement of the intersection of closed sets is the union of the complement of closed sets. That is, it is the union of open sets . Since union of open set is open, the complement of intersection of closed set is open. Hence, the intersection of closed set is closed. Again by De Morgan's law, complement of union of closed set is the intersection of open set, which is open. Hence, union of closed set is closed. □

5 Continuity in Hypergraph Topology

Let us define the continuity in hypergraph topological space.
Let (H_1, T_1) and (H_2, T_2) be hypergraph topological spaces. A function $f : H_1 \to H_2$ is said to be continuous if for each neighbourhood $N(v_i)$ of H_2, the set $f^{-1}(N(v_i))$ is a neighbourhood of H_1. By this definition, we derive the following theorem using closed sets.

Theorem 3 *Let* (H_1, T_1) *and* (H_2, T_2) *be hypergraph topological spaces. Then* $f : H_1 \to H_2$ *is continuous if and only if for every closed set W in* H_2*, the set* $f^{-1}(W)$ *is closed in* H_1.

Proof Assume that f is continuous. By definition, for each neighbourhood $N(v_i)$ in H_2, the set $f^{-1}N(v_i)$ is a neighbourhood of H_1. Let W be a closed set in H_2. Then by definition of closed set W^c is a neighbourhood in H_2. By continuity, $f^{-1}(W^c)$ is a neighbourhood in H_1. $f^{-1}(W^c) = (f^{-1}(W))^c$

Let $v_i \in f^{-1}(W^c)$
$\Longleftrightarrow f(v_i) \in W^c$
$\Longleftrightarrow f(v_i) \notin W$
$\Longleftrightarrow v_i \notin f^{-1}(W)$,since f is continuous
$\Longleftrightarrow v_i \in (f^{-1}(W^))^c$.
Therefore $f^{-1}(W^c) = (f^{-1}(W^))^c$

$(f^{-1}(W^))^c$ is a neighbourhood in H_1 . This implies that $f^{-1}(W)$ is closed in H_1.
Conversely assume that for every closed set W in H_2, $f^{-1}(W)$ is closed in H_1.
W is closed in $H_2 \implies W^c$ is open in H_2
$\implies W^c$ is a neighbourhood in H_2
$f^{-1}(W)$ is closed in $H_1 \implies (f^{-1}(W))^c$ is a neighbourhood in H_1
$\implies f^{-1}(W^c)$ is a neighbourhood in H_1. Thus, for every neighbourhood W^c in H_2 there exist $f^{-1}(W^c)$, neighbourhood in $H_1 \implies f$ is continuous. □

Using continuous mapping the connectedness can be proved by the next theorem.

Theorem 4 *Let* (H_1, T_1) *and* (H_2, T_2) *be two hypergraph topological spaces. If* $f : H_1 \to H_2$ *is continuous and* H_1 *is a connected, then* $f(H_1)$ *is a connected hypergraph topological space.*

Proof Since H_1 is connected, there exists a hyperpath between every pair of vertices.
$\implies \cap_{i=1}^n N(v_i) \neq \phi$ for every i
claim: $f(\cap_{i=1}^n N(v_i)) = \cap_{i=1}^n f(N(v_i))$
$w \epsilon f(\cap_{i=1}^n N(v_i)) \Longleftrightarrow f^{-1}(w) \epsilon \cap_{i=1}^n N(v_i))$
$\Longleftrightarrow f^{-1}(w) \epsilon N(v_i))$ for every i
$\Longleftrightarrow w \epsilon f(N(v_i))$ for every i
$\Longleftrightarrow w \epsilon \cap_{i=1}^n f(N(v_i))$
Hence, $f(\cap_{i=1}^n N(v_i)) = \cap_{i=1}^n f(N(v_i)) \neq \cap_{i=1}^n f(\phi) \neq \phi$
Thus, there is a hyperpath between every pair of vertices in $f(H_1)$
Thus, $f(H_1)$ is connected. □

6 Metric in Hypergraph Topology

The function $d(v_i, v_j) = k$, where k is the length of shortest hyperpath in a hypergraph topology, follows all the axioms of the metric. So with this metric defined by $d(v_i, v_j) = k$, where k is the length of shortest hyperpath in a hypergraph topology is a function $d : V \times V \rightarrow R$, which satisfies the following conditions

- $d(v_i, v_j) \geq 0$
- $d(v_i, v_j) = 0 \iff v_i = v_j$
- $d(v_i, v_j) = d(v_j, v_i)$
- $d(v_i, v_k) \leq d(v_i, v_j) + d(v_j, v_k)$

This hypergraph topology induced by this metric is called *metric hypergraph topology*. For example, by Fig. 1, $d(v_1, v_1) = 0, d(v_1, v_4) = 1, d(v_1, v_5) = 2, d(v_1, v_7) = 3$. That is the first three conditions are satisfied. Also $d(v_1, v_7) = 3$ and $d(v_1, v_4) + d(v_4, v_7) = 1 + 2 = 3$. From Fig. 4, $d(v_1, v_8) = 4$ and $d(v_1, v_3) + d(v_3, v_8) = 2 + 3 = 5$. Thus, $d(v_i, v_k) \leq d(v_i, v_j) + d(v_j, v_k)$. Hence, the fourth condition is also satisfied. With this metric (H, d) is a *metric hypergraph space*.

7 Homeomorphism in Hypergraphs

Homeomorphism is a bijective correspondence that preserves the topological structure, it gives the connection between the neighbourhoods of H_1 and H_2. Two hypergraphs H_1 and H_2 are *homeomorphic* if one of the hypergraph is obtained from the other by subdivision or smoothing out of the vertices.

7.1 Subdivision

In subdivision, edges are subdivided into edges by introducing suitable number of vertices. Using Fig. 1,consider the edge $e_1 = \{v_1, v_2, v_3, v_4\}$
Add two new vertices as common.Let the new vertex be w_1 and w_2.Hence by subdivision, the new edges are $\{v_1, v_2, w_1, w_2\}$ and $\{w_1, w_2, v_3, v_4\}$ (Fig. 4).

7.2 Smoothing Out

In smoothing out the edge is smoothed out to an edge by deleting suitable number of vertices in common. For example, in Fig. 4 consider the edges $\{v_1, v_2, w_1, w_2\}$ and $\{w_1, w_2, v_3, v_4\}$. They have the intersection $\{w_1, w_2\}$. By smoothing out w_1 and w_2, the edge $\{v_1, v_2, v_3, v_4\}$ is obtained .

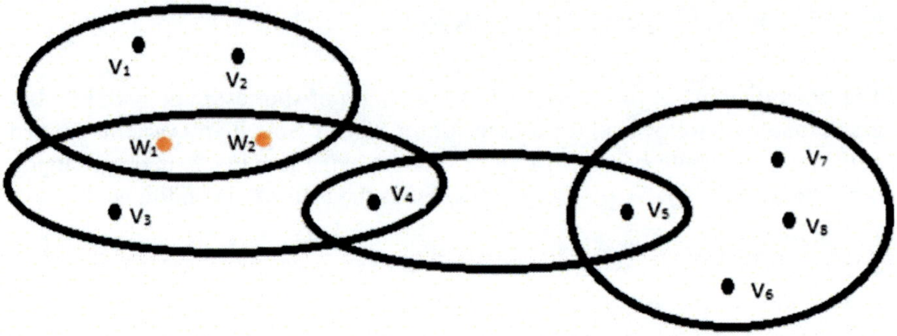

Fig. 4 Hypergraph obtained by subdivision

8 Conclusion

By defining hypergraph topological space with hypergraph topology using neighbourhood of a vertex a new branch of topology is evolved. Thus, hypergraph is a topological space in which vertices are points, and each edge is a region in a plane. The concept of closed set, continuity, connectedness in hypergrah topological space are discussed. Using hyperpath metric is defined by giving metric hypergraph space. The homeomorphism in hypergraph is defined. Further, investigation is being done on the homeomorphism and to homology of hypergraph topological space.

References

1. Antoine Vella, *A fundamentally topological perspective on Graph Theory* (Combinatorics and Optimization Waterloo, Ontario, Canada, 2005)
2. P. Srinivasan, P. Palanichamy, *Applications of Topology in Automobile Engineering* (Kamban Engineering College, Thiruvannamalai, Anna University, India)
3. D. Archdeacon, *Variations on a theme of Kuratowski* (Department of Mathematics and Statistics, University of Vermont, Burlington, USA), Received 7 December 2001, received in revised form 15 October 2002, accepted 22 July 2004
4. D. Zhou, J. Huang, B. Scholkopf, *Learning with Hypergraphs, Clustering,Classification, and Embedding* (2006)
5. J. Conan, *Boolean formulae, Hypergraphs and combinatorial topology*
6. R. James, *Munkres, Topology, Second Edition by* (Massachusetts Institute of Technology, Prentice Hall of India, 2004)
7. K. Došen , Z. Petric, *Hypergraph polytopes* (Mathematical Institute, SANU, Knez Mihailova 36, p.f. 367, Belgrade, Serbia)
8. S. Buzaglo R. Pinchasi Günter Rote *Topological Hypergraph* (2007)
9. S. Bressan, J. Li, S. Ren, J. Wu, *The Embedded Homology of Hypergraphs and Applications* (2018)